Advances in Computer Recognition, Image Processing and Communications, Selected Papers from CORES 2021 and IP&C 2021

Advances in Computer Recognition, Image Processing and Communications, Selected Papers from CORES 2021 and IP&C 2021

Editors

Michał Choraś
Robert Burduk
Agata Giełczyk
Rafał Kozik
Tomasz Marciniak

MDPI • Basel • Beijing • Wuhan • Barcelona • Belgrade • Manchester • Tokyo • Cluj • Tianjin

Editors
Michał Choraś
Bydgoszcz University of
Science and Technology
Poland

Robert Burduk
Wroclaw University of
Science and Technology
Poland

Agata Giełczyk
Bydgoszcz University of
Science and Technology
Poland

Rafał Kozik
Bydgoszcz University of
Science and Technology
Poland

Tomasz Marciniak
Bydgoszcz University of
Science and Technology
Poland

Editorial Office
MDPI
St. Alban-Anlage 66
4052 Basel, Switzerland

This is a reprint of articles from the Special Issue published online in the open access journal *Entropy* (ISSN 1099-4300) (available at: https://www.mdpi.com/journal/entropy/special_issues/CR_IPC).

For citation purposes, cite each article independently as indicated on the article page online and as indicated below:

LastName, A.A.; LastName, B.B.; LastName, C.C. Article Title. *Journal Name* **Year**, *Volume Number*, Page Range.

ISBN 978-3-0365-5313-9 (Hbk)
ISBN 978-3-0365-5314-6 (PDF)

Contents

entropy

MDPI

Editorial

Advances in Computer Recognition, Image Processing and Communications

Michał Choraś [1,*], Robert Burduk [2], Agata Giełczyk [1], Rafał Kozik [1] and Tomasz Marciniak [1]

[1] Faculty of Telecommunications, Computer Science and Electrical Engineering, Bydgoszcz University of Science and Technology, 85-796 Bydgoszcz, Poland; agata.gielczyk@pbs.edu.pl (A.G.); rkozik@pbs.edu.pl (R.K.); tommar@pbs.edu.pl (T.M.)

[2] Department of Systems and Computer Networks, Wroclaw University of Science and Technology, 50-370 Wroclaw, Poland; robert.burduk@pwr.edu.pl

* Correspondence: chorasm@utp.edu.pl

Citation: Choraś, M.; Burduk, R.; Giełczyk, A.; Kozik, R.; Marciniak, T. Advances in Computer Recognition, Image Processing and Communications. *Entropy* **2022**, *24*, 108. https://doi.org/10.3390/e24010108

Received: 4 January 2022
Accepted: 7 January 2022
Published: 10 January 2022

This Special Issue aimed to gather high-quality advancements in theoretical and practical aspects of computer recognition, pattern recognition, image processing and machine learning (shallow and deep), including, in particular, novel implementations of these techniques in the areas of modern telecommunications and cybersecurity. As almost all human activities have been moved online due to the pandemic, novel robust and efficient approaches and further research have been in higher demand in the field of computer science and telecommunication.

The selected authors of the contributions presented on 12th International Conference on Computer Recognition Systems (CORES) and the 12th International Conference on Image Processing and Communications (IP&C), held jointly with the 22nd International Conference on Advanced Computer Systems (ACS), were invited to submit extended versions of their original papers for this Special Issue. Our multi-conference took place in June 2021. This Special Issue was open to other submissions outside of the conference, too.

Finally, after a rigorous review process, we accepted 13 interesting papers in various domains and applications of computer recognition, image processing and communications.

Image processing can be applied in numerous areas of our daily lives: for entertainment, security and also improving medicine. This type of computer vision implementation was presented in [1]. In this paper, the authors proposed a method for detecting veins. They used digital image correlation (DIC) for detecting the micro-shifts in the skin caused by pulsation of the underlying veins. This method enabled observing the thickness of the veins up to a certain length. This also allowed for an initial, quick determination of whether the patient should be referred for diagnostic tests for venous embolism and thrombosis.

In [2], the authors investigated whether it is possible to use data pre-processing methods to robustify an ANN-based classifier against an adversarial evasion attack in the person re-identification problem. They examined the set of following methods: JPEG compression, Gaussian noise, Local Spatial Smoothing, Total Variance Minimisation and Block-Matching Convolutional Neural Network (BMCNN) for image denoising. Furthermore, they proposed a pre-processing pipeline that can robustly defend the ANN-based classifier against the adversarial attack without re-training the classifier. The proposed method is especially valuable since in the computer vision domain, the training process can be a huge computational endeavor.

The authors in [3] focused on the problem of quality assessment of stitched images. The process of stitching is an important element of many virtual reality and remote sensing applications where the panoramic images may be used as a background. As it is a complex issue (stitching quality can be affected by geometric distortions, ghosting, blurring and color distortion), the authors aimed to develop a new objective image quality metric based on the entropy measure. The metric proposed in the paper ensured achieving a considerably higher correlation of the designed objective metrics with subjective quality scores of the stitched images delivered in the ISIQA database, which was used in the research.

In the article [4], a novel approach to pointer instrument reading was presented. As the digitalization of the industry has become widely implemented, the automatization of instrument reading has become an emerging issue. The authors proposed the following pipeline: instrument detection from an image (VGG-16), pointer position extraction (Hough transform) and, finally, the pointer position reading (analysis of the angle between the center line and the extracted pointer). The presented experiments proved that the proposed method can read the pointer instruments comparably to human vision (1.354% of relative error).

The papers [5,6] addressed different problems of modeling a smooth monotone curve. Modeling is an effective tool for investigating objects, phenomena and processes. It can also be applied in reverse engineering.

In [7], a novel approach to remote sensing image segmentation is proposed. The authors implemented their method in order to distinguish the crop and vacancy fields on the farmland images obtained by UAV. In this approach, the ResNet architecture was analyzed, improved and then implemented as a part of the encoder–decoder structure. The authors also claimed that the proposed method can be easily adapted to handle other similar issues, e.g., crack segmentation.

In [8], the problem of selecting an optical channel for transporting the double sideband radio-frequency-over-fiber (DSB-RFoF) radio signal over the optical fronthaul path, avoiding the dispersion-induced power penalty (DIPP) phenomenon, is addressed. The presented method complements the possibilities of a short-range optical network working in the flexible dense wavelength division multiplexing (DWDM) format, where chromatic dispersion compensation is not applied. In [9], the authors describe their algorithm for dynamic routing with dedicated path protection. They proposed the algorithm in the context of optical networks, but it can be applicable to other networks, where services have to be protected and the network resources are finite and discrete, e.g., wireless radio or networks capable of advance resource reservation.

In [10], the authors presented a deep neural network in order to automatically classify ECG signals. Three neural network architectures were proposed: the first based on the convolutional network, the second on SincNet and the third on the convolutional network but with additional entropy-based features. The author of [11] touches on a new approach in the last-mile network structural solutions for smart grid networks and suggests a new method for finding the optimal SM localization, which can also work as a data concentrator.

In [12], the authors proposed Hfinger, a novel effective malware HTTP request fingerprinting tool. It extracts information from the parts of the request, such as URI, protocol information, headers and payload, providing a concise request representation that preserves the extracted information in a form interpretable by a human analyst.

The authors of [13] establish the minimal amount of data that is sufficient to efficiently train machine learning algorithms in intrusion detection. The authors also identify the most valuable NetFlow-based features that facilitate effective, machine-learning-based network intrusion detection in the real world. Their objectives are reached in a series of experiments with the use of several feature selection techniques, machine learning algorithms and intrusion detection benchmark datasets. The paper [13] is the result of the EU Horizon 2020 SIMARGL project (simargl.eu).

Acknowledgments: The Guest Editors are grateful to all authors for their contributions to this Special Issue and to the anonymous peer-reviewers and editors for carefully reading the submissions and their constructive feedback. We would also like to thank the *Entropy* editorial team, especially Vincent Shang.

Conflicts of Interest: The authors declare no conflict of interest.

References

1. Lutowski, Z.; Bujnowski, S.; Marciniak, B.; Kloska, S.; Marciniak, A.; Lech, P. A Novel Method of Vein Detection with the Use of Digital Image Correlation. *Entropy* **2021**, *23*, 401. [CrossRef] [PubMed]
2. Pawlicki, M.; Choraś, R.S. Preprocessing Pipelines including Block-Matching Convolutional Neural Network for Image Denoising to Robustify Deep Reidentification against Evasion Attacks. *Entropy* **2021**, *23*, 1304. [CrossRef] [PubMed]
3. Okarma, K.; Chlewicki, W.; Kopytek, M.; Marciniak, B.; Lukin, V. Entropy-Based Combined Metric for Automatic Objective Quality Assessment of Stitched Panoramic Images. *Entropy* **2021**, *23*, 1525. [CrossRef] [PubMed]
4. Wang, L.; Wang, P.; Wu, L.; Xu, L.; Huang, P.; Kang, Z. Computer Vision Based Automatic Recognition of Pointer Instruments: Data Set Optimization and Reading. *Entropy* **2021**, *23*, 272. [CrossRef] [PubMed]
5. Havrylenko, Y.; Kholodniak, Y.; Halko, S.; Vershkov, O.; Bondarenko, L.; Suprun, O.; Miroshnyk, O.; Shchur, T.; Śrutek, M.; Gackowska, M. Interpolation with Specified Error of a Point Series Belonging to a Monotone Curve. *Entropy* **2021**, *23*, 493. [CrossRef] [PubMed]
6. Havrylenko, Y.; Kholodniak, Y.; Halko, S.; Vershkov, O.; Miroshnyk, O.; Suprun, O.; Dereza, O.; Shchur, T.; Śrutek, M. Representation of a Monotone Curve by a Contour with Regular Change in Curvature. *Entropy* **2021**, *23*, 923. [CrossRef] [PubMed]
7. Zhang, X.; Yang, Y.; Li, Z.; Ning, X.; Qin, Y.; Cai, W. An improved encoder-decoder network based on strip pool method applied to segmentation of farmland vacancy field. *Entropy* **2021**, *23*, 435. [CrossRef]
8. Zakrzewski, Z. Optical Channel Selection Avoiding DIPP in DSB-RFoF Fronthaul Interface. *Entropy* **2021**, *23*, 1554. [CrossRef] [PubMed]
9. Szcześniak, I.; Olszewski, I.; Woźna-Szcześniak, B. Towards an Efficient and Exact Algorithm for Dynamic Dedicated Path Protection. *Entropy* **2021**, *23*, 1116. [CrossRef] [PubMed]
10. Śmigiel, S.; Pałczyński, K.; Ledziński, D. ECG Signal Classification Using Deep Learning Techniques Based on the PTB-XL Dataset. *Entropy* **2021**, *23*, 1121. [CrossRef] [PubMed]
11. Kiedrowski, P. Selection of the Optimal Smart Meter to Act as a Data Concentrator with the Use of Graph Theory. *Entropy* **2021**, *23*, 658. [CrossRef] [PubMed]
12. Białczak, P.; Mazurczyk, W. Hfinger: Malware HTTP Request Fingerprinting. *Entropy* **2021**, *23*, 507. [CrossRef] [PubMed]
13. Komisarek, M.; Pawlicki, M.; Kozik, R.; Hołubowicz, W.; Choraś, M. How to Effectively Collect and Process Network Data for Intrusion Detection? *Entropy* **2021**, *23*, 1532. [CrossRef] [PubMed]

entropy

Article

A Novel Method of Vein Detection with the Use of Digital Image Correlation

Zbigniew Lutowski [1,*,†], Sławomir Bujnowski [1,†], Beata Marciniak [1,†], Sylwester Kloska [2,†], Anna Marciniak [2,†] and Piotr Lech [3,†]

1 Faculty of Telecommunications, Computer Science and Electrical Engineering, UTP University of Science and Technology, 85-796 Bydgoszcz, Poland; slawomir.bujnowski@utp.edu.pl (S.B.); beata.marciniak@utp.edu.pl (B.M.)
2 Faculty of Medicine Ludwik Rydygier Collegium Medicum in Bydgoszcz, Nicolaus Copernicus University in Torun, 85-067 Bydgoszcz, Poland; mkloska93@gmail.com (S.K.); marciniak.anna1993@gmail.com (A.M.)
3 Department of Signal Processing and Multimedia Engineering, West Pomeranian University of Technology, 70-313 Szczecin, Poland; Piotr.Lech@zut.edu.pl
* Correspondence: zbigniew.lutowski@utp.edu.pl
† These authors contributed equally to this work.

Abstract: Digital image correlation may be useful in many different fields of science, one of which is medicine. In this paper, the authors present the results of research aimed at detecting skin micro-shifts caused by pulsation of the veins. A novel technique using digital image correlation (DIC) and filtering the resulting shifts map to detect pulsating veins was proposed. After applying the proposed method, the veins in the forearm were visualized. The proposed technique may be used in the diagnosis of venous stenosis and may also contribute to reducing the number of adverse events during blood collection. The great advantage of the proposed method is the lack of the need to have specialized equipment, only a typical mobile phone camera is needed to perform the test.

Keywords: vein detection; digital image processing; correlation; displacement measurement

Citation: Lutowski, Z.; Bujnowski, S.; Marciniak, B.; Kloska, S.; Marciniak, A.; Lech, P. A Novel Method of Vein Detection with the Use of Digital Image Correlation. *Entropy* **2021**, *23*, 401. https://doi.org/10.3390/e23040401

Academic Editor: Amelia Carolina Sparavigna and Tae Sun Choi

Received: 16 February 2021
Accepted: 25 March 2021
Published: 28 March 2021

1. Introduction

Measurement of displacements may be useful in many various fields. Because it is based on point-wise measurements, they offer solutions to relatively simple problems. One of the uses of the method could be in the medical field. The new application of digital image correlation (DIC), proposed in this paper, is to detect micro-shifts in the skin caused by pulsation of the underlying veins. The most important methods currently used for precise vein imaging are—Magnetic Resonance Angiography (MRA) methods, hyperspectral imaging methods (data captured in the range 900–1700 nm) and ultrasound imaging methods. However, mainly due to cost reasons, near infrared (NIR) with a wavelength from 740 nm to 760 nm imaging is the most popular method [1]. Use of vein finders minimalizes the risk of pre-analytical error, reduces potential pain of patient and is fast to perform. This method is based on the fact that hemoglobin absorbs the light emitted by diodes. Hemoglobin is composed of protein and iron containing compounds. The main function of hemoglobin is to transport blood rich in oxygen (O_2) from lungs to all tissues of the body and blood rich in carbon dioxide (CO_2) in the opposite direction [2]. It is also worth to mention that blood rich in oxygen will be transparent for NIR and will not visualize arteries. With the help of vein detector, it is possible to observe veins under the skin, as it appears in darker colors than background.

Medical diagnostics play a key role in the process of disease recognition. To provide the patient with the best quality medical care, it is important to plan the diagnostic procedure properly, starting with the sample. Blood is an extremely important source of information about the patient; therefore, it is most often used in diagnostic tests as a sample. Blood collection is an invasive method, so it is extremely important to choose veins accurately, so the patient experiences the least discomfort during the procedure. To obtain reliable information about a patient, samples should be taken in an appropriate manner, in order to reflect actual state of the patient. Many previous studies reported that there is a high percentage of error in blood sample collection. That is why it is extremely important to ensure proper preparation of the pre-analytical phase [3,4]. Peripheral venous access is very frequently performed procedure in hospitals. There are over 1 billion venipunctures performed every day, to obtain blood sample for testing. This procedure may be time-consuming and difficult in various groups of patients, such as neonates, children, obese patients, or patients after difficult, long treatment for example, chemotherapy. Venipuncture involves piercing a vein with a needle that makes it possible to draw it. This procedure can lead to patient stress, pain, and other unwanted reactions. Failures during blood collection can cause various types of complications, including thrombosis [5,6]. Such events can lead to court hearings in extreme cases because patients can hold people responsible for health damage. To locate a vein, a compression band and a gentle knock are usually used around the elbow joint. Locating vein is not easy due to various factors such as skin color, obesity, and dehydration of patients.

First attempts to use vein detectors were back in early 2000 [7]. A device, the Luminetx VeinViewer (LVV), produced by Luminetx Corp was tested. The aim of the work was to test whether LVV would be helpful in detecting feeder veins in various conditions and situations, especially in patients with telangiectasia. Device allowed localization of bigger number of veins than naked eye would, because feeder veins are too deep. Treating this disease was reliant on experience of physician, which is not a pleasant situation. LVV was compared to another device and showed a higher resolution and sensitivity. LVV proved to be an excellent device, however, the research was conducted on a relatively small group of 23 patients.

Along with globally increasing health awareness, the authors of [8] had the idea of developing an application that would allow mobile devices, such as smartphones or tablets, to act as vein detectors with the use of NIR. Considering the importance of portable devices in everyday life, it had a good chance of being successful. Nevertheless, it should be considered that the use of devices such as telephones is not a fairly reliable reference point from a medical point of view. Most applications that are now used to take care of health have not been tested for their effectiveness and reliable value of the results obtained. During their work they used vein detectors available on the market and had valuable clinical reviews. High cost and efficacy of studied vein detectors were not always as good as would be required. There were also several other problems with vein detectors and their evaluation. Most of the research was performed on small, homogenous groups which did not allow them to draw clear conclusions in the context of a larger, diverse population. Cost-effectiveness is one of the biggest problems for vein detectors. Those devices are not cheap themselves. They may even cost a few hundred dollars and they require trained personnel to use them. Preparations and training are also time consuming and cost money.

On the contrary, the authors of [9] created a device characterized by high-intensity and low-leak light source, near-infrared CMOS camera and a small and light one-eye head mounted display. With the use of these technologies, they could visualize veins at a depth of 67 mm. This allowed the performance of several venous interventions with a low chance of failure.

The authors of [10] also developed their version of vein detectors with the use of NIR. They used a CCD camera but were obliged to eliminate the cut-off filter to gain access to the infrared part of radiation spectrum. Source of light used to develop their detector were Infrared LEDs. Obtained images in resolution of 640 × 480 were then transferred to computer. Images were then processed with the use of several algorithms and programs, including LabVIEW. Proposed system managed to obtain vein images; however, they were not at the highest quality and indicated the need of future improvements in both software and hardware. Authors also pointed out that the vein detecting systems in general have difficulties with vein detection in people with many tattoos that blocked the transmission of light.

Despite many advantages, vein detectors also have disadvantages. One of which is the cost of devices and their portability. These facts can encourage researchers to continue their work to improve the devices and create reliable cost-effective and efficient vein finders for medical applications. It is also an area where roads of medics and biomedical engineers may cross. The concept of vein finders may seem to be simple, however, there are a lot of factors which hinder the task of creating the best one—to combine high quality devices with attractive price. The growth of interest in vein detectors led various producers to create their own devices.

Most of the vein detection methods based on image processing algorithms relate to biometrics. The use of the finger venous system due to its individual uniqueness. Biometric security features are much more effective than passwords or Personal Identification Numbers (PINs). Therefore, the authors of [11] proposed a method recognizing the venous pattern of the finger. The authors used algorithm based on gradient correlation. The proposed method consists in obtaining a high-quality image, which is then processed and stored in the database. The result of such operations is then compared with the pattern. Based on the arrangement of the finger veins, access can therefore be allocated or denied.

The authors of [12] used infrared images to detect finger veins. The method they proposed used the Gabor filter and the SIFT feature. It is characterized by a low error rate (below 0.5%). However, this method only focuses on the detection of the finger veins.

This is why the field needs to be more deeply explored and novel solutions need to be proposed. Most of the known solutions are based on the use of NIR, which automatically makes it necessary to have an illuminator and a camera working with such a wavelength of light. Looking for the cheapest solution that does not require any additional hardware components and is easily accessible to as many potential users as possible, our team decided to use a technique unique in this field—the correlation of images obtained with a camera of a typical mobile phone. The main idea of the method is based on the comparison of two images taken within a small amount of time, to capture skin deformation caused by blood flow. It is possible to increase the resolution of this method with the use of sub-pixel information calculations. There are many algorithms that allow this kind of calculation. One of the methods used for many years is the basic theory of digital image correlation (DIC), which enables the determination of image motion, distortions, or cracks. The theoretical accuracy of sub-pixel methods is well known; however, little is known about its real accuracy [13]. The goal of the research is to locate veins with the use of a camera directed at the examined limb. This allows observation of skin pulsations during blood flow through the veins. However, we should take into consideration the fact that it is impossible to keep a limb perfectly immobilized—the human body naturally shakes, which influences the quality of the obtained pictures. Regardless, it is possible to indicate pictures on which there are visible veins.

2. Methodology

2.1. Digital Image Correlation

The technique of digital image correlation is used for measuring displacements of textured objects. One image from the series is selected and it represents a reference picture for all subsequent analyses. Then it is divided into small rectangular regions (called subsets) containing $N \times N$ pixels (Figure 1). Subset size depends on the size of the random pattern, but also on its quality. Position of subsets is tracked from the reference picture to all other pictures from the measurement series. Cross-correlation coefficient calculation is used to search subsets positions. The displacement vectors U and V are calculated for every subset (Figure 1). Subpixel accuracy is achieved with the use of advanced interpolation methods, including Newton-Raphson method. This method is used to increase the accuracy of tracking, but also to speed up the process. At the beginning, it had extremely big computational costs, however, in recent years, scientists have been working on improving this algorithm, by removing its unnecessary parts. Thanks to their endeavors Newton-Raphson algorithm has become faster, and at the same time requires less computing power. This approach is a gold standard for detection of subpixel movement. The use of the Newton-Raphson algorithm during analysis improves the efficiency of the entire process. Set of displacement maps is obtained as a result, and therefore it can be used for deformation maps computation [14,15].

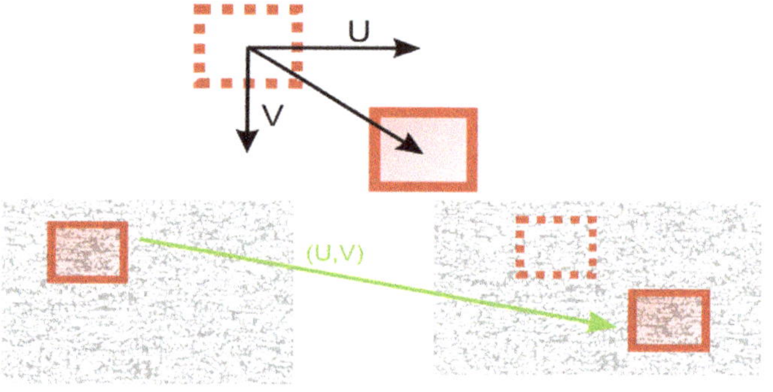

Figure 1. The principle of the digital image correlation (DIC) method.

The DIC algorithm uses the correlation function to search for its maximum value. There are two different correlation criteria, which are used for the initial guess finding and its subsequent refinement. By computing at integer locations, the normalized cross correlation (NCC) (1) the initial guess is found.

$$C_{cc} = \frac{\sum_{(i,j)\in S}(f(\tilde{x}_{refi}, \tilde{y}_{refj}) - f_m)(g(\tilde{x}_{curi}, \tilde{y}_{curj}) - g_m)}{\sqrt{\sum_{(i,j)\in S}[f(\tilde{x}_{refi}, \tilde{y}_{refj}) - f_m]^2 [g(\tilde{x}_{curi}, \tilde{y}_{curj}) - g_m]^2}}, \tag{1}$$

where f and g are respectively the reference and current image grayscale intensity functions at a specified location (x, y). Functions f_m (2) and g_m (3) correspond to the mean grayscale values of reference and current subset.

$$f_m = \frac{\sum_{(i,j)\in S}(f(\tilde{x}_{refi}, \tilde{y}_{refj})}{n(S)} \tag{2}$$

$$g_m = \frac{\sum_{(i,j)\in S}(g(\tilde{x}_{refi}, \tilde{y}_{refj})}{n(S)}, \tag{3}$$

where $n(S)$ is the number of data points in subset S. The initial guess yields u and v with integer (pixel) accuracy. To refine these results with sub-pixel resolution the next step is to use a nonlinear optimizer by finding the minimum of Equation (4).

$$C_{LS} = \sum_{(i,j)\in S}\left[\frac{f(\tilde{x}_{refi}, \tilde{y}_{refj}) - f_m}{\sqrt{\sum_{(i,j)\in S}[f(\tilde{x}_{refi}, \tilde{y}_{refj}) - fm]^2}} - \frac{g(\tilde{x}_{refi}, \tilde{y}_{refj}) - g_m}{\sqrt{\sum_{(i,j)\in S}[g(\tilde{x}_{refi}, \tilde{y}_{refj}) - g_m]^2}}\right]^2. \tag{4}$$

In conclusion, in our research we performed the calculation of displacement maps by computing the NCC with sub-pixel accuracy obtained by applying the Newton-Raphson algorithm. At a later stage of the research, it turned out that it was necessary to develop additional methods of filtering the results, which were proposed in the section 'Methodology improvements and results'.

2.2. Data and Algorithm Parameters

The research work was carried out on data obtained with an ordinary mobile phone camera (Samsung Galaxy S10). Movies were recorded with 1920 × 1080 pixels resolution and 30 frames per second (FPS) speed. The resulting movie with a 23 s duration consisted of 675 frames (pictures). Since the physical dimensions of the area observed by the camera were 100 mm × 56 mm, a single pixel corresponded to a square with a side of 52 µm. The first frame was set as a base frame (BF) for the DIC algorithm, so the shifts were always calculated between BF and subsequent movie frames. To limit the calculation time, the area of interest has been defined as a grid designating points for which shifts were calculated. The grid consisted of 130 columns and 200 rows spaced 4 picture pixels apart (Figure 2). Each intersection of a column and a row marked a single pixel in the image for which the shift was calculated, resulting in total 26,000 points calculation for each pair of BF and n-th frame.

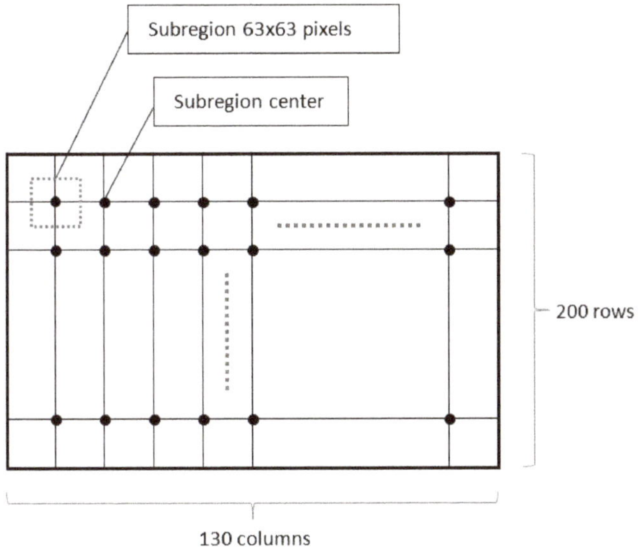

Figure 2. Grid of analyzed points.

The use of DIC method for single pixel displacement calculation requires an analysis of its neighborhood. This neighborhood was defined as a square-shaped area (subregion) of 63 pixels side size.

3. Methodology Improvements and Results

The direct use of the methodology described in the previous chapter gave completely illegible shift maps. This was due to the fact that the limb vibrations were greater several orders of magnitude than the skin movements caused by the change in pressure during the blood flow through the blood vessel (Figure 3). Therefore, the resulting movements map was filtered using average shift calculated from 100 × 300 pixel area without visible veins. Then the calculated vector was subtracted from the shift map (Figure 4). The shifts maps on Figures 4, 6 and 8 are colored according to the absolute shift value, where the minimum shift value is blue and the maximum shift is red. Due to the poor visibility of the vessels at the edges of the hands, filtering using one, fixed shift vector turned out to be insufficient.

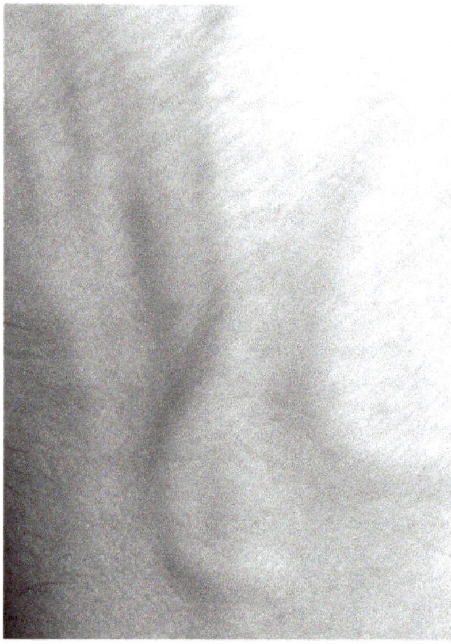

Figure 3. Base frame—No. 1 from a sequence of 675 frames (30 fps).

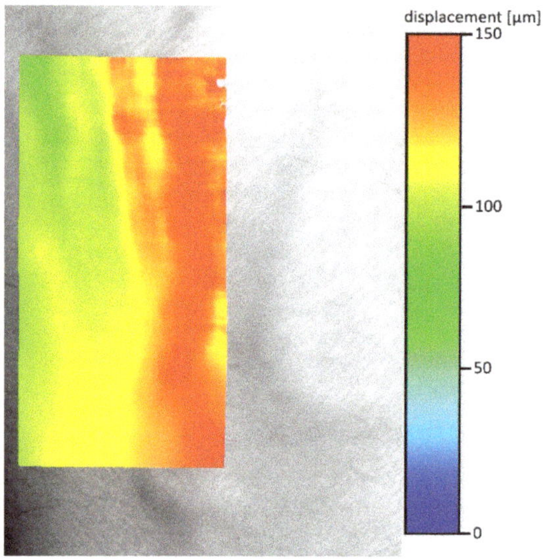

Figure 4. Shift between frames 1 and 206 with removal of the component designated by the area without visible veins of 100 × 300 pixels (the same value subtracted from all shifts).

The next improvement was to use the subtraction of the average calculated not once for the entire shift map, but the average determined by a sliding window, as shown on Figure 5. From each point in the shift map, the local average was subtracted. This local average was calculated from 30 adjacent shift points (15 from left and 15 from right), so each point in shift map was corrected by its own local average value. The resulting shifts map is shown on Figure 6. Comparison of Figures 4 and 6 indicates better separation of vein edges in Figure 6. The optimal size of the averaging window, in terms of filtration, depends on the ratio of the distance between individual points of the offset map to the expected size of the vein. Thanks to the use of the window mechanism to determine the local mean, the effect of automatic adjustment of the algorithm to the non-uniform vector of the entire hand shift was achieved. These heterogeneities result from both the complicated nature of forehand vibrations and the distortions introduced by its curvature. Too small averaging window will "blur" the most interesting information about the location of the edges of the veins. On the other hand, too large will not consider the local changes in the optical shift of the entire hand area. The proposed value was selected experimentally.

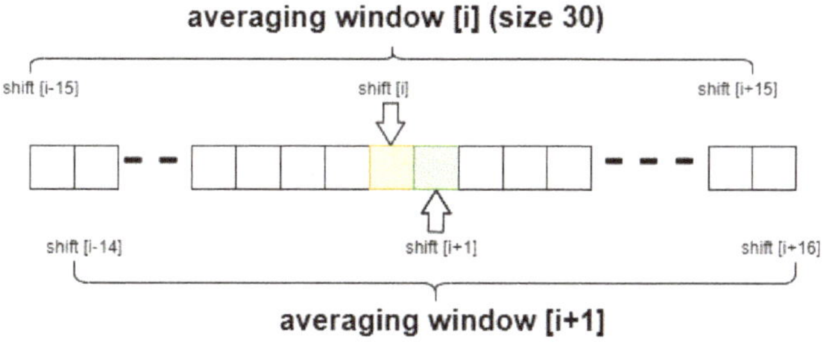

Figure 5. Schema of moving average calculation for exemplary *i* and *i*+1 shift.

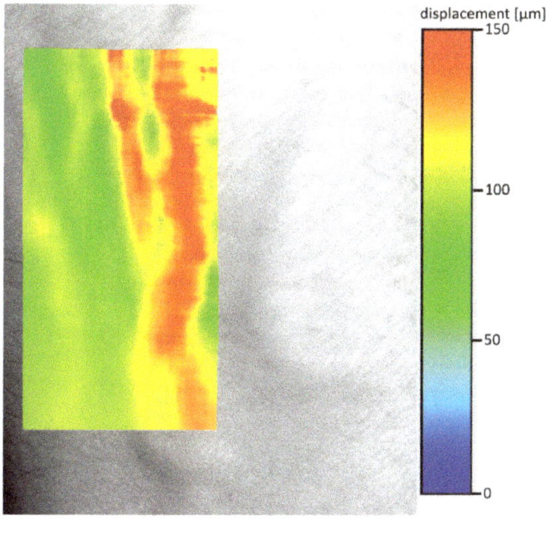

Figure 6. Shift between frames 1 and 206 with the removal of the component determined by a moving window (moving average) with a size of 30 pixels (15 on each side) (the window includes a single line).

Because the averaging window is determined only based on a single row of shifts, the presented approach will not work for veins running parallel to the x-axis of the image (Figure 7). Since we focused on analyzing the images of the arm, we considered the above disadvantage to be insignificant at this stage of the research, but in the case of an unpredictable direction of the veins, it would probably be more appropriate to choose a filtering window with the size of not a single line, but the entire square area around the analyzed point. The resulting shifts map shows one more interesting effect—as it is clearly visible on Figure 8—map shows only edges of veins.

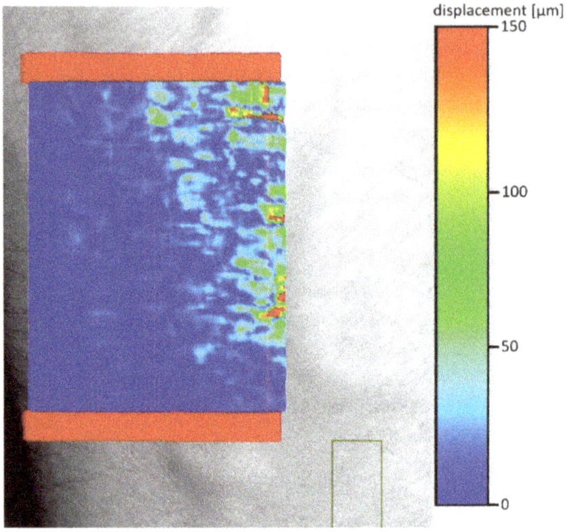

Figure 7. Detection of veins and arteries, image rotated 90 degrees (moving average determined from rows is not effective).

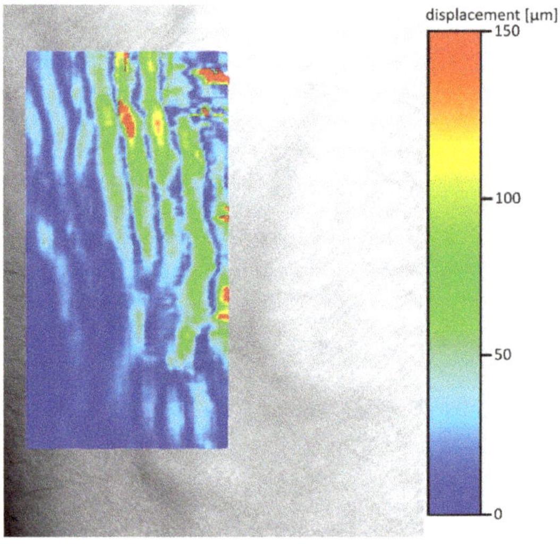

Figure 8. Shift between frames 1 and 423, moving average, visible edges of veins and arteries.

The vein edge detection mechanism causing visualization of two edges per single vein is explained in Figure 9. The method detects only shifts in axes parallel to the image plane, so the movement of point 6 has greater axis X delta value than the movement of point 5.

In the next step, we tested the presented method on a group of 31 people. Veins were detected in each of the participants using the proposed method on both hands (Figure 10). The images of veins obtained with the proposed method were compared with the images obtained with the 'Infrared vein finder Medcaptain NAVI-60' device. Some examples of vein detection in one randomly selected participant is described below.

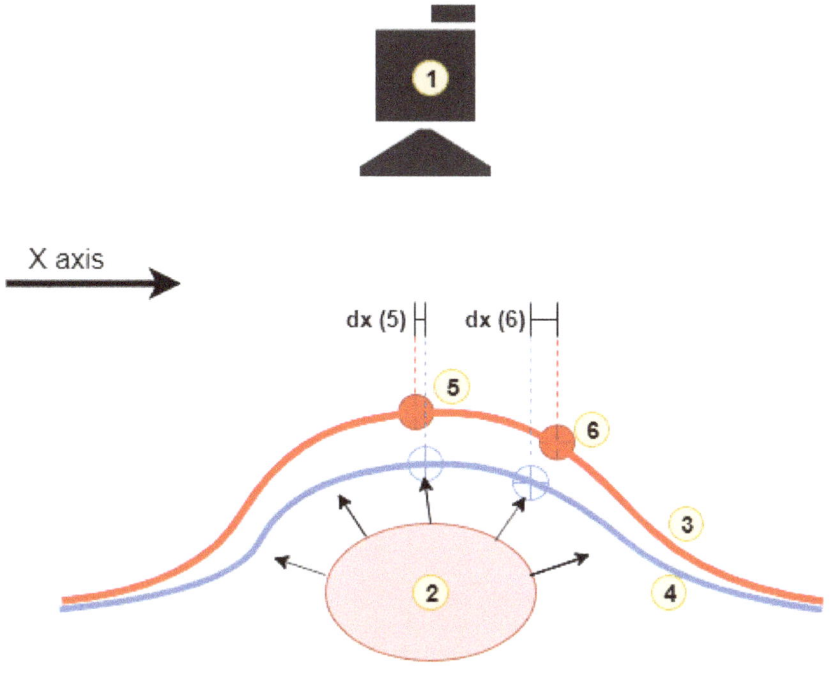

Figure 9. Explaining the mechanism causing visualization of the edges of the arteries (1—camera, 2—artery, 3—skin surface in the max pressure phase, 4—skin surface in the pressure min phase, 5 and 6 selected two points on the skin and their position in the maximum/minimum phase).

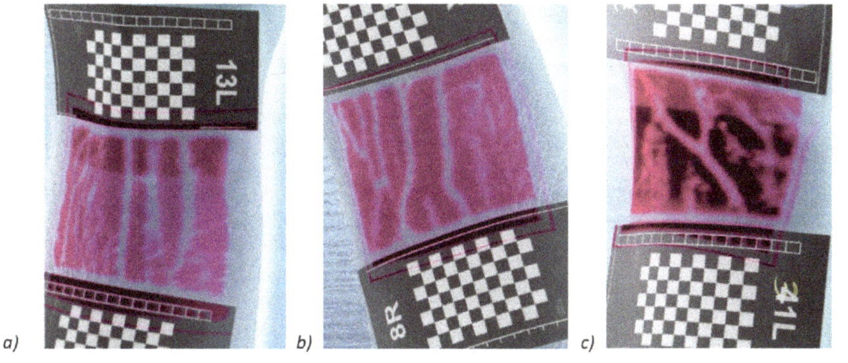

Figure 10. Negatives of the reference images (better readability when applied). (**a**) 13L, (**b**) 8R, (**c**) 31L describes the number of participant and which hand was tested, L—left, R—right.

Presented method shows the image of the pulsation of the veins at a selected moment in time, in order to obtain a complete image of the veins, the frames of shifts from all time moments had to be summed up into one. Examples of images from different time moments are shown in Figure 11, which shows the shift maps making up the Figure 12.

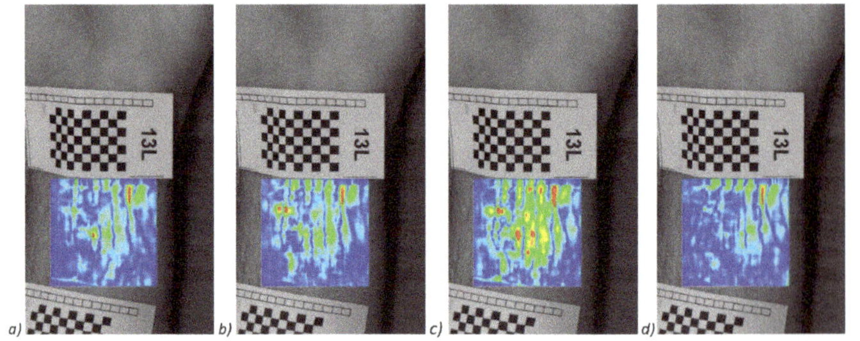

Figure 11. Exemplary images that makes up to the Figure 12—shift maps for frames #35,78,101,119 (**a**–**d**) of 120.

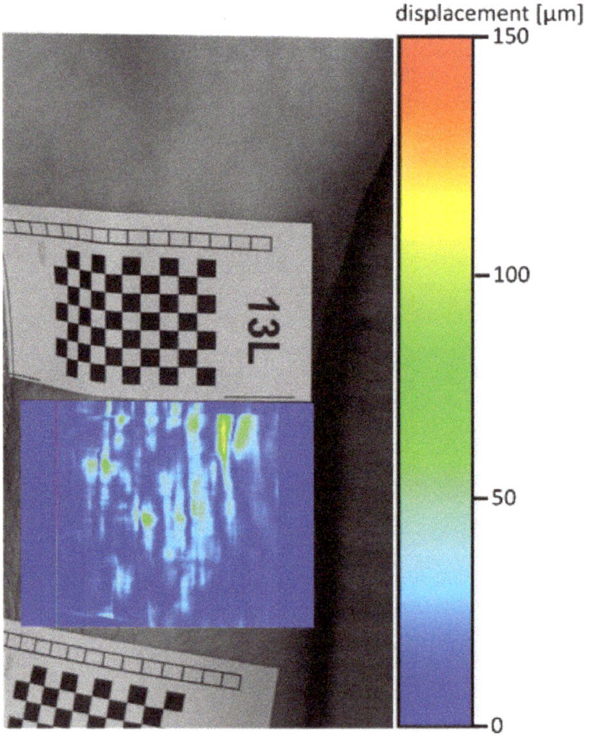

Figure 12. Cumulative shifts for 13L participant.

Figure 12 shows cumulative (sum) shifts for 120 frames (movie with 30 FPS, that is, about 4 s of the movie) for 13L participant. Calculation parameters of the map was: a grid of points 400 × 310, the matching area for one point of the grid: 97 pixels. Dark blue places of the image indicate no shift, which means that they indicate the edges of the veins (where the skin does not pulsate). Figure 13 shows reference image with overlaid cumulative shifts from Figure 12. In our opinion, this is the least accurate and legible image of the veins, so based on this image, we created the manually traced edges of the veins detected—Figure 14. Some areas detected as 'pulsating' do not coincide with the base (1), and edges of veins not visible in the base were also detected (2).

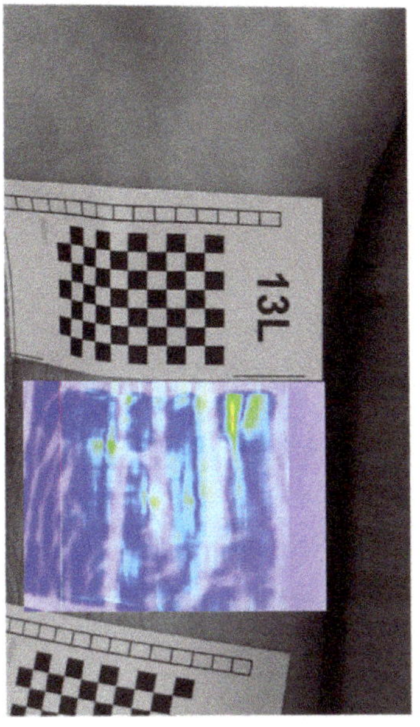

Figure 13. Cumulative shifts overlaid on the negative reference image.

Figure 14. The manually traced edges of the veins detected.

The research carried out on a test group of 31 people confirmed that the proposed method can be used to detect pulsating veins, however, determining the qualitative parameters (accuracy of determining the location of veins, maximum depth of detected veins) of the method requires even more detailed tests. Our method allowed to detect the pulsation of about 70% of the veins visualized with the NAVI-60 device. Typically, compared to the reference device, the smallest veins were hardly visible. The calculations were made using our own software based on the OpenCV library. The average computation time for 120 measurement frames was 10 s (on the computer with Intel i5 8265 processor and 8GB RAM). According to the processor specifications and benchmark results, similar computation times should be possible with the Exynos 9820 processor used in the mobile device used for image acquisition. Therefore, it seems that the proposed algorithms can be fully implemented on mobile devices.

4. Discussion

The application of this method is to improve the patient's quality of life during all types of medical procedures that were mentioned before. Considering the fact that there are a lot of these procedures performed daily it is highly important.

This method allows us to observe the thickness of the veins for a certain length. This allows an initial, quick determination of whether the patient should be referred for diagnostic tests for venous embolism and thrombosis. A completely different principle of operation compared to the competing NIR method allows for the visualization of subsequent phases of blood flow through the veins. In addition, unlike the NIR-based methods, the proposed method allows us to check if the blood flow in the veins is not slowed down or if there is a blood-returning disorder.

5. Conclusions

The above-described method allows the finding of veins. It is characterized by high accuracy and efficiency. However, despite many advantages, it is not an ideal method. One of the biggest issues of the proposed method is the fact that it depends on camera position. During the study, we noticed that during horizontal positioning of the hand relative to the camera, it failed to effectively visualize the vein. No problems were encountered when the hand was vertically positioned towards the camera. One of the biggest advantages of the proposed method is complete independence from specialized equipment; it can be performed with the use of an ordinary mobile phone using its camera and its computing power. The authors believe that this method is promising, so further optimization tests should be carried out.

Author Contributions: Data curation, S.B.; Formal analysis, Z.L., B.M. and P.L.; Investigation, Z.L., S.K. and A.M.; Methodology, Z.L., S.B. and P.L.; Resources, B.M., A.M., S.K. and P.L.; Writing—review & editing, Z.L., B.M., S.B., S.K., A.M. and P.L. All authors have read and agreed to the published version of the manuscript.

Funding: This research received no external funding.

Institutional Review Board Statement: According to our best knowledge the experiment we conducted, even though it required human participants was not invasive or intimate, so it did not require Ethics Committee consent.

Informed Consent Statement: Informed consent was obtained from all subjects involved in the study.

Data Availability Statement: https://github.com/UTP-WTIiE/NovelMethodOfVeinDetection, accessed on 24 March 2021.

Conflicts of Interest: The authors declare no conflict of interest.

References

1. Timm, U.; McGrath, D.; Lewis, E.; Kraitl, J.; Ewald, H. Sensor system for non-invasive optical hemoglobin determination. In Proceedings of the SENSORS, New Orleans, LA, USA, 17–19 February 2009; pp. 1975–1978.
2. Thom, C.S.; Dickson, C.F.; Gell, D.A.; Weiss, M.J. Hemoglobin variants: biochemical properties and clinical correlates. *Cold Spring Harb. Perspect. Med.* **2013**, *3*, a011858. [CrossRef] [PubMed]
3. Lippi, G.; Baird, G.S.; Banfi, G.; Bölenius, K.; Cadamuro, J.; Church, S.; Cornes, M.P.; Dacey, A.; Guillon, A.; Hoffmann, G.; et al. Improving quality in the preanalytical phase through innovation, on behalf of the European Federation for Clinical Chemistry

and Laboratory Medicine (EFLM) Working Group for Preanalytical Phase (WG-PRE). *Clin. Chem. Lab. Med.* **2017**, *55*, 489–500. [CrossRef] [PubMed]
4. Giavarina, D.; Lippi, G. Blood venous sample collection: Recommendations overview and a checklist to improve quality. *Clin. Biochem.* **2017**, *50*, 568–573. [CrossRef] [PubMed]
5. Lamperti, M.; Pittiruti, M. II. Difficult Peripheral Veins: Turn on the Lights. *Br. J. Anaest.* **2013**, *110*, 888–891. [CrossRef] [PubMed]
6. Buowari, O.Y. Complications of venepuncture. *Adv. Biosci. Biotechnol.* **2013**, *4*, 38. [CrossRef]
7. Miyake, R.K.; Zeman, H.D.; Duarte, F.H.; Kikuchi, R.; Ramacciotti, E.; Lovhoiden, G.; Vrancken, C. Vein imaging: a new method of near infrared imaging, where a processed image is projected onto the skin for the enhancement of vein treatment. *Dermatol. Surg.* **2006**, *32*, 1031–1038. [CrossRef] [PubMed]
8. Juric, S.; Flis, V.; Debevc, M.; Holzinger, A.; Zalik, B. Towards a low-cost mobile subcutaneous vein detection solution using near-infrared spectroscopy. *Sci. World J.* **2014**, *2014*, 365902. [CrossRef] [PubMed]
9. Tobisawa, N.; Namita, T.; Kato, Y.; Shimizu, K. Injection assist system with surface and transillumination images. In Proceedings of the 2011 5th International Conference on Bioinformatics and Biomedical Engineering, Wuhan, China, 13–15 May 2011; pp. 1–4.
10. Wadhwani, M.; Sharma, A.D.; Pillai, A.; Pisal, N.; Bhowmick, M. Vein detection system using infrared light. *Int. J. Sci. Eng. Res.* **2015**, *6*, 780–786.
11. Lin, C.; Li, M.; Sun, X. A finger vein recognition algorithm based on gradient correlation. *Aasri Procedia* **2012**, *1*, 40–45.
12. Peng, J.; Wang, N.; Abd El-Latif, A.A.; Li, Q.; Niu, X. Finger-vein verification using Gabor filter and SIFT feature matching. In Proceedings of the 2012 Eighth International Conference on Intelligent Information Hiding and Multimedia Signal Processing, Athens, Greece, 18–20 July 2012; pp. 45–48.
13. Lutowski, Z.; Marciniak, B.; Marciniak, T.; Bujnowski, S. Precision of sub-pixel image displacement measurements. *J. Mach. Constr. Maint. Probl. Eksploat.* **2017**, *4*, 21–26.
14. Pan, B.; Li, K. A fast digital image correlation method for deformation measurement. *Opt. Lasers Eng.* **2011**, *49*, 841–847. [CrossRef]
15. Khoo, S.W.; Karuppanan, S.; Tan, C.S. A review of surface deformation and strain measurement using two-dimensional digital image correlation. *Metrol. Meas. Syst.* **2016**, *23*, 461–480. [CrossRef]

entropy

Article

Preprocessing Pipelines Including Block-Matching Convolutional Neural Network for Image Denoising to Robustify Deep Reidentification against Evasion Attacks

Marek Pawlicki [1,*] and Ryszard S. Choraś [2]

[1] ITTI Sp. z o.o., 61-612 Poznań, Poland
[2] Institute of Telecommunications and Computer Science, Bydgoszcz University of Science and Technology, 85-796 Bydgoszcz, Poland; Ryszard.Choras@utp.edu.pl
* Correspondence: mpawlicki@itti.com.pl

Abstract: Artificial neural networks have become the go-to solution for computer vision tasks, including problems of the security domain. One such example comes in the form of reidentification, where deep learning can be part of the surveillance pipeline. The use case necessitates considering an adversarial setting—and neural networks have been shown to be vulnerable to a range of attacks. In this paper, the preprocessing defences against adversarial attacks are evaluated, including block-matching convolutional neural network for image denoising used as an adversarial defence. The benefit of using preprocessing defences comes from the fact that it does not require the effort of retraining the classifier, which, in computer vision problems, is a computationally heavy task. The defences are tested in a real-life-like scenario of using a pre-trained, widely available neural network architecture adapted to a specific task with the use of transfer learning. Multiple preprocessing pipelines are tested and the results are promising.

Keywords: deep learning; computer vision; adversarial attacks; adversarial defences

Citation: Pawlicki, M.; Choraś, R.S. Preprocessing Pipelines Including Block-Matching Convolutional Neural Network for Image Denoising to Robustify Deep Reidentification against Evasion Attacks. *Entropy* **2021**, 23, 1304. https://doi.org/10.3390/e23101304

Academic Editor: Ernestina Menasalvas

Received: 17 August 2021
Accepted: 30 September 2021
Published: 3 October 2021

1. Introduction

Artificial neural networks offer a collection of benefits which have proved useful in image processing, especially in tasks including artificial-intelligence-based computer aided detection [1]. The progress of the last decade allowed to push the results obtained by artificial neural networks to levels surpassing human performance, in select tasks [2]. In computer vision, deep neural networks became the go-to solution for a wide variety of problems [3], capable of producing an impressive result in a sensible time frame [4]. Recently, artificial neural networks found success in person reidentification [5–7].

In general, reidentification (reID) refers to the process of re-attaching publicly available data to an anonymised record in order to discover the identity of an individual. In the context of computer vision, the phrase refers to the ability of an image recognition system to spot an individual across different cameras, and different angles [7]. ReID is a challenging task which stirred up a significant amount of research recently, particularly due to the significant benefits it could bring for public safety [7]. The use in the context of intelligent surveillance systems forces the consideration of adversarial behaviour against the artificial intelligence (AI) technologies used for reID. In a real-world scenario, impressive detection metrics are not the only thing that matters [8]. The current trend in reID involves the use of deep neural networks, which have been proven to be susceptible to a novel kind of attacks [9–12].

Deep neural networks, particularly convolutional neural networks (CNN), are widely used for the CV tasks [13]; some of the best-performing ImageNet contest architectures were based on the premise of utilising convolutional layers. The network architectures tend to be very deep: Inception features over 6 million trainable parameters [14,15], ResNet18

(Residual neural network) over 11 million [16], AlexNet over 60 million [17], VGG16 (Visual Geometry Group) over 138 million [18], etc. Therefore, training a top-tier deep neural network is a huge computational endeavour [19]. In order not to repeat this effort for each task, transfer learning can be employed [20]. Transfer learning leverages pre-trained networks, essentially using them as feature extractors with frozen weights, feeding samples to the network and only training the added dense layers at the output end of the topology. However, the use of openly-available, pre-trained networks poses a security problem in an adversarial setting, as it raises the capability of the attacker [21,22].

The idea of attacking deep neural networks has focused the attention of the deep learning community over the last few years [23–25]. A range of adversarial attacks effective against AI were discovered, uncovering the vulnerabilities of data driven technologies [25]. In this work, the attacks performed at test-time are considered, which are known as evasion attacks [26,27].

The goal of an Evasion attack is to force the AI-based system to misclassify a particular sample. This is achieved by adding a specifically crafted noise to the tested sample. This added noise, in case of images, is imperceptible to humans, but leverages the 'intriguing properties of neural networks' to fool the AI algorithm [28]. The issue of defending against those attacks is a fierce arms race and the satisfactory defence has not yet emerged [29].

The algorithms and technologies presented in this paper were used to form a submission to the reidentification defences track of the H2020 SPARTA SAFAIR contest. The task was formulated around the CelebA face recognition dataset [30,31]. The dataset, as used in the task, featured 5304 classes, with 85,612 samples in the training subset and 28,523 samples in the testing set. The objective of the defensive track was to propose ways of preventing adversarial samples from lowering the accuracy of the face recognition model. The following sections describe the specific technologies used for defining the submission of the contest, the rationale behind those choices, the formulated defences, and provide the results of the experiments.

As such, the research and, thus, the paper is conducted and formulated to answer the following research question:

- **RQ1** Is it possible to use data preprocessing methods to robustify an ANN-based classifier against adversarial evasion attack in computer vision (CV)?
- **RQ2** Does using all the identified defensive preprocessing methods provide a better protection than using just a selection of those?

Thus, the innovative contribution of this paper comes in the formulation and evaluation of a plug-and-play preprocessing pipeline for robustification of already-existing or pre-trained CV classifiers, easily deployable in a real-world situation and saving on the cost of re-training the classifier

The paper is structured as follows: In Section 2, the related works are introduced and the most important categories of defences are described. Section 3 lists the setup of the used reidentification pipeline, showcases the effects of the adversarial attacks and introduces the specific defences, including the block-matching convolutional neural network (BMCNN) for image denoising, which, to the best of our knowledge, has never before been used to counter adversarial attacks. Section 4 contains the experimental setups and the results obtained by specific pipelines. Section 5 encompasses the conclusions along with the impact the defensive pipeline has over a clean dataset.

2. Related Works

The advent of adversarial perturbations revealed the vulnerabilities of contemporary AI-based technologies. There is a considerable body of research into both the attacks and the defences. However, as noted by [32], the construction of a theoretical model of crafting adversarial perturbations is problematic, as it is a sophisticated optimisation procedure for most machine learning models. This absence of a theoretical baseline makes it troublesome to verify whether administering a certain defence can proof a system against a certain set of attacks. This situation finds its expression in the fact that whenever a new defence is proposed, a new attack capable of breaking through that defence appears [33–38].

Against this canvas, the authors of [38] propose a set of guidelines for research into the defensive mechanisms against adversarial attacks, listing common pitfalls and a range of best practices. There is a substantial body of work gathering both the available attacks and possible defences geared towards machine and deep learning [27,29,32,39–42] and even specifically deep learning in computer vision [23,43,44].

A thorough analysis of the sources allows one to roughly divide the adversarial defences into these categories:

1. Gradient masking;
2. Input reconstruction;
3. Detectors.

According to [29], the category of gradient masking encompasses defences which fit either intentionally or unintentionally. This category of defences relies on making the gradient unfit for the operation of the attack algorithms. Some defences do not aim at gradient masking specifically, but achieve it as a by-product of defensive procedures. One of the most popular approaches, adversarial training, frequently has a gradient masking effect, even though it is not the goal of the process.

Adversarial (re)training is considered as the brute-force approach [32]. The procedure relies on crafting adversarial samples and including them in the training set. The problem with retraining the whole classifier is the computational cost of such course of conduct. This problem will be touched upon later in this paper.

The defences in the input reconstruction category perform various forms of input pre-processing. Although it might be possible to circumvent those methods in a scenario where the attacker has full knowledge of the system, in a real-world setting the defences from this category can be very effective, and computationally much cheaper in use than retraining. The detection approaches are effective as long as the adversary is not aware of the existence of the detector. For an attacker of sufficient capability it is possible to build an adversarial sample which, at the same time, circumvents the detector and fools the classifier, as proven by [33].

3. Materials and Methods

3.1. Classifier Setup

In this work, the VGG-face network was used [45] with the pre-trained 'resnet-50' [46] architecture. VGGFace is trained on a dataset containing 2.6 million face images of over 2.6k people. The resnet50 network is a CNN assembled of 50 layers. The detailed hyperparameter setup of the entire network can be found in [45]. The final layer of the pre-trained network is AveragePooling2D with the shape of (None, 1, 1, 2048). To perform transfer learning, a dense layer of 2048 neurons is added to the the pre-trained network, followed by a dropout layer, and wrapped up by the softmax layer set with the number of neurons equal to the number of classes. The added dense layer uses the rectified linear unit (ReLU) activation function. The weights between the AveragePooling layer and the dense layer along with the weights between the dense layer and the output layer constitute the part of the network that is trained on the CelebA dataset, with the weights of the remainder of the network frozen. The batch size used for training was set to 1, while early stopping was used to find the optimal number of epochs, which capped at 32. Multiple different hyperparameters setup were tested, and learning rate scheduling was also tested. For the reduced dataset used in the experiments the default learning rate of straight 0.01 proved optimal.

The trainable part of the model contains 15,064,248 parameters when it is prepared to recognise all the 5304 identities found in the CelebA dataset. To allow fast prototyping, a toy model was built on fourteen most populated classes in the CelebA dataset. The most populated classes were chosen to avoid having to deal with the data imbalance problem, allowing the research to focus on adversarial defences. Changing just the number of classes allowed to reduce the number of trainable parameters to just over 4 million; a reduction of over 70%. The prior probability of the occurrence of each of the classes is displayed in Table 1

Table 1. The priors of the classes

Class	1757	2114	2820	3227	3699	3745	3782	4262
Prior Probability (%)	6.72	6.72	7.84	7.84	7.56	7.56	7.84	6.72
	4740	4978	6568	8968	9152	9256		
Prior Probability (%)	6.72	6.72	6.72	7.00	7.00	7.00		

Multi-task cascaded convolutional neural networks (MTCNN) is a technique capable of spotting faces and extracting them for later processing by other networks. A state-of-the-art face recognition processing pipeline consists of MTCNN for face detection and landmark placement, and a CNN used for placing the extracted face in adequate categories [47–49]. In this work, MTCNN is used for preprocessing the CelebA images for both training and testing. The CelebA subset selected for the formulation of the model was further split into the training set and the testing set. The classifier performance on the test set containing the 14 most populated classes is presented in Table 2.

Table 2. Classifier performance on the test set containing the 14 most populated classes.

Label	Precision	Recall	f1-Score
1757.0	1.00	1.00	1.00
2114.0	1.00	1.00	1.00
2820.0	0.88	1.00	0.93
3227.0	1.00	0.86	0.92
3699.0	0.88	1.00	0.93
3745.0	1.00	1.00	1.00
3782.0	1.00	1.00	1.00
4262.0	0.88	1.00	0.93
4740.0	1.00	1.00	1.00
4978.0	1.00	1.00	1.00
6568.0	1.00	1.00	1.00
8968.0	1.00	1.00	1.00
9152.0	1.00	1.00	1.00
9256.0	1.00	0.71	0.83
macro avg	0.97	0.97	0.97
weighted avg	0.97	0.97	0.97
accuracy	0.9693877551020408		
balanced accuracy	0.9693877551020408		

For better evaluation of the effects of adversarial perturbations and adversarial defences, the misclassified samples were removed from the set, manually pushing the performance to 100% accuracy. That way, any adversarial perturbations are registered as drops in performance, avoiding a situation where an attack pushes the misclassified sample to the correct class. Furthermore, the way the defences affect the classifier performance is more clearly readable.

3.2. Adversarial Attacks

The testing set was then subjected to the procedure of creating the adversarial samples. To produce the adversarial attacks, the projected gradient descent (PGD) method was used, considering PGD as the universal first-order adversary, following [50]. The maximum number of iterations was set to 100, the epsilon step to 0.1. The value of epsilon determines the maximum size of perturbation allowed for the attack. Along with the number of iterations, multiple values of epsilon were tested to simulate different strengths of attack. The effect different strengths of the attacks have on the image can be seen in Figure 1. The pictures are reformatted to fit the vgg-face input shape. The effects of PGD eps = 4 on the performance of the classifier can be seen in Table 3.

Figure 1. The effects of different strengths of the attacks on the image.

Table 3. The effects of PGD eps=4 on the performance of the classifier.

Label	Precision	Recall	f1-Score
1757.0	1.00	0.14	0.25
2114.0	0.33	0.14	0.20
2820.0	0.00	0.00	0.00
3227.0	1.00	0.17	0.29
3699.0	0.32	1.00	0.48
3745.0	0.00	0.00	0.00
3782.0	0.00	0.00	0.00
4262.0	0.33	0.71	0.45
4740.0	0.08	0.14	0.11
4978.0	0.00	0.00	0.00
6568.0	1.00	0.14	0.25
8968.0	0.00	0.00	0.00
macro avg	0.40	0.21	0.20
weighted avg	0.38	0.21	0.19
accuracy	0.21052631578947367		
balanced accuracy	0.2139455782312925		

3.3. *Defences*

There have been a number of defences proposed by the research community [51]. The task is to design robust AI tools that are resilient to adversarial attacks. Some methods rely on retraining the entire classifier using attacks generated with the known attack methods [52]. This method, called adversarial training, not only impacts the effectiveness of the classifier, but also requires an immense computational effort. The proposition contained in this section utilises the idea of using pre-processing methods to robustify existing AI-based classifiers, so as the users do not need to re-train their models. The proposed methods are accompanied by an assessment of how the defensive measures affect the classifier performance, which helps optimise the resiliency of AI against the loss of performance some defences introduce.

3.3.1. JPEG Compression

The Joint Photographic Experts Group (JPEG) compression used as adversarial defence relies on the fact that JPEG-compressed images are very prevalent in contemporary usage. Following the authors of [53], who noted that JPEG compression often has the ability to reverse the effects of small adversarial perturbations, the technique is evaluated here for the use as a purely pre-processing defence against adversarial attacks. The compression has the effect of removing additive artefacts in square blocks of an image, effectively working as a filter removing adversarial perturbations [54]. The effect of different magnitudes of compression (20, 40, 80) can be seen in Figure 2. The results of the classifier using JPEG compression with quality set to 20 on PGD attacks with epsilon = 4 can be found in Table 4

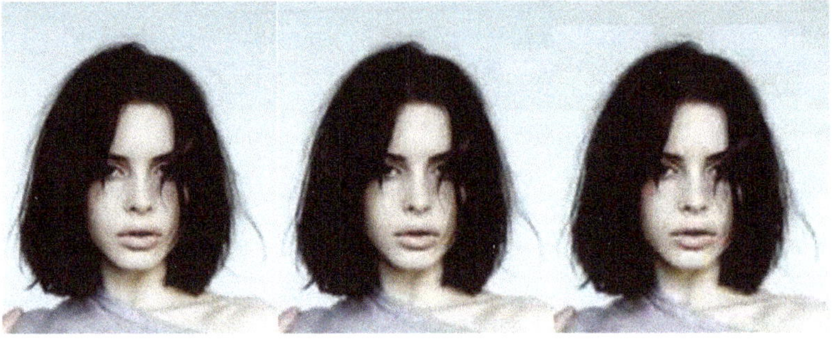

Figure 2. JPEG compression.

Table 4. The results of the classifier using JPEG compression with quality set to 20 on PGD attacks with epsilon = 4.

Label	Precision	Recall	f1-Score
1757.0	1.00	1.00	1.00
2114.0	1.00	1.00	1.00
2820.0	1.00	1.00	1.00
3227.0	1.00	0.83	0.91
3699.0	0.88	1.00	0.93
3745.0	0.86	0.86	0.86
3782.0	0.86	0.86	0.86
4262.0	0.78	1.00	0.88
4740.0	1.00	1.00	1.00
4978.0	0.86	0.86	0.86
6568.0	1.00	1.00	1.00
8968.0	1.00	0.86	0.92
9152.0	1.00	0.86	0.92
9256.0	0.80	0.80	0.80
macro avg	0.93	0.92	0.92
weighted avg	0.93	0.93	0.93
accuracy	0.9263157894736842		
balanced accuracy	0.9227891156462587		

3.3.2. Gaussian Data Augmentation

Gaussian data augmentation [55] is a process of adding Gaussian noise to a sample. This method is proven not to produce adversarial samples and can reverse the effects of known adversarial attacks. Image samples with different sigma settings can be seen in Figure 3. The value of sigma expresses the variance.

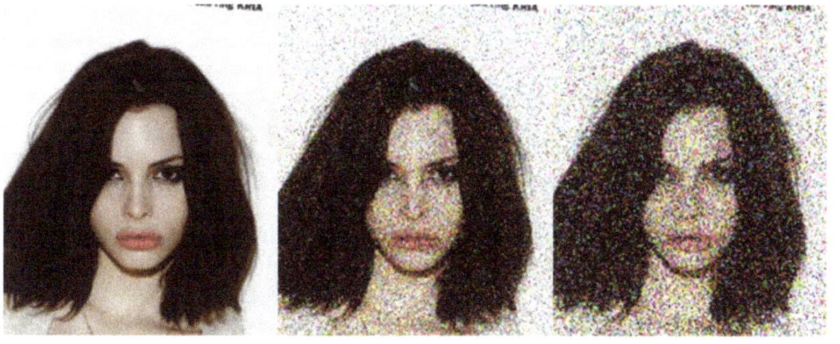

Figure 3. Gaussian augmentation—sigma 255.0/5, 255.0/17, 255.0/3.

3.3.3. Local Spatial Smoothing

Following the research of [56], spatial smoothing can be used to reduce the effects of added adversarial noise. The algorithm uses local blurring filters to remove the effects of adversarial noise. The approach is one of the feature squeezing methods and can be effectively applied as a pre-processor-based defence. The image before and after spatial smoothing can be seen in Figure 4.

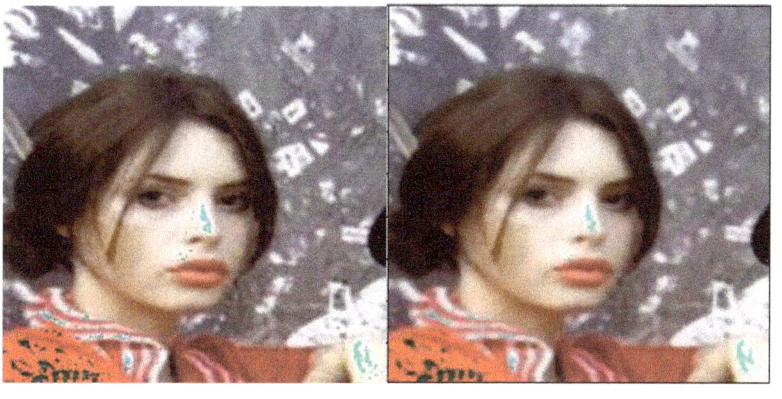

Figure 4. The image before and after spatial smoothing.

3.3.4. Total Variance Minimisation

Total variance minimisation is a model-agnostic preprocessor approach. In the original paper [57], the defence is used for retraining the model and then the inputs are also preprocessed at test time. The method reassembles the image by rebuilding a randomly chosen set of pixels with the plainest depiction of these pixels. The image before and after total variance minimisation can be seen in Figure 5.

Figure 5. The image before and after total variance minimisation.

3.3.5. Block-Matching Convolutional Neural Network (BMCNN) for Image Denoising as an Adversarial Defence

Following the work in image denoising presented in [58], and extending the idea of applying autoencoders as adversarial defences [59], the BMCNN is proposed for the a method of robustifying the image recognition system against adversarial attacks. BMCNN is an attempt to merge two leading approaches to image denoising: non-local self-similarity prior based methods [60] and feed-forward denoising with the use of convolutional neural networks [61]. The method is applied as a pre-processor to remove adversarial noise before the sample is fed to the classifier. The results of the BMCNN with sigma set to 20 used on adversarial samples created with PGD with epsilon set to four can be seen in Table 5. The value of sigma has been chosen experimentally.

Table 5. The results of the classifier using BMCNN with sigma set to 20 used on adversarial samples created with PGD with epsilon set to four.

Label	Precision	Recall	f1-Score
1757.0	1.00	1.00	1.00
2114.0	1.00	1.00	1.00
2820.0	1.00	1.00	1.00
3227.0	0.83	0.83	0.83
3699.0	0.70	1.00	0.82
3745.0	1.00	0.71	0.83
3782.0	0.88	1.00	0.93
4262.0	0.78	1.00	0.88
4740.0	1.00	1.00	1.00
4978.0	0.88	1.00	0.93
6568.0	1.00	0.86	0.92
8968.0	1.00	0.86	0.92
9152.0	0.80	0.57	0.67
9256.0	1.00	0.8	0.89
macro avg	0.92	0.90	0.90
weighted avg	0.92	0.91	0.90
accuracy	0.9052631578947369		
balanced accuracy	0.9023809523809525		

4. Results

The low computational cost of the preprocessors in comparison with re-training the classifier allows to mix and match the defences. The experiments show that some pipelines are more effective than others. An example of a defensive pipeline which utilises all the researched defences is displayed in Figure 6.

Figure 6. A defensive pipeline which utilises all the researched defences.

The pipeline makes intuitive sense, as blurring the image should remove some of the artefacts added by PGD, same for JPEG compression, then adding Gaussian noise and removing it with BMCNN denoising has the potential of removing both the Gaussian and the adversarial noise at the same time. The results of this particular pipeline are shown in Table 6.

Table 6. The results of the classifier using spatial smoothing with JPEG compression, Gaussian augmentation, total variance minimisation and BMCNN with sigma set to 20 on PGD images with epsilon set to four.

Label	Precision	Recall	f1-Score
1757.0	0.50	0.71	0.59
2114.0	0.50	0.43	0.46
2820.0	0.00	0.00	0.00
3227.0	0.40	0.33	0.36
3699.0	0.37	1.00	0.54
3745.0	0.25	0.14	0.18
3782.0	0.25	0.86	0.39
4262.0	0.25	0.14	0.18
4740.0	0.50	0.57	0.53
4978.0	0.67	0.29	0.40
6568.0	1.00	0.14	0.25
8968.0	0.50	0.14	0.22
9152.0	0.67	0.29	0.40
9256.0	0.00	0.00	0.00
macro avg	0.42	0.36	0.32
weighted avg	0.43	0.37	0.33
accuracy	0.3684210526315789		
balanced accuracy	0.36054421768707484		

As showcased by the results of the experiment in Table 6, the mix of defences improved the detection metrics as compared to the undefended model; however it did not perform as well as, for example, BMCNN denoising alone (Table 5). For the next experiment, the total variance minimisation preprocessor was removed, as it has a similar filtering effect as localised spatial smoothing. The pipeline is shown in Figure 7. The results of the experiment are contained in Table 7.

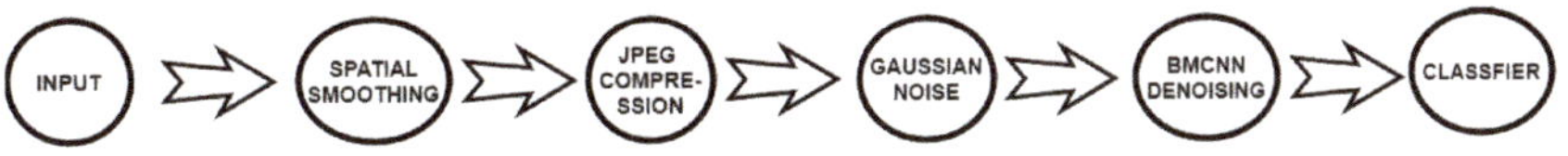

Figure 7. A defensive pipeline which utilises all the researched defences, except total variance minimisation.

Table 7. The results of the classifier using spatial smoothing with JPEG compression, Gaussian augmentation, and BMCNN with sigma set to 20 on PGD images with epsilon set to four, without total variance minimisation.

Label	Precision	Recall	f1-Score
1757.0	1.00	1.00	1.00
2114.0	1.00	1.00	1.00
2820.0	1.00	1.00	1.00
3227.0	0.83	0.83	0.83
3699.0	0.78	1.00	0.88
3745.0	1.00	0.86	0.92
3782.0	0.75	0.86	0.80
4262.0	0.78	1.00	0.88
4740.0	1.00	1.00	1.00
4978.0	0.86	0.86	0.86
6568.0	1.00	0.86	0.92
8968.0	1.00	0.86	0.92
9152.0	1.00	0.57	0.73
9256.0	0.83	1.00	0.91
macro avg	0.92	0.91	0.90
weighted avg	0.92	0.91	0.90
accuracy	0.9052631578947369		
balanced accuracy	0.9064625850340137		

To find the optimal mix of preprocessors that would minimise or eliminate the effect of adversarial perturbations without significantly deteriorating the classifier results, a range of experiments was performed. The results of some of those tests are contained in Tables 8 and 9. To assess the results of the preprocessing defences, the best performing preprocessing pipeline was tested on a clean, unperturbed set. The results of this experiment can be found in Table 10. The best performing pipeline is illustrated in Figure 8.

Table 8. The results of the classifier using spatial smoothing with JPEG compression on PGD images with epsilon set to four.

Label	Precision	Recall	f1-Score
1757.0	1.00	1.00	1.00
2114.0	1.00	1.00	1.00
2820.0	1.00	1.00	1.00
3227.0	1.00	0.83	0.91
3699.0	0.78	1.00	0.88
3745.0	0.86	0.86	0.86
3782.0	0.86	0.86	0.86
4262.0	0.78	1.00	0.88
4740.0	1.00	1.00	1.00
4978.0	0.86	0.86	0.86
6568.0	1.00	1.00	1.00
8968.0	1.00	0.86	0.92
9152.0	1.00	0.71	0.83
9256.0	0.80	0.80	0.80
macro avg	0.92	0.91	0.91
weighted avg	0.93	0.92	0.92
accuracy	0.9157894736842105		
balanced accuracy	0.9125850340136055		

Figure 8. A defensive pipeline with JPEG compression, Gaussian augmentation, and BMCNN.

Table 9. The results of the classifier using JPEG compression, Gaussian augmentation, and BMCNN on PGD images with epsilon set to four.

Label	Precision	Recall	f1-Score
1757.0	0.88	1.00	0.93
2114.0	1.00	1.00	1.00
2820.0	1.00	1.00	1.00
3227.0	1.00	0.83	0.91
3699.0	0.78	1.00	0.88
3745.0	0.86	0.86	0.86
3782.0	0.86	0.86	0.86
4262.0	0.88	1.00	0.93
4740.0	1.00	1.00	1.00
4978.0	0.86	0.86	0.86
6568.0	1.00	1.00	1.00
8968.0	1.00	0.86	0.92
9152.0	1.00	0.71	0.83
9256.0	1.00	1.00	1.00
macro avg	0.94	0.93	0.93
weighted avg	0.93	0.93	0.93
accuracy	0.9263157894736842		
balanced accuracy	0.9268707482993197		

Table 10. Results of classification with preprocessing defences on a clean dataset.

Label	Precision	Recall	f1-Score
1757.0	1.00	1.00	1.00
2114.0	1.00	1.00	1.00
2820.0	1.00	1.00	1.00
3227.0	1.00	0.83	0.91
3699.0	0.88	1.00	0.93
3745.0	0.83	0.71	0.77
3782.0	0.75	0.86	0.80
4262.0	0.78	1.00	0.88
4740.0	1.00	1.00	1.00
4978.0	1.00	1.00	1.00
8968.0	1.00	1.00	1.00
9152.0	1.00	1.00	1.00
9256.0	1.00	0.60	0.75
macro avg	0.95	0.93	0.93
weighted avg	0.94	0.94	0.94
accuracy	0.9368421052631579		
balanced accuracy	0.9289115646258503		

5. Conclusions

The classifier performance indicates that using preprocessing defences causes a drop in the measured metrics; at the same time, the achieved robustness is considerable. The results of the experiments prove that input transformations are an effective weapon against adversarial attacks, though the robustness comes at a cost. The utility of the proposed preprocessing pipeline solution comes in the fact that it can be used as a plug-and-play quick-fix, granting a measure of robustness against adversarial attacks without having to incur the costs of re-training the classifier. This answers RQ1 affirmatively, using preprocessing defensive methods is feasible for robustification of ANN-based classifiers against adversarial evasion attacks in computer vision tasks. The results of the experiments also point out that passing the images through a series of filters can have adverse effects on the accuracy of protected classifier. Joining all the researched preprocessing methods in one pipeline did alleviate some of the effects of the adversarial attacks. However, the accumulative distortion introduced by those methods hindered the effectiveness of the classifier to a considerable extent. Extensive experimentation made it possible to answer RQ2—some combinations are more effective than others and more effective than using all the preprocessors together.

Additionally, data augmentation is a booming area of research [62], and mixing preprocessing adversarial defences with novel approaches to data augmentation could potentially offset the performance loss of the researched defensive techniques, an approach which is part of future research.

Author Contributions: The authors contributed equally. All authors have read and agreed to the published version of the manuscript.

Funding: This work is funded under the SPARTA project, which has received funding from the European Union's Horizon 2020 research and innovation programme under grant agreement No. 830892.

Institutional Review Board Statement: Not applicable.

Informed Consent Statement: Not applicable.

Data Availability Statement: The dataset used in this study is the Large-Scale CebelFaces Attributes (CelebA) Dataset, available here: https://mmlab.ie.cuhk.edu.hk/projects/CelebA.html (accessed on 30 September 2021).

Acknowledgments: This work is funded under the SPARTA project, which has received funding from the European Union's Horizon 2020 research and innovation programme under grant agreement No. 830892.

Conflicts of Interest: The authors declare no conflicts of interest.

References

1. Fazal, M.I.; Patel, M.E.; Tye, J.; Gupta, Y. The past, present and future role of artificial intelligence in imaging. *Eur. J. Radiol.* **2018**, *105*, 246–250. doi:10.1016/j.ejrad.2018.06.020.
2. Ho-Phuoc, T. CIFAR10 to compare visual recognition performance between deep neural networks and humans. *arXiv* **2018**, arXiv:1811.07270.
3. Huang, G.; Liu, Z.; Van Der Maaten, L.; Weinberger, K.Q. Densely Connected Convolutional Networks. In Proceedings of the 2017 IEEE Conference on Computer Vision and Pattern Recognition (CVPR), Honolulu, HI, USA, 21–26 July 2017; pp. 2261–2269. doi:10.1109/CVPR.2017.243.
4. LeCun, Y.; Bengio, Y.; Hinton, G. Deep learning. *Nature* **2015**, *521*, 436–444.
5. Wang, G.; Yuan, Y.; Chen, X.; Li, J.; Zhou, X. Learning discriminative features with multiple granularities for person re-identification. In Proceedings of the 26th ACM international conference on Multimedia, Seoul, Korea, 22–26 October 2018; pp. 274–282.
6. Luo, H.; Gu, Y.; Liao, X.; Lai, S.; Jiang, W. Bag of tricks and a strong baseline for deep person re-identification. In Proceedings of the IEEE/CVF Conference on Computer Vision and Pattern Recognition Workshops, Long Beach, CA, USA, 16–17 June 2019; pp. 1487–1495.
7. Ye, M.; Shen, J.; Lin, G.; Xiang, T.; Shao, L.; Hoi, S.C. Deep Learning for Person Re-identification: A Survey and Outlook. *IEEE Trans. Pattern Anal. Mach. Intell.* **2021**, 1. doi:10.1109/TPAMI.2021.3054775.
8. Choraś, M.; Pawlicki, M.; Puchalski, D.; Kozik, R. Machine learning–the results are not the only thing that matters! what about security, explainability and fairness? In *International Conference on Computational Science*; Springer: Cham, Switzerland, 2020; pp. 615–628.
9. Grosse, K.; Manoharan, P.; Papernot, N.; Backes, M.; McDaniel, P. On the (Statistical) Detection of Adversarial Examples. *arXiv* **2017**, arXiv:1702.06280.

10. Papernot, N.; McDaniel, P.; Wu, X.; Jha, S.; Swami, A. Distillation as a Defense to Adversarial Perturbations Against Deep Neural Networks. In Proceedings of the 2016 IEEE Symposium on Security and Privacy (SP), San Jose, CA, USA, 22–26 May 2016. doi:10.1109/sp.2016.41.
11. Papernot, N.; McDaniel, P.; Jha, S.; Fredrikson, M.; Celik, Z.B.; Swami, A. The Limitations of Deep Learning in Adversarial Settings. In Proceedings of the 2016 IEEE European Symposium on Security and Privacy (EuroS&P), Saarbruecken, Germany, 21–24 March 2016. doi:10.1109/eurosp.2016.36.
12. Papernot, N.; McDaniel, P.; Goodfellow, I.; Jha, S.; Celik, Z.B.; Swami, A. Practical black-box attacks against machine learning. In Proceedings of the 2017 ACM on Asia Conference on Computer and Communications Security, Abu Dhabi, United Arab Emirates, 2–6 April 2017; pp. 506–519.
13. Srinivas, S.; Sarvadevabhatla, R.K.; Mopuri, K.R.; Prabhu, N.; Kruthiventi, S.S.; Babu, R.V. An Introduction to Deep Convolutional Neural Nets for Computer Vision. In *Deep Learning for Medical Image Analysis;* Elsevier: Cambridge, MA, USA, 2017; pp. 25–52. doi:10.1016/B978-0-12-810408-8.00003-1.
14. Anwar, A. Difference between AlexNet, VGGNet, ResNet, and Inception. 2019. Available online: https://towardsdatascience.com/the-w3h-of-alexnet-vggnet-resnet-and-inception-7baaaecccc96 (accessed on 10 February 2021).
15. Szegedy, C.; Liu, W.; Jia, Y.; Sermanet, P.; Reed, S.; Anguelov, D.; Erhan, D.; Vanhoucke, V.; Rabinovich, A. Going Deeper with Convolutions. *arXiv* **2014**, arXiv:1409.4842.
16. He, K.; Zhang, X.; Ren, S.; Sun, J. Deep residual learning for image recognition. In Proceedings of the IEEE Conference on Computer Vision and Pattern Recognition, Las Vegas, NV, USA, 27–30 June 2016; pp. 770–778.
17. Krizhevsky, A.; Sutskever, I.; Hinton, G.E. ImageNet classification with deep convolutional neural networks. *Commun. ACM* **2017**, *60*, 84–90. doi:10.1145/3065386.
18. Simonyan, K.; Zisserman, A. Very Deep Convolutional Networks for Large-Scale Image Recognition. *arXiv* **2014**, arXiv:1409.1556.
19. Thompson, N.C.; Greenewald, K.; Lee, K.; Manso, G.F. The Computational Limits of Deep Learning. *arXiv* **2020**, arXiv:2007.05558.
20. Kaya, A.; Keceli, A.S.; Catal, C.; Yalic, H.Y.; Temucin, H.; Tekinerdogan, B. Analysis of transfer learning for deep neural network based plant classification models. *Comput. Electron. Agric.* **2019**, *158*, 20–29. doi:10.1016/j.compag.2019.01.041.
21. Wang, B.; Yao, Y.; Viswanath, B.; Zheng, H.; Zhao, B.Y. With great training comes great vulnerability: Practical attacks against transfer learning. In Proceedings of the 27th {USENIX} Security Symposium ({USENIX} Security 18), Baltimore, MD, USA, 15–17 August 2018; pp. 1281–1297.
22. Davchev, T.; Korres, T.; Fotiadis, S.; Antonopoulos, N.; Ramamoorthy, S. An empirical evaluation of adversarial robustness under transfer learning. *arXiv* **2019**, arXiv:1905.02675.
23. Akhtar, N.; Mian, A. Threat of Adversarial Attacks on Deep Learning in Computer Vision: A Survey. *IEEE Access* **2018**, *6*, 14410–14430. doi:10.1109/ACCESS.2018.2807385.
24. Choraś, M.; Pawlicki, M.; Kozik, R. The feasibility of deep learning use for adversarial model extraction in the cybersecurity domain. In *International Conference on Intelligent Data Engineering and Automated Learning;* Springer: Cham, Switzerland, 2019; pp. 353–360.
25. Chakraborty, A.; Alam, M.; Dey, V.; Chattopadhyay, A.; Mukhopadhyay, D. Adversarial Attacks and Defences: A Survey. *arXiv* **2018**, arXiv:1810.00069.
26. Pawlicki, M.; Choraś, M.; Kozik, R. Defending network intrusion detection systems against adversarial evasion attacks. *Future Gener. Comput. Syst.* **2020**, *110*, 148–154.
27. Pitropakis, N.; Panaousis, E.; Giannetsos, T.; Anastasiadis, E.; Loukas, G. A taxonomy and survey of attacks against machine learning. *Comput. Sci. Rev.* **2019**, *34*, 100199.
28. Szegedy, C.; Zaremba, W.; Sutskever, I.; Bruna, J.; Erhan, D.; Goodfellow, I.; Fergus, R. Intriguing properties of neural networks. *arXiv* **2013**, arXiv:1312.6199.
29. Kurakin, A.; Goodfellow, I.; Bengio, S.; Dong, Y.; Liao, F.; Liang, M.; Pang, T.; Zhu, J.; Hu, X.; Xie, C.; et al. Adversarial Attacks and Defences Competition. In *The NIPS '17 Competition: Building Intelligent Systems;* Escalera, S., Weimer, M., Eds.; Springer International Publishing: Cham, Switzerland, 2018; pp. 195–231.
30. Liu, Z.; Luo, P.; Wang, X.; Tang, X. Deep Learning Face Attributes in the Wild. *arXiv* **2014**, arXiv:1411.7766.
31. Liu, Z.; Luo, P.; Wang, X.; Tang, X. Large-scale CelebFaces Attributes (CelebA) Dataset. 2016. Available online: https://mmlab.ie.cuhk.edu.hk/projects/CelebA.html (accessed on 10 February 2021).
32. Chakraborty, A.; Alam, M.; Dey, V.; Chattopadhyay, A.; Mukhopadhyay, D. A survey on adversarial attacks and defences. *CAAI Trans. Intell. Technol.* **2021**, *6*, 25–45.
33. Carlini, N.; Wagner, D. Adversarial examples are not easily detected: Bypassing ten detection methods. In Proceedings of the 10th ACM Workshop on Artificial Intelligence and Security, Dallas, TX, USA, 3 November 2017; pp. 3–14.
34. Uesato, J.; O'donoghue, B.; Kohli, P.; Oord, A. Adversarial risk and the dangers of evaluating against weak attacks. In Proceedings of the International Conference on Machine Learning. PMLR, 2018, Stockholm, Sweden, 10–15 July 2018; pp. 5025–5034.
35. Shafahi, A.; Huang, W.R.; Studer, C.; Feizi, S.; Goldstein, T. Are adversarial examples inevitable? *arXiv* **2018**, arXiv:1809.02104.
36. Athalye, A.; Carlini, N.; Wagner, D. Obfuscated Gradients Give a False Sense of Security: Circumventing Defenses to Adversarial Examples. In *Proceedings of Machine Learning Research, Proceedings of the 35th International Conference on Machine Learning, PMLR 2018, Stockholm, Sweden, 10-15.07.2018;* Dy, J., Krause, A., Eds.; JMLR: Cambridge, MA, USA, 2018; Volume 80, pp. 274–283.

37. Carlini, N.; Wagner, D. Towards Evaluating the Robustness of Neural Networks. In Proceedings of the 2017 IEEE Symposium on Security and Privacy (SP), San Jose, CA, USA, 22–26 May 2017; pp. 39–57. doi:10.1109/SP.2017.49.
38. Carlini, N.; Athalye, A.; Papernot, N.; Brendel, W.; Rauber, J.; Tsipras, D.; Goodfellow, I.; Madry, A.; Kurakin, A. On evaluating adversarial robustness. *arXiv* **2019**, arXiv:1902.06705.
39. de Mello, F.L. A survey on machine learning adversarial attacks. *J. Inf. Secur. Cryptogr. (Enigm.)* **2020**, *7*, 1–7.
40. Yuan, X.; He, P.; Zhu, Q.; Li, X. Adversarial Examples: Attacks and Defenses for Deep Learning. *IEEE Trans. Neural Netw. Learn. Syst.* **2019**, *30*, 2805–2824. doi:10.1109/TNNLS.2018.2886017.
41. Sadeghi, K.; Banerjee, A.; Gupta, S.K. A system-driven taxonomy of attacks and defenses in adversarial machine learning. *IEEE Trans. Emerg. Top. Comput. Intell.* **2020**, *4*, 450–467.
42. Wiyatno, R.R.; Xu, A.; Dia, O.; de Berker, A. Adversarial examples in modern machine learning: A review. *arXiv* **2019**, arXiv:1911.05268.
43. Serban, A.; Poll, E.; Visser, J. Adversarial examples on object recognition: A comprehensive survey. *ACM Comput. Surv. (CSUR)* **2020**, *53*, 1–38.
44. Xu, H.; Ma, Y.; Liu, H.C.; Deb, D.; Liu, H.; Tang, J.L.; Jain, A.K. Adversarial attacks and defenses in images, graphs and text: A review. *Int. J. Autom. Comput.* **2020**, *17*, 151–178.
45. Parkhi, O.M.; Vedaldi, A.; Zisserman, A. Deep Face Recognition. In Procedings of the British Machine Vision Conference 2015, Swansea, UK, 7–10 September 2015; British Machine Vision Association: Durham, UK, 2015; pp. 41.1–41.12. doi:10.5244/C.29.41.
46. ResNet-50. ResNet-50 Pre-trained Model for Keras. 2017. Available online: https://www.kaggle.com/keras/resnet50 (accessed on 10 February 2021).
47. Du, J. High-Precision Portrait Classification Based on MTCNN and Its Application on Similarity Judgement. *J. Phys. Conf. Ser.* **2020**, *1518*, 12066. doi:10.1088/1742-6596/1518/1/012066.
48. Xiang, J.; Zhu, G. Joint Face Detection and Facial Expression Recognition with MTCNN. In Proceedings of the 2017 4th International Conference on Information Science and Control Engineering (ICISCE), Changsha, China, 21–23 July 2017; pp. 424–427. doi:10.1109/ICISCE.2017.95.
49. Zhang, K.; Zhang, Z.; Li, Z.; Qiao, Y. Joint Face Detection and Alignment using Multi-task Cascaded Convolutional Networks. *IEEE Signal Process. Lett.* **2016**, *23*, 1499–1503. doi:10.1109/LSP.2016.2603342.
50. Madry, A.; Makelov, A.; Schmidt, L.; Tsipras, D.; Vladu, A. Towards Deep Learning Models Resistant to Adversarial Attacks. *arXiv* **2017**, arXiv:1706.06083.
51. Qiu, S.; Liu, Q.; Zhou, S.; Wu, C. Review of Artificial Intelligence Adversarial Attack and Defense Technologies. *Appl. Sci.* **2019**, *9*, 909. doi:10.3390/app9050909.
52. Bai, T.; Luo, J.; Zhao, J.; Wen, B.; Wang, Q. Recent Advances in Adversarial Training for Adversarial Robustness. *arXiv* **2021**, arXiv:2102.01356.
53. Dziugaite, G.K.; Ghahramani, Z.; Roy, D.M. A study of the effect of JPG compression on adversarial images. *arXiv* **2016**, arXiv:1608.00853.
54. Das, N.; Shanbhogue, M.; Chen, S.T.; Hohman, F.; Chen, L.; Kounavis, M.E.; Chau, D.H. Keeping the Bad Guys Out: Protecting and Vaccinating Deep Learning with JPEG Compression. *arXiv* **2017**, arXiv:1705.02900.
55. Zantedeschi, V.; Nicolae, M.I.; Rawat, A. Efficient Defenses Against Adversarial Attacks. *arXiv* **2017**, arXiv:1707.06728.
56. Xu, W.; Evans, D.; Qi, Y. Feature Squeezing: Detecting Adversarial Examples in Deep Neural Networks. *arXiv* **2017**, arXiv:1704.01155. doi:10.14722/ndss.2018.23198.
57. Guo, C.; Rana, M.; Cisse, M.; van der Maaten, L. Countering Adversarial Images using Input Transformations. *arXiv* **2017**, arXiv:1711.00117.
58. Ahn, B.; Kim, Y.; Park, G.; Cho, N.I. Block-Matching Convolutional Neural Network (BMCNN): Improving CNN-Based Denoising by Block-Matched Inputs. In Proceedings of the 2018 Asia-Pacific Signal and Information Processing Association Annual Summit and Conference (APSIPA ASC), Honolulu, HI, USA, 12–15 November 2018; pp. 516–525, doi:10.23919/APSIPA.2018.8659548.
59. Creswell, A.; Bharath, A.A. Denoising Adversarial Autoencoders. *arXiv* **2017**, arXiv:1703.01220.
60. Xu, J.; Zhang, L.; Zuo, W.; Zhang, D.; Feng, X. Patch Group Based Nonlocal Self-Similarity Prior Learning for Image Denoising. In Proceedings of the 2015 IEEE International Conference on Computer Vision (ICCV), Santiago, Chile, 1–13 December 2015; pp. 244–252, doi:10.1109/ICCV.2015.36.
61. Zhang, K.; Zuo, W.; Chen, Y.; Meng, D.; Zhang, L. Beyond a Gaussian Denoiser: Residual Learning of Deep CNN for Image Denoising. *IEEE Trans. Image Process.* **2016**, *26*, 3142–3155. doi:10.1109/TIP.2017.2662206.
62. Salazar, A.; Vergara, L.; Safont, G. Generative Adversarial Networks and Markov Random Fields for oversampling very small training sets. *Expert Syst. Appl.* **2021**, *163*, 113819. doi:10.1016/j.eswa.2020.113819.

entropy

MDPI

Article

Entropy-Based Combined Metric for Automatic Objective Quality Assessment of Stitched Panoramic Images

Krzysztof Okarma [1,*], Wojciech Chlewicki [1], Mateusz Kopytek [1], Beata Marciniak [2] and Vladimir Lukin [3]

[1] Department of Signal Processing and Multimedia Engineering, West Pomeranian University of Technology in Szczecin, 70-313 Szczecin, Poland; wojciech.chlewicki@zut.edu.pl (W.C.); km46880@zut.edu.pl (M.K.)

[2] Faculty of Telecommunications, Computer Science and Electrical Engineering, Bydgoszcz University of Science and Technology, 85-796 Bydgoszcz, Poland; beata.marciniak@pbs.edu.pl

[3] Department of Information and Communication Technologies, National Aerospace University, 61070 Kharkov, Ukraine; lukin@ai.kharkov.com

* Correspondence: okarma@zut.edu.pl

Abstract: Quality assessment of stitched images is an important element of many virtual reality and remote sensing applications where the panoramic images may be used as a background as well as for navigation purposes. The quality of stitched images may be decreased by several factors, including geometric distortions, ghosting, blurring, and color distortions. Nevertheless, the specificity of such distortions is different than those typical for general-purpose image quality assessment. Therefore, the necessity of the development of new objective image quality metrics for such type of emerging applications becomes obvious. The method proposed in the paper is based on the combination of features used in some recently proposed metrics with the results of the local and global image entropy analysis. The results obtained applying the proposed combined metric have been verified using the ISIQA database, containing 264 stitched images of 26 scenes together with the respective subjective Mean Opinion Scores, leading to a significant increase of its correlation with subjective evaluation results.

Keywords: image quality assessment; stitched images; panoramic images; image analysis; image entropy

Citation: Okarma, K.; Chlewicki, W.; Kopytek, M.; Marciniak, B.; Lukin, V. Entropy-Based Combined Metric for Automatic Objective Quality Assessment of Stitched Panoramic Images. *Entropy* **2021**, *23*, 1525. https://doi.org/10.3390/e23111525

Academic Editors: Michal Choras, Robert Burduk, Agata Giełczyk, Rafal Kozik and Tomasz Marciniak

Received: 8 October 2021
Accepted: 15 November 2021
Published: 17 November 2021

1. Introduction

Panoramic images, constructed as a result of image stitching operation conducted for a series of constituent images with partially overlapping regions, may suffer from various distortions, including blur, ghosting artifacts, and quite well visible geometric and color distortions. The presence of such issues decreases the perceived image quality and in some cases may be unacceptable from an aesthetic point of view. Although modern cameras and smartphones are usually equipped with software functions making it possible to properly register the overlapping areas of individual photos to create panoramic images, some additional requirements should be fulfilled by users during the recording to prevent such problems. Nevertheless, the growing availability of software and hardware solutions causes higher popularity of panoramic images which may be useful, e.g., as wide background images, in virtual reality scenarios, as well as in mobile robotics for the Visual Simultaneous Localization and Mapping (VSLAM) applications.

Considering the modern applications of image stitching and image registration algorithms, related to the use of cameras mounted on mobile robots, the quality of obtained panoramic images is very important due to potential errors in vision-based control of their motion. In the case of decreased image quality, such images might be removed from the analysis to prevent their influence on the robot's control. Another interesting direction of such research in mobile robotics concerns the fusion of images acquired by unmanned aerial vehicles (UAVs) [1,2].

One of the most relevant factors, influencing the final quality of the panoramic images, is the appropriate choice of distinctive image features used to match the same regions visible on the "neighboring" constituent images. Nevertheless, some additional post-processing operations conducted after the assignment, such as blending and interpolation, may also have a significant impact on the quality of stitched images. Some obvious examples might be related to different lighting conditions and background changes visible on the constituent images which may cause some easily noticeable seams. Another factor, related to geometric distortions, is the influence of lens imperfections and a too low number of detected keypoints used for further image matching, particularly for constituent images with overlapping areas less than 15–20% of the image area. Although some corrections, e.g., calibration, chromatic aberration or vignetting corrections, may be conducted using both freeware and commercial software for image stitching, even after the final blending some imperfections may still be visible. Since a synchronous acquisition of constituent images using multiple cameras may be troublesome in many practical applications, some problems may also occur for moving objects, particularly leading to motion blur and ghosting artifacts.

Although during several recent years a great progress has been made in general-purpose image quality assessment (IQA), the direct application of those methods proposed by various researches for an automatic objective evaluation of stitched images is troublesome, or even impossible. This situation is caused by significant differences between the most common types of distortions and those which may be found in stitched images. Therefore, the development of stitched images quality assessment methods is limited by the availability of the image databases containing panoramic images subject to different types of distortions typical for image stitching together with subjective quality scores. Since the first attempts to such quality metrics have not been verified using such datasets, there is a need of their additional verification, as well as the analysis of their usefulness in the combination with some other approaches.

Such experiments are possible with the use of the Indian Institute of Science Stitched Image Quality Assessment (ISIQA) dataset consisting of 264 stitched images and 6600 human quality ratings. One of the methods recently proposed for quality assessment of stitched panoramic images, verified using the ISIQA database, is the Stitched Image Quality Evaluator (SIQE) proposed by the authors of this dataset [3]. This method utilized a comparison of 36 features calculated for the constituent and stitched images, namely the eigenvalues determined for the covariance matrix where the covariances are computed for each patch for the pair of wavelet coefficients for a bivariate distribution determined from the Gaussian Mixture Model (GMM) and shape parameters of the Generalized Gaussian Distribution (GGD). The application of the bivariate statistics for the GMM is useful for detection of the correlation caused by ghosting artifacts, whereas the shape parameters of the GGD represent features sensitive to geometric distortions caused by presence of additional edges as well as blur [3]. A more detailed description of the SIQE metric is presented in Section 2.2. Nevertheless, the authors of the SIQE method have used only randomly selected 20% of stitched images for testing whereas 80% of the images have been used for training. Therefore, the reported relatively high correlation results should be considered as harder to obtain for the whole database due to the higher number of images and therefore such overall correlation is significantly decreased [4].

One of the methods for the increase of the correlation of objective metrics with subjective quality evaluation results is the application of the combined metrics, successfully applied for general-purpose IQA [5,6], multiply distorted images, remote sensing [7], as well as for the quality evaluation of the 3D printed surfaces [8]. Although such methods cannot be directly applied for the stitched images, the general idea of the combination of various metrics is worth investigating, leading to promising results as presented in the further sections of the paper.

The motivation for the combination of the entropy-based features with some existing metrics has been related to the observed increase of the local image entropy for the regions containing some kinds of distortions typical for the stitched images. According to expectation, an increase of the global image entropy for lower quality images may also be observed. Nevertheless, as the image entropy is highly dependent on the image contents, a more useful approach is the comparison of the entropy-based features calculated for the constituent and the stitched images in a similar way as for 36 features extracted in the SIQE framework [3]. Hence, the additional entropy-based features may be added after the SVR step and combined with the SIQE values, as well as with some additional features or sub-metrics. The additional use of the variance of the local entropy and two additional features originating from the MIQM metric [9,10], leading to the extension of the idea initially verified in the paper [4], makes it possible to increase the correlations with subjective scores significantly, as presented in the further part of the paper.

2. Materials and Methods

2.1. Overview of Methods for Stitched Image Quality Assessment

Objective image quality assessment methods may be generally divided into full-reference (FR) and no-reference (NR) methods. The latter group also referred to as "blind" metrics, seems to be more attractive for many applications since FR metrics require the full knowledge of the reference (undistorted) images. Since such "pristine" images are often unavailable, a partial solution for this problem might be the use of the reduced-reference methods where the knowledge of some characteristic parameters or features of the original image is necessary.

Nevertheless, the FR quality assessment of the stitched images should be considered in another way since perfect quality panoramas are usually unavailable, however, there is still a possibility of some comparisons with constituent images that are typically at the disposal. Therefore, the stitched image quality assessment may be considered as an indirect assessment of the quality of the applied stitching method. In view of these assumptions, these methods cannot be directly classified as "purely" FR or NR IQA algorithms, since they use the data from additional (constituent) images but do not utilize any direct comparisons of the distorted panoramas with the "pristine" stitched images.

One of the first interesting approaches to stitched IQA is based on the attempt of using the well-known Structural Similarity (SSIM) method [11] examined by Qureshi et al. [12]. In this method, the SSIM has been used for comparisons of the high-frequency data, e.g., concerning the edges, in overlapping regions of constituent and stitched images, leading to the HFI_SSIM metric. Additionally low-frequency information is used in this metric to assess the photometric quality of the panorama image using the spectral angle mapper (based on the angle between two vectors representing pixels' colors in the RGB color space) and intensity magnitude ratio measures.

Color correction and balancing in the image and video stitching has also been investigated in the paper [13], whereas the mosaicking performance has been examined by Paalanen et al. [14]. A classification of color correction methods for image stitching can be made using the framework proposed by Bellavia and Colombo [15] who utilize well-known Feature Similarity (FSIM) metric [16] together with the improved Color Image Difference (iCID) measure [17] to assess the quality. Another idea, useful for the analysis of color inconsistency, has been proposed by Niu et al. [18] and is based on the calculation of the color contrast similarity and the color value difference.

The application of the local variance of optical flow field energy between the distorted and reference images has been combined with the intensity and chrominance gradient calculations in highly-structured patches by Cheung et al. [19], allowing mainly for the measurements of the geometric and structure errors.

Unfortunately, regardless of their popularity and good results obtained in some other applications, some data-driven quality assessment methods cannot be successfully applied for the quality assessment of stitched images due to the necessity of training with the use of a great number of images [20]. Some recent examples might be Generated Image Quality Assessment (GIQA) [21] or Face Forensics in the Wild [22] but these methods focus on the evaluation of generated images with distortions different than typical for stitched images or face forgery detection being related to classification rather than quality assessment.

Although similar methods might be successfully applied for the general-purpose IQA, it should be kept in mind that for the general-purpose IQA several large-scale databases containing the subjective quality scores are available, differently than for the evaluation of stitched images limited to the use of the ISIQA dataset [3].

One of the methods partially utilized in this paper has been proposed by Solh and AlRegib [9,10] who have developed the Multi-view Image Quality Measure (MIQM) consisting of luminance, contrast, spatial motion, and edge-based structure components. A more detailed description of the application of its simplified version used in our experiments is provided in the further part of the paper (Section 2.3).

2.2. The SIQE Metric

As mentioned earlier, one of the most interesting approaches to quality assessment of panoramic stitched images has been recently developed by the inventors of the ISIQA database [3]. The main assumption of the SIQE framework is the use of 36 features divided into two sets. The first set is sensitive to structural changes visible as blurring or changes of edges, whereas the second one captures the distortions caused by variations of the spatial correlation caused by ghosting artifacts. The image-level features are determined as the weighted average of the local patch-level features (calculated for 100×100 pixels patches), further used to predict the final quality score using the Support Vector Machine (SVM) regression.

The detection of ghosting artifacts observed as some additional edges or replications of some image regions, caused by imprecise aligning of the overlapping regions of constituent images during the stitching procedure, is based on the use of the multi-scale multi-orientation decomposition. For this purpose, the use of the steerable pyramids has been proposed by the authors of the paper [3], who have used 2 scales and 6 orientations to decompose the image into 12 subbands. Then, three groups of features may be determined for these subbands both for constituent and stitched images. Fitting a GGD model to subband coefficients their shape coefficients may be determined as the first group of 12 pairs of features. Using the bivariate statistics based on the GMM model, the additional 24 features, representing the covariance values for pairs of wavelet coefficients for the horizontal and vertical neighborhood, may be calculated for the stitched and constituent images, accordingly. These features may be expressed as the eigenvalues of the bivariate distribution [3]. Finally, the differences of all 36 features extracted from the stitched image and the corresponding constituent images (denoted as f^s_{1-36} and f^c_{1-36}) are calculated, being the input for the SVM regression procedure.

Although the Pearson's Linear Correlation Coefficient (PLCC) for the ISIQA dataset is equal to 0.8395 with Spearman Rank Order Correlation SROCC = 0.8318 reported in [3], these results have been obtained for 1000 iterations of randomly chosen train and test sets, using only 20% of images for testing. Unfortunately, applying this metric for the whole ISIQA dataset, significantly lower values of the PLCC = 0.7488 and SROCC = 0.7057 may be achieved [4].

Considering the necessity of the use of a large amount of the ground truth data for training to avoid overfitting of the trained model, there is a limited possibility of application of the deep CNN-based methods, as stated by Hou et al. [20]. For this reason, regardless of the popularity of the deep learning methods, an interesting direction of research seems to be the development of combined metrics, utilizing the SIQE method and some other approaches based on handcrafted features.

2.3. The Simplified MIQM Implementation

Although the fundamental element for our research is the SIQE metric, its extension towards a combined metric requires an implementation of some additional metrics and calculation of additional features, as well as their further optimization making it possible to increase the correlation with subjective quality scores.

Two additional sub-measures have been incorporated from [9] for this study, i.e., luminance and contrast index, and edge-based structural index, being the elements of the MIQM. The first one is focused on recognizing sharp local changes in luminance and contrast around structured regions. The computation formula was derived and adjusted in [10] to provide higher weights for structured regions. It is mathematically expressed in the following way

$$k_{x,y}(I,J) = \frac{4(\sigma_I \sigma_J) \cdot (\mu_I \mu_J) + C}{(\sigma_I^2 + \sigma_J^2) \cdot (\mu_I^2 + \mu_J^2) + C} \tag{1}$$

where (x, y) is the coordinate of the upper left corner of the macroblock. The mean intensity is denoted as μ, and the standard deviation as σ, respectively. Both σ and μ are computed for the macroblocks of the dimension $s \times s$. In our study, we have set s to 21, which is a trade-off between a reasonable computation time and accuracy. The subscript I denotes the reference image whereas J stands for the distorted image, and C is a constant added to prevent instability in case of the denominator value being close to zero.

To compute the overall quality index a weighted average of luminance and contrast index of each macroblock should be used. The weights' values are obtained based on the reference image in the following way. First, the texture randomness index at macroblock (x, y) of the image I has to be computed using the formula [23]

$$t(x,y) = E_I(x,y) \times M_I(x,y) \tag{2}$$

where E_I is an edge intensity binary image with values equal to 1 where the function recognizes edges, and values 0 elsewhere, with M_I being the mean intensity of I. These values have been computed for the same non-overlapping macroblocks as previously. Finally, the texture randomness index has been mapped to the object index in the following way

$$T(x,y) = \begin{cases} K_1 + \left(0.5 \times K_1 \times \frac{\log_2 t(x,y)}{\log_2 \beta_1}\right) & \beta_1 \leq t(x,y) < \beta_2 \\ K_2 + \left(0.5 \times K_2 \times 2^{-(t(x,y)-\beta_2)}\right) & t(x,y) \geq \beta_2 \\ K_1 & \text{otherwise} \end{cases} \tag{3}$$

where K_1 and K_2 are the constant parameters that control the weights assigned to the structured regions and randomly assigned regions, accordingly. If K_1 is much larger than K_2, then higher weights are assigned to the structured regions. Parameters β_1 and β_2 are the edge detector thresholds. Such computed $T(x, y)$ is employed to compute both sub-measures according to the Formulas (4) and (5).

The luminance and contrast index for $M \times N$ macroblocks may be calculated as

$$K(I,J) = \frac{\sum_{x=1}^{M} \sum_{y=1}^{N} k_{x,y}(I,J) \times T(x,y)}{\sum_{x=1}^{M} \sum_{y=1}^{N} T(x,y)} \tag{4}$$

whereas the edge-based structural index for $M \times N$ macroblocks is defined as

$$E(I,J) = \frac{\sum_{x=1}^{M} \sum_{y=1}^{N} \left(1 - \left\| \frac{T_{x,y}(I) - T_{x,y}(J))}{T_{x,y}(I)} \right\|\right)}{M \times N}. \tag{5}$$

The values of $K(I, J)$ and $E(I, J)$ are close to 1 for minimum distortions and consequently values almost 0 for maximum distortions.

In our study, the reference image has been a region of interest (ROI) selected from each constituent image and the corresponding ROI found in the evaluated stitched image. All the formulas have been implemented as MATLAB functions. Instead of the third MIQM term, namely spatial motion index, partially utilizing the local entropy, we have used the additional global and local entropy-based features, leading to the increase of the proposed metric's correlation with subjective MOS values.

2.4. The Proposed Entropy-Based Approach

The initial experiments, conducted using 264 stitched images obtained for 26 scenes that are included in the ISIQA dataset (sample images are shown in Figure 1) as well as some additional stitched images generated using the freeware Hugin software with various parameters, have demonstrated the potential improvements of existing metrics caused by their diversity.

Figure 1. Sample constituent and stitched images with various distortions from the ISIQA dataset.

Assuming the usefulness of the image entropy, reflecting the amount of information in an image, the first experiments have been made utilizing the global entropy values calculated for the image X according to the well-known formula:

$$ent^X_{global} = -\sum(p \cdot \log_2(p)) \; , \qquad (6)$$

where p contains the normalized histogram counts determined for the image X using 256 bins, as well as the local entropy values calculated using the same Formula (6) for the 9×9 pixels neighborhood of the specified pixel from the image X.

Therefore, the additional entropy-based features (added after the SVR step and further combined with the SIQE values) are defined as [4]:

$$\overline{ent}_{local} = \overline{ent^c}_{local} - \overline{ent^s}_{local} \; , \qquad (7)$$

and

$$ent_{global} = ent^c_{global} - ent^s_{global} \; , \qquad (8)$$

where the average local entropy values $\overline{ent}_{local}$ and the global entropy values ent_{global} for constituent (c) and stitched (s) images are subtracted, respectively. Regardless of these two differential features, their equivalents for the constituent and stitched images may be independently analyzed as well.

After the experimental verification of the possible combinations, the initially considered combined metrics, referred to as EntSIQE, have been defined in two variants based on the weighted sum and weighted product [4]:

$$\text{EntSIQE}_1 = \alpha \cdot \text{SIQE} + \beta \cdot ent_{global} + \gamma \cdot \overline{ent^s}_{local} \; , \qquad (9)$$

and

$$\text{EntSIQE}_2 = (\text{SIQE})^{\alpha} \cdot \left(ent_{global}\right)^{\beta} \cdot \left(\overline{ent^s}_{local}\right)^{\gamma} \; , \qquad (10)$$

where the parameters α, β, and γ may be optimized (independently for each of the above formulas) to provide the highest correlation with the MOS values for the specified database, e.g., the ISIQA database as used in this paper. Since the use of the averaged $\overline{ent}_{local}$ features, calculated only for the stitched images, has provided better results than the use of the differences for the constituent and stitched images, only the ent_{global} features have been calculated as the difference of features for the constituent and stitched images.

The additional extension of these metrics with the use of two indexes, originating from the MIQM approach (Equations (4) and (5)), described in Section 2.3, may be conducted in the same way, leading to the finally proposed metrics referred to as EntSIQE_1^+ and EntSIQE_2^+. Nevertheless, applying the Formulas (4) and (5), two vectors containing the similarity indexes between the ROIs extracted from the stitched image and each corresponding constituent image may be obtained. Therefore, for the images from the ISIQA database, depending on the number of constituent images (four in two sets and five in the other cases), 8 or 10 values may be obtained in aggregate. Considering the use of the maximum, minimum, average and median values for $K(I,J)$ and $E(I,J)$ indexes, the best results have been obtained using the median values.

The additional useful feature, leading to a further increase of the correlation of the proposed metric with subjective scores is the variance of the local entropy that may be calculated subtracting the averaged variances determined for the constituent and stitched images according to:

$$\overline{var}_{ent} = \overline{var}(ent^c{}_{local}) - \overline{var}(ent^s{}_{local}) \; . \qquad (11)$$

Hence, the final formulas may be expressed as:

$$\begin{aligned}\mathrm{EntSIQE}_1^+ &= a_1 \cdot (\mathrm{SIQE})^{w_1} + a_2 \cdot \left(ent_{global}\right)^{w_2} + a_3 \cdot \left(\overline{ent^s}_{local}\right)^{w_3} \\ &+ a_4 \cdot \left(\overline{var}_{ent}\right)^{w_4} + a_5 \cdot (med(K(I,J)))^{w_5} + a_6 \cdot (med(E(I,J)))^{w_6}\end{aligned}, \tag{12}$$

and

$$\begin{aligned}\mathrm{EntSIQE}_2^+ &= (\mathrm{SIQE})^{w_1} \cdot \left(ent_{global}\right)^{w_2} \cdot \left(\overline{ent^s}_{local}\right)^{w_3} \\ &\cdot \left(\overline{var}_{ent}\right)^{w_4} \cdot (med(K(I,J)))^{w_5} \cdot (med(E(I,J)))^{w_6}\end{aligned}. \tag{13}$$

It is worth to note that the necessity of the use of additional weighting exponents in Formula (12) in comparison to the Formula (9) is caused by different dynamic ranges of individual features used in the proposed metrics. Nevertheless, the application based on the weighted product might be a better choice due to the use of only six weighting coefficients (w_1–w_6).

3. Results and Discussion

To verify the correlation between the objective and subjective quality scores for the 264 images from the ISIQA database, three correlation metrics being the most typical in the IQA research, have been used.

Pearson's Linear Correlation Coefficient (PLCC) between the objective metric Q the Mean Opinion Score (MOS) values, illustrating the prediction accuracy of the image quality, is defined as the ratio of the covariance to the product of the standard deviations:

$$r = \frac{\mathrm{cov}(Q, MOS)}{\sigma_Q \cdot \sigma_{MOS}}. \tag{14}$$

It should be noted that in many IQA related papers, the additional nonlinear regression is applied, usually with the use of the logistic function, according to the recommendations of the Video Quality Experts Group (VQEG). Nevertheless, in the case of the combined metrics, it does not lead to meaningful changes of the correlation coefficients (differences are typically below 0.003) due to the nonlinear combination of various features. This has also been verified experimentally both for the original SIQE metric as well as for all the proposed combinations.

To verify the prediction monotonicity, two rank-order correlations may be applied. Spearman's Rank Order Correlation Coefficient (SROCC) is given as:

$$\rho = 1 - \frac{6 \cdot \sum d_i^2}{n \cdot (n^2 - 1)}, \tag{15}$$

where d_i stands for the difference between the ranks of corresponding images in two sets sorted according to objective (Q) and subjective (MOS) quality scores and n denotes the number of images. The second one is Kendall Rank Order Correlation Coefficient (KROCC) expressed as:

$$\tau = 2 \cdot \frac{n_c - n_d}{n \cdot (n - 1)}, \tag{16}$$

where n_c and n_d are the number of concordant and discordant, respectively, that are considered as the pairs of images ordered in the same way and reversely.

The calculations of all parameters as well as the optimizations have been conducted in MATLAB environment. For the optimization of exponential parameters w_i as well as the multipliers a_i the derivative-free method without constraints based on the Nelder–Mead simplex method has been used in the version implemented in MATLAB's *fminsearch* function with additional verification of the local minima.

The obtained results for the original SIQE metric as well as for the initially considered and finally proposed combined metrics are presented in Tables 1 and 2 as well as on the scatter plots shown in Figure 2. Since some comparisons of the original SIQE with the older metrics (i.a. NIQE, BRISQUE, DIVIINE or HFI_SSIM [12]), presented in the paper [3], have demonstrated its significantly better performance with correlation's increase by over 0.2 for the same ISIQA dataset, the analysis in this paper is limited to the comparison to the state-of-the-art SIQE metric to increase the clarity of presented results.

As can be easily noticed much more linear relation between the proposed objective metrics and MOS values can be observed in comparison to the original SIQE metric. Analyzing the number and location of outliers, most of them are located closer to the linear trend visible on the scatter plots for the proposed metrics. The values of the parameters obtained for the proposed combined metrics are presented in Table 3.

Table 1. Correlations with subjective scores obtained for the ISIQA database for the initially considered metrics and their parameters [4].

Metric	Parameters			Correlation with MOS		
	α	β	γ	PLCC	SROCC	KROCC
SIQE	-	-	-	0.7488	0.7057	0.5308
$EntSIQE_1$	0.3592	0.7938	2.3176	0.8012	0.7920	0.5971
$EntSIQE_2$	0.8597	−0.0036	0.4579	0.8101	0.7945	0.5990

Table 2. Correlations with subjective scores obtained for the ISIQA database using the proposed combined metrics and during the ablation study assuming the removal of one of the elementary metrics or features.

Metric	Correlation with MOS		
	PLCC	SROCC	KROCC
$EntSIQE_1^+$ (proposed)	0.8338	0.8338	0.6418
$EntSIQE_1^+$ with $a_1 = 0$	0.2869	0.2859	0.1833
$EntSIQE_1^+$ with $a_2 = 0$	0.8319	0.8326	0.6401
$EntSIQE_1^+$ with $a_3 = 0$	0.8283	0.8267	0.6341
$EntSIQE_1^+$ with $a_4 = 0$	0.8288	0.8264	0.6334
$EntSIQE_1^+$ with $a_5 = 0$	0.8326	0.8335	0.6417
$EntSIQE_1^+$ with $a_6 = 0$	0.8295	0.8250	0.6312
$EntSIQE_2^+$ (proposed)	0.8337	0.8341	0.6432
$EntSIQE_2^+$ with $w_1 = 0$	0.2652	0.2870	0.1885
$EntSIQE_2^+$ with $w_2 = 0$	0.8336	0.8336	0.6424
$EntSIQE_2^+$ with $w_3 = 0$	0.8128	0.8037	0.6130
$EntSIQE_2^+$ with $w_4 = 0$	0.8282	0.8225	0.6296
$EntSIQE_2^+$ with $w_5 = 0$	0.8333	0.8311	0.6399
$EntSIQE_2^+$ with $w_6 = 0$	0.8273	0.8249	0.6309

Table 3. Parameters obtained for the ISIQA database for the newly proposed metrics.

Metric	Parameters					
	w_1	w_2	w_3	w_4	w_5	w_6
$EntSIQE_1^+$	0.9010	1.8909	1.534×10^{-4}	1.4566	9.359×10^{-5}	2.438×10^{-4}
$EntSIQE_2^+$	1.1574	−0.0037	0.5554	0.0262	0.1271	1.1078
	a_1	a_2	a_3	a_4	a_5	a_6
$EntSIQE_1^+$	1.391×10^{-5}	1.582×10^{-4}	3.1237	2.079×10^{-4}	0.9362	1.6689

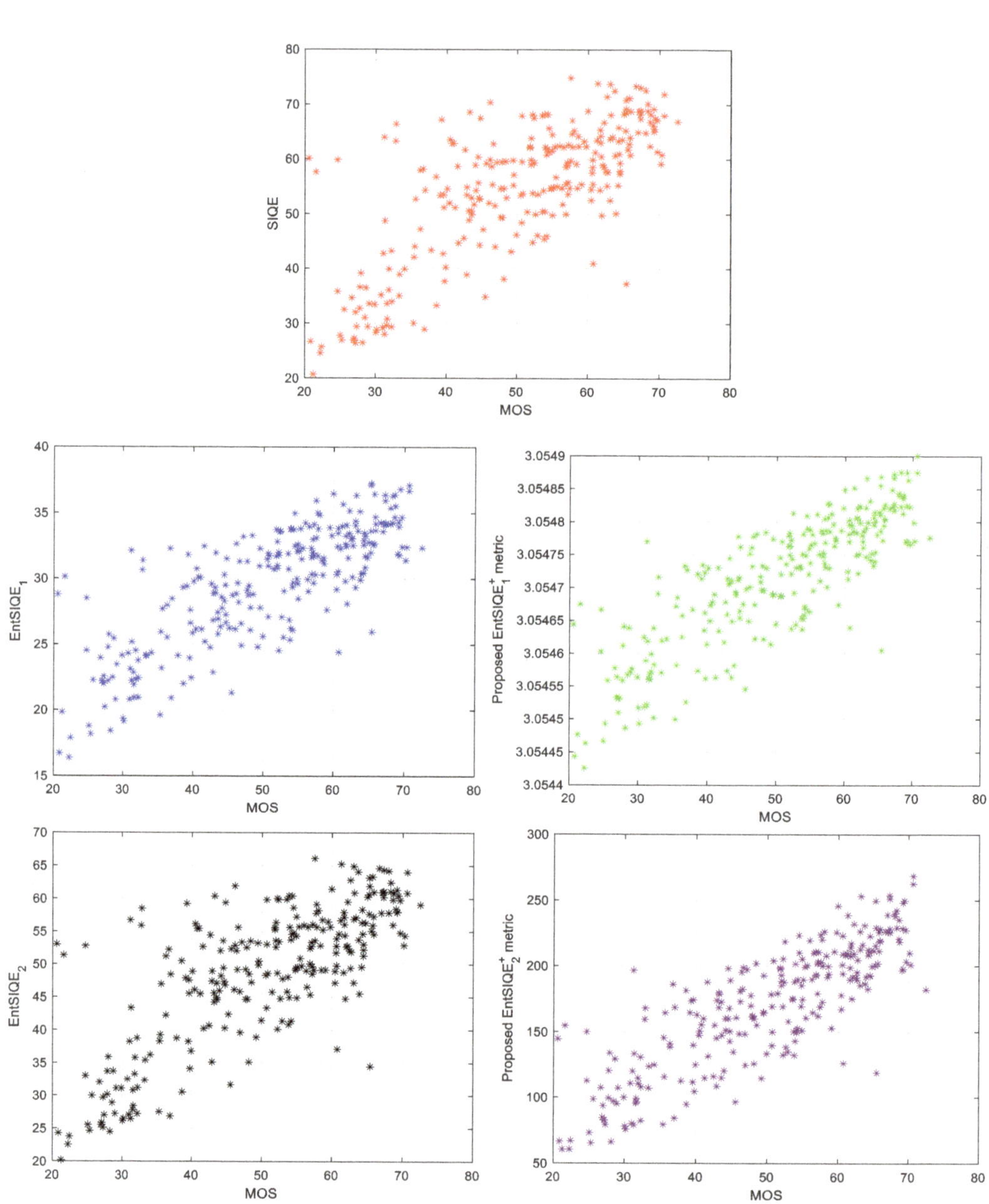

Figure 2. Scatter plots for SIQE (red points), two variants of the EntSIQE metric (blue and black points), and the proposed two variants of the EntSIQE$^+$ metric (green and violet points).

Analyzing the obtained results, a significant increase of the correlation with subjective scores may be observed for the proposed approach. Since the values of the parameters used for all six metrics or features are not close to zero (for the EntSIQE$_1^+$ metric there is no pair of parameters a_i and w_i being close to zero), a removal of any of the parts of the combined metric would decrease the correlation of the combined metric with the MOS values. To illustrate this, the results of the ablation study (with independent optimization) are presented in Table 2 for the six versions of the 5-element combined metrics (with removed one of the components). As it may be easily noticed, in both cases the most relevant element is undoubtedly the original SIQE metric. Nevertheless, only a slightly lower correlation with subjective scores may be achieved without the use of the global entropy or median values of $(K(I,J))$ calculated according to Formula (4).

4. Conclusions

The extensions of the recently proposed SIQE metric towards the combined metric proposed in the paper make it possible to achieve considerably higher correlation of the designed objective metrics with subjective quality scores of the stitched images delivered in the ISIQA database. The obtained results are promising and confirm the usefulness of the combined metrics also for the automatic quality assessment of the stitched panoramic images. As shown in the ablation study, the application of the additional entropy-based features utilizing the local image entropy and its variance is one of the crucial elements increasing the correlation with the MOS values, since their removal causes the most significant decrease of the PLCC, SROCC and KROCC values, obviously with the exception of the original SIQE metric.

One of the potential directions of further research might be the application of the proposed approach for quality assessment of parallax-tolerant image stitching methods [24] as well as the validation of the proposed approach for some other types of images and video sequences, containing similar types of distortions, also using some other combination models.

Author Contributions: Conceptualization, K.O. and W.C.; methodology, K.O. and V.L.; software, K.O. and W.C.; validation, K.O., B.M. and V.L.; formal analysis, K.O. and W.C.; investigation, K.O. and W.C.; resources, K.O. and M.K.; data curation, K.O., M.K. and B.M.; writing—original draft preparation, K.O. and W.C.; writing—review and editing, K.O., W.C. and V.L.; visualization, K.O.; project administration, K.O. and V.L.; funding acquisition, K.O., V.L. and B.M. All authors have read and agreed to the published version of the manuscript.

Funding: The research is partially co-financed by the Polish National Agency for Academic Exchange (NAWA) and the Ministry of Education and Science of Ukraine under the project no. PPN/BUA/2019/1/00074 entitled "Methods of intelligent image and video processing based on visual quality metrics for emerging applications". This research is also supported by ZUT Highfliers School (Szkoła Orłów ZUT) project within the framework of the program of the Minister of Education and Science (Grant No. MNiSW/2019/391/DIR/KH, POWR.03.01.00-00-P015/18), co-financed by European Social Fund.

Institutional Review Board Statement: Not applicable.

Informed Consent Statement: Not applicable.

Data Availability Statement: Not applicable.

Conflicts of Interest: The authors declare no conflict of interest. The funders had no role in the design of the study; in the collection, analyses, or interpretation of data; in the writing of the manuscript, or in the decision to publish the results.

Abbreviations

The following abbreviations are used in this manuscript:

CNN	Convolutional Neural Network
FR IQA	Full-Reference Image Quality Assessment
FSIM	Feature Similarity
GIQA	Generated Image Quality Assessment
GGD	Generalized Gaussian Distribution
GMM	Gaussian Mixture Model
iCID	improved Color Image Difference
ISIQA	Indian Institute of Science Stitched Image Quality Assessment (dataset)

KROCC	Kendall Rank Order Correlation Coefficient
MIQM	Multi-view Image Quality Measure
MOS	Mean Opinion Scores
NR IQA	No-Reference Image Quality Assessment
PLCC	Pearson's Linear Correlation Coefficient
ROI	region of interest
SIQE	Stitched Image Quality Evaluation
SROCC	Spearman Rank Order Correlation Coefficient
SVM	Support Vector Machine
UAV	Unmanned Aerial Vehicle
VQEG	Video Quality Experts Group
VSLAM	Visual Simultaneous Localization and Mapping

References

1. Duan, H.; Liu, Y.; Huang, H.; Wang, Z.; Zhao, H. Image Stitching Algorithm for Drones Based on SURF-GHT. *IOP Conf. Ser. Mater. Sci. Eng.* **2019**, *569*, 052025. [CrossRef]
2. Xiong, P.; Liu, X.; Gao, C.; Zhou, Z.; Gao, C.; Liu, Q. A Real-time Stitching Algorithm for UAV Aerial Images. In Proceedings of the 2nd International Conference on Computer Science and Electronics Engineering (ICCSEE 2013), Hangzhou, China, 22–23 March 2013; Atlantis Press: Amsterdam, The Netherlands, 2013. [CrossRef]
3. Madhusudana, P.C.; Soundararajan, R. Subjective and Objective Quality Assessment of Stitched Images for Virtual Reality. *IEEE Trans. Image Process.* **2019**, *28*, 5620–5635. [CrossRef] [PubMed]
4. Okarma, K.; Kopytek, M. Application of Image Entropy Analysis for the Quality Assessment of Stitched Images. In *Progress in Image Processing, Pattern Recognition and Communication Systems*; Choraś, M., Choraś, R.S., Kurzyński, M., Trajdos, P., Pejaś, J., Hyla, T., Eds.; Springer International Publishing: Cham, Switzerland, 2021; pp. 125–132. [CrossRef]
5. Liu, T.J.; Lin, W.; Kuo, C.C.J. Image quality assessment using multi-method fusion. *IEEE Trans. Image Process.* **2013**, *22*, 1793–1807. [CrossRef] [PubMed]
6. Oszust, M. Decision Fusion for Image Quality Assessment using an Optimization Approach. *IEEE Signal Process. Lett.* **2016**, *23*, 65–69. [CrossRef]
7. Ieremeiev, O.; Lukin, V.; Okarma, K.; Egiazarian, K. Full-Reference Quality Metric Based on Neural Network to Assess the Visual Quality of Remote Sensing Images. *Remote Sens.* **2020**, *12*, 2349. [CrossRef]
8. Okarma, K.; Fastowicz, J.; Lech, P.; Lukin, V. Quality Assessment of 3D Printed Surfaces Using Combined Metrics Based on Mutual Structural Similarity Approach Correlated with Subjective Aesthetic Evaluation. *Appl. Sci.* **2020**, *10*, 6248. [CrossRef]
9. Solh, M.; AlRegib, G. MIQM: A novel Multi-view Images Quality Measure. In Proceedings of the 2009 International Workshop on Quality of Multimedia Experience, San Diego, CA, USA, 29–31 July 2009;
10. Solh, M.; AlRegib, G. MIQM: A Multicamera Image Quality Measure. *IEEE Trans. Image Process.* **2012**, *21*, 3902–3914. [CrossRef] [PubMed]
11. Wang, Z.; Bovik, A.; Sheikh, H.; Simoncelli, E. Image Quality Assessment: From Error Visibility to Structural Similarity. *IEEE Trans. Image Process.* **2004**, *13*, 600–612. [CrossRef] [PubMed]
12. Qureshi, H.; Khan, M.; Hafiz, R.; Cho, Y.; Cha, J. Quantitative quality assessment of stitched panoramic images. *IET Image Process.* **2012**, *6*, 1348–1358. [CrossRef]
13. Xu, W.; Mulligan, J. Performance evaluation of color correction approaches for automatic multi-view image and video stitching. In Proceedings of the 2010 IEEE Computer Society Conference on Computer Vision and Pattern Recognition (CVPR), San Francisco, CA, USA, 13–18 June 2010. [CrossRef]
14. Paalanen, P.; Kämäräinen, J.K.; Kälviäinen, H. Image Based Quantitative Mosaic Evaluation with Artificial Video. In *Image Analysis, Proceedings of the 16th Scandinavian Conference, SCIA 2009, Oslo, Norway, 15–18 June 2009*; Salberg, A.B., Hardeberg, J.Y., Jenssen, R., Eds.; Springer: Berlin/Heidelberg, Germany, 2009; LNCS Volume 5575, pp. 470–479. [CrossRef]
15. Bellavia, F.; Colombo, C. Dissecting and Reassembling Color Correction Algorithms for Image Stitching. *IEEE Trans. Image Process.* **2018**, *27*, 735–748. [CrossRef] [PubMed]
16. Zhang, L.; Zhang, L.; Mou, X.; Zhang, D. FSIM: A Feature Similarity Index for Image Quality Assessment. *IEEE Trans. Image Process.* **2011**, *20*, 2378–2386. [CrossRef] [PubMed]
17. Preiss, J.; Fernandes, F.; Urban, P. Color-Image Quality Assessment: From Prediction to Optimization. *IEEE Trans. Image Process.* **2014**, *23*, 1366–1378. [CrossRef] [PubMed]
18. Niu, Y.; Zhang, H.; Guo, W.; Ji, R. Image Quality Assessment for Color Correction Based on Color Contrast Similarity and Color Value Difference. *IEEE Trans. Circuits Syst. Video Technol.* **2018**, *28*, 849–862. [CrossRef]
19. Cheung, G.; Yang, L.; Tan, Z.; Huang, Z. A Content-Aware Metric for Stitched Panoramic Image Quality Assessment. In Proceedings of the 2017 IEEE International Conference on Computer Vision Workshops (ICCVW), Venice, Italy, 22–29 October 2017. [CrossRef]

20. Hou, J.; Lin, W.; Zhao, B. Content-Dependency Reduction With Multi-Task Learning In Blind Stitched Panoramic Image Quality Assessment. In Proceedings of the 2020 IEEE International Conference on Image Processing (ICIP), Abu Dhabi, United Arab Emirates, 25–28 October 2020. [CrossRef]
21. Gu, S.; Bao, J.; Chen, D.; Wen, F. GIQA: Generated Image Quality Assessment. In Proceedings of the European Conference on Computer Vision—ECCV 2020, Glasgow, UK, 23–28 August 2020; Vedaldi, A., Bischof, H., Brox, T., Frahm, J.M., Eds.; Springer International Publishing: Cham, Switzerland, 2020; pp. 369–385. [CrossRef]
22. Zhou, T.; Wang, W.; Liang, Z.; Shen, J. Face Forensics in the Wild. In Proceedings of the 2021 IEEE/CVF Conference on Computer Vision and Pattern Recognition (CVPR), Nashville, TN, USA, 20–25 June 2021; pp. 5774–5784. [CrossRef]
23. Tang, C.W.; Chen, C.H.; Yu, Y.H.; Tsai, C.J. Visual sensitivity guided bit allocation for video coding. *IEEE Trans. Multimed.* **2006**, *8*, 11–18. [CrossRef]
24. Jung, K.; Hong, J. Quantitative Assessment Method of Image Stitching Performance Based on Estimation of Planar Parallax. *IEEE Access* **2021**, *9*, 6152–6163. [CrossRef]

entropy

Article

Computer Vision Based Automatic Recognition of Pointer Instruments: Data Set Optimization and Reading

Lu Wang †, Peng Wang †, Linhai Wu, Lijia Xu, Peng Huang and Zhiliang Kang *

College of Mechanical and Electrical Engineering, Sichuan Agricultural University, Ya'an 625000, China; 201800656@stu.sicau.edu.cn (L.W.); 2019317016@stu.sicau.edu.cn (P.W.); 201800655@stu.sicau.edu.cn (L.W.); 10887@stu.sicau.edu.cn (L.X.); 14130@stu.sicau.edu.cn (P.H.)

* Correspondence: 12200@stu.sicau.edu.cn; Tel.: +86-186-0835-1703

† These authors contributed to the work equally and should be regarded as co-first authors.

Abstract: With the promotion of intelligent substations, more and more robots have been used in industrial sites. However, most of the meter reading methods are interfered with by the complex background environment, which makes it difficult to extract the meter area and pointer centerline, which is difficult to meet the actual needs of the substation. To solve the current problems of pointer meter reading for industrial use, this paper studies the automatic reading method of pointer instruments by putting forward the Faster Region-based Convolutional Network (Faster-RCNN) based object detection integrating with traditional computer vision. Firstly, the Faster-RCNN is used to detect the target instrument panel region. At the same time, the Poisson fusion method is proposed to expand the data set. The K-fold verification algorithm is used to optimize the quality of the data set, which solves the lack of quantity and low quality of the data set, and the accuracy of target detection is improved. Then, through some image processing methods, the image is preprocessed. Finally, the position of the centerline of the pointer is detected by the Hough transform, and the reading can be obtained. The evaluation of the algorithm performance shows that the method proposed in this paper is suitable for automatic reading of pointer meters in the substation environment, and provides a feasible idea for the target detection and reading of pointer meters.

Keywords: pointer instrumentation; image processing; object detection; K-fold cross-validation; Faster-RCNN

Citation: Wang, L.; Wang, P.; Wu, L.; Xu, L.; Huang, P.; Kang, Z. Computer Vision Based Automatic Recognition of Pointer Instruments: Data Set Optimization and Reading. *Entropy* **2021**, *23*, 272. https://doi.org/10.3390/e23030272

Academic Editors: Michal Choras, Robert Burduk, Agata Giełczyk, Rafal Kozik and Tomasz Marciniak

Received: 12 January 2021
Accepted: 20 February 2021
Published: 25 February 2021

1. Introduction

There are many instruments in substations, such as pressure gauges, ammeters, oil temperature gauges, and so on. There are two types of current instruments, i.e., pointer type and digital type. Figure 1 shows the working conditions of pointer instruments and digital display instruments in substations.

Different working principles of such two types lead to different characteristics of them: the share of digital display instruments in the instrument industry is increasing for their advantages of high accuracy and convenient reading, while they are also not applicable to some occasions such as harsh oily or dusty environments, or this type of instrument won't be applicable when the field instrument input variables change over fast [1]; In this regard, the pointer instrument has many advantages over the digital display instrument including simple structure, low price, strong anti-interference ability, dustproof, waterproof, anti-freeze, anti-interference, oil resistance, and so on [2].

Based on the above reasons, pointer instruments have been widely using in the industry, while recognition of their numerical value has always been a hot spot in the industrial study. In the early days of industrial development, the operators read the pointer instruments mainly by visual recognition, that is, manual interpretation, so it is inevitably subject to interferences caused by various artificial factors. Due to a large number of meters, the complex external environment, the dial observation angle, visual fatigue, the error caused by the observation distance and the deviation caused by the influence of light on the pointer artificial, and other factors, the recognition process is boring, troublesome and easily affected by subjective factors, which will inevitably cause inaccurate readings [1,2]. As technology develops, inspection robots have been introduced into more and more substations to replace manual inspections over recent years. Thereby the automation degree of substations has been greatly improved. The solution in this article is implemented offline. The robot takes photos and stores them, and then sends them to the computer for processing.

(**a**) (**b**)

Figure 1. Working status of different instruments: (**a**) Digital display instrument; (**b**) pointer instrument.

At present, there has been a lot of research on the automatic recognition of pointer meters. Peng et al. obtained pointer connection by converting Red, Green, Blue (RGB) space to Hue, Saturation, Value (HSV) space and utilizing color features to detect the beginning and end scales of oil level gauges. However, this method relies highly on environmental factors. Hua et al. [3] proposed a recognition algorithm for the readings of pointer instruments. It uses equipment images of the instrument to establish instrument templates and utilizes Scale-Invariant Feature Transform (SIFT), Oriented Fast and Rotated Brief (ORB), and other feature point detection algorithms, matching and extracting sub-images of the instrument dial area from the input image to realize instrument positioning. However, the template matching algorithm relies strictly on corner calibration, and corner detection is susceptible to deformation, occlusion, and other environmental factors, and a large number of false matches are likely to occur in complex scenes. To detect the instrument dial, Wei et al. [4] adopted Support Vector Machine (SVM) training, but the training accuracy of large sample data is not satisfactory [5]. Haoqiang et al. performed target detection over dial plate with the Single Shot multibox Detector (SSD) deep learning model to extract the meter area and remove irrelevant background [6]. The above methods are not ideal for object detection.

In recent years, the deep learning-based neural network method has been widely applied in the image processing of electric power meters. However, little research describes the data set. In the actual operation, the quality of the data set is critical to the accuracy of object detection. To ensure that the intelligent meter reading system forms a complex scene to obtain high-quality images of the dial area, this paper proposes to expand the data set with Poisson fusion algorithms to achieve its diversity, and then use the K-fold verification method to preprocess and optimize the training data aiming at fully training the model and suppressing the sample over-fitting caused by the insufficient quantity and uneven distribution. The utilization rate of the data set and the accuracy of model detection are greatly optimized. These data set optimization methods are unprecedented. On this basis, combined with the optimized classifier, the high-quality dial region image is extracted for image processing. The main contributions are as follows:

(1) Establishing a data set optimized by data fusion expansion and K-fold verification algorithm and getting it applied to industrial production, which optimized the data set quality and greatly reduced the workload of data set collection.

(2) An intelligent meter reading system whose accuracy rate up to 98.65% is obtained by utilizing Faster-RCNN and Hough transform straight line detection. The evaluation of the algorithm performance shows that the method proposed in this paper is suitable for the extraction of pointer meters in the substation environment, and a feasible thought for object detection and reading of pointer meters is provided thereof.

The overall scheme of the method proposed in this paper is shown in Figure 2. All experiments are performed on Visual Studio 2019. Visual Studio 2019 is a product developed by American Microsoft Corporation.

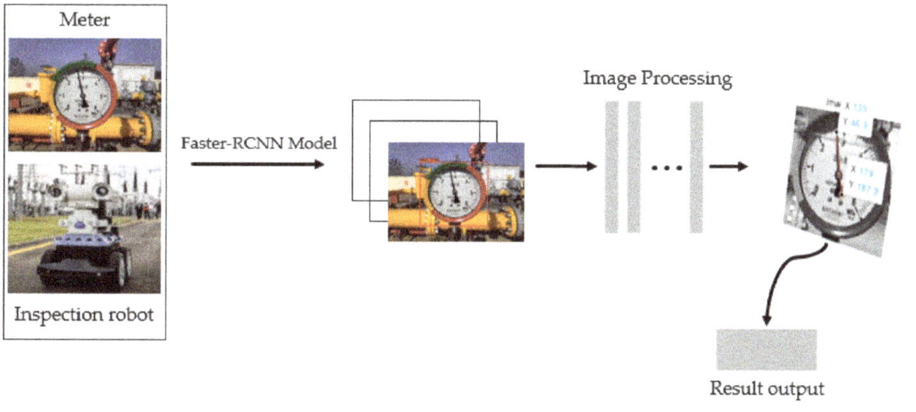

Figure 2. The overall flow of the solution in this paper is that: the image obtained by the robot is passed through the Faster-RCNN network to obtain the image of the dial area, and then preprocessed, the pointer position is detected by the Hough transform line, and finally, the indication is obtained.

This paper is organized into six sections, including the present one. Section 2 introduces the one-stage algorithm and the two-stage algorithm in deep learning compares the pros and cons of the two algorithms and explains the reasons why the Faster-RCNN algorithm is chosen. Section 3 describes the optimization of the data set and introduces how to improve the quantity and quality of data sets in detail. And the experimental results of the optimized data set are discussed therein. Section 4 explains the Faster-RCNN algorithm-based object detection network and the relevant network structure used in this paper. And it also compares and analyzes various classic classification network models. Section 5 elaborately describes dial image preprocessing and identifying the reading of pointer meter with Hough transform [7] and discusses the feasibility and practicability of the experimental results. Section 6 emphasizes some significant features of this scheme.

2. Deep Learning-Based Object Detection

2.1. Object Detection Algorithm

In the substation and factory environment, the accuracy of the reading of the pointer instrument is of primary importance to improving production efficiency and avoiding excessive losses. When an inspection robot shoots the images of a pointer instrument, it will inevitably include the complex background into the images. Accurately obtaining high-quality dial images from complex environmental backgrounds is the basis of accurate readings for the intelligent meter reading system. With regard to extraction and determination of shallow surface features of specific objects in the early development of object detection, Hough transforms [7–9], circle detection algorithm, and SIFT [10] feature matching algorithm are more taken among traditional computer vision methods. These two algorithms carry out detection based on certain features concluded from a large amount of research and summary work done by professionals, but this method has poor recognition accuracy, and the detection effect is quite limited. Aiming at this shortcoming, this paper uses a deep learning-based object detection algorithm, and by establishing a deep neural network model and using a large amount of data for training, the network model can more accurately identify the target from a complex background with high reliability and wider scope of adaptation.

The neural network-based object detection algorithms can be roughly divided into two-stage and one-stage. The two-stage detection algorithm divides the object detection process into two steps. First, it generates region proposals [11] relying on the Convolutional Neural Network (CNN to obtain the possible positions of the targets in the entire picture, and then it classifies and refines the proposed regions before carrying out subtle target recognition. The typical algorithms of the two-stage model include Region-based Convolutional Neural Network (R-CNN) [12,13], Fast Region-based Convolutional Neural Network (Fast R-CNN) [14], Faster R-CNN [15], and so on. One-stage does not require the region proposal stage. The position coordinates and category probability of the target can be generated through the CNN to obtain the position coordinates and corresponding confidence of the target directly from the picture. Typical algorithms of the one-stage model include You Only Look Once (YOLO) [16] and SSD [17].

2.2. Algorithm Selection

To select the best model, several commonly used models are tested, and the speed and accuracy are compared, as shown in Table 1. The Faster R-CNN algorithm has the best performance among the R-CNN series. This algorithm integrates the regional recommendation network and the convolutional neural network to classify and locate the target. Comparing with the other R-CNN series algorithms, it greatly reduces the time and space for training and testing, improves the detection speed while maintaining superiority over small object detection [12–15]. The YOLO series pursues detection speed at the expense of accuracy. The SSD algorithm strikes a balance between speed and accuracy, gaining a greater increase in speed at the expense of less decrease in accuracy [16,17]. However, in principle, the meter reading system in the substation does not allow false negatives, so this paper selects the Faster R-CNN model with higher detection accuracy and faster rate.

Table 1. Test results of five models.

Model	R-CNN	Fast-RCNN	Faster-RCNN	YOLO	SSD
mAP (%)	63.20	70.21	90.27	81.32	79.20
Time (s)	22.1	4.1	0.69	0.48	0.57

3. Preparation and Optimization of Data Set

The pictures in this paper were taken at a substation in Nanchong City, Sichuan Province. This data set was taken by Canon EOS90D. The camera comes from Canon Inc., Japan. The image resolution is 3472 × 2320. There are both long-distance shots and short-distance shots, and the shooting angles are both top-down and parallel. The data sets are dominated by top-down and short-distance shooting since it is more in line with the angular range of the inspection robot camera and is more practical. There is a great variety of pointer instruments in substations and factories, of which common types of instruments for measuring electrical quantities include ammeters, voltmeters, power meters, etc., while non-electrical quantities are mainly measured by pressure gauges, thermometers, and oil level gauges. To achieve better detection results, it is necessary to shoot tens of thousands of images in substations and factories for preparing a large number of data sets, but currently, there is no public instrument data set in substations, so we have to collect images by ourselves. This also results in a substantial increase in the workload of data set acquisition. Given the labor cost, this study obtained almost 1000 images through field shooting and from the Internet. To enrich the data set, the following process had been performed on the images for achieving high-quality data sets.

3.1. Data Set Processing

Only a part of the region in any given image contains meters. As shown in Figure 3. The open-source software labelImg was employed to mark and generate corresponding label information, and then randomly cut such information according to the location of the meter.

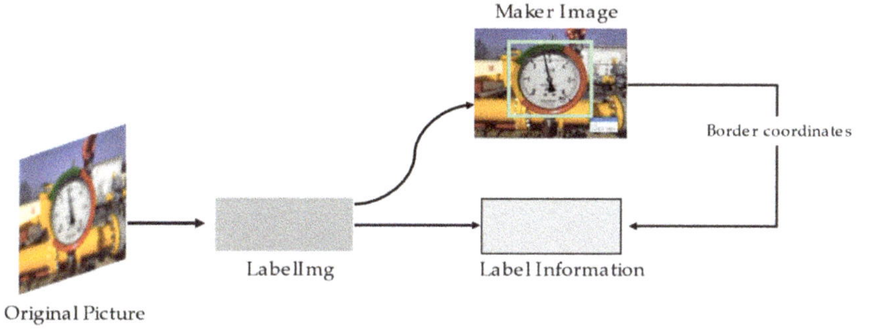

Figure 3. Preliminary meter image processing: Obtaining the label information of the dial position from original images.

3.2. Data Set Expansion

Because the shooting distance and the shooting time significantly affect the pixels and the quality of the images with the instrument on them, and the data set is not large enough, we expand the data set by randomly flipping and cropping the images. Two images of different objects were merged to obtain a higher recall rate. This method mainly merges training images with different targets through pixel-by-pixel blending, thus improving the generalization performance of image classification and object detection.

Data Fusion

The real captured image scenes are much more complicated than the cropped image contained in the data set. The background of the image may include complex objects that interfere with dial recognition, such as buildings, pipes, electric boxes, and electric poles. The complex background in the image causes a high false detection rate. To adapt to the complex background environment, it should compose images with different complex backgrounds by means of data fusion to add negative samples into the training data set and reduce false alarms by expanding the data set.

This paper uses the Poisson fusion method for image fusion when expanding data samples. The Poisson fusion method, as a well-known image editing algorithm, was proposed by R. Patrick Pérez in 2003 [18–20], which can achieve a more natural fusion effect and has been widely applied in image fusion and image restoration fields [21].

The Poisson fusion method regards the image fusion problem as a solution to the minimization problem of Formula (1).

$$\min_f \iint | \nabla f - v |^2, f|_{\partial\Omega} = f^* |_{\partial\Omega} \tag{1}$$

where Ω is the fusion area corresponding to the foreground image; $\partial\Omega$ is the boundary of such area; represents the fused image; f^* represents fusing the known background image outside the fusion area; v is called guide field, and the gradient field of the original foreground image is taken. The minimization problem is to minimize the gap between the gradient field of the fusion area and the guide field while guaranteeing the boundary value of the fusion area is consistent with the background image, to preserve the gradient field of the original foreground image to the greatest extent. This problem can be transformed into the solution of Poisson's equation with Dirichlet boundary conditions [22].

Figure 4 shows a composing example. According to the gradient information of the source image and the boundary information of the target image, the image pixels within the composing area are reconstructed by means of interpolation. This method is based on Poisson image fusion to fuse background and dial samples to enrich the data set and form new samples.

(a) (b) (c)

Figure 4. Composing example: (**a**) original image; (**b**) background image; (**c**) blended image.

It greatly reduced the workload of data acquisition and improved the object detection accuracy that the amount of data set was expanded through data fusion.

3.3. K-Fold Verification Algorithm

Overfitting, caused by an inadequate data set, is used in this paper. To improve the detection accuracy, this paper chooses the k-fold method [23] to preprocess the data set. With regard to this data set, the training set and the validation set are divided by 4:1. In this experiment, the value of k is 5, i.e., the training data is randomly divided into 5 portions, one of them is taken as the validation set and the other 4 portions are used as the training set. The original Faster-RCNN model is trained for 5 rounds in total, and we obtained the mean Average Precision (mAP) of 5 sets of model parameters. They are respectively used for testing the test set, and the original MobileNet V2 network framework is uniformly used in this experiment. Finally, the average of 5 test results is used as an evaluation of model accuracy. The test results of the 5 rounds of training in this experiment and the comparative experimental results of the final model evaluation are shown in Tables 2 and 3.

Table 2. K-fold verification test results.

Target Detection	Epoch1 (%)	Epoch2 (%)	Epoch3 (%)	Epoch4 (%)	Epoch5 (%)
mAP	95.45	95.35	95.21	95.61	95.52

Table 3. Comparison of 5-fold data preprocessing based experimental results.

Target Detection	mAP (Faster-RCNN)	mAP (K-fold)
mAP	92.67	95.43

By employing the K-fold verification data preprocessing method, it is significantly improved comparing with the original Faster-RCNN model. Moreover, the mAP of the model is increased by 2.76% after data preprocessing. This data set shows that the K-fold verification data preprocessing method has a good effect on inhibiting overfitting, which is beneficial to optimize model parameters and improve model detection performance.

4. Object Detection Results and Experimental Process

4.1. Faster-RCNN Based Dial Object Detection

The RPN-based Faster-RCNN network structure is shown in Figure 5. The algorithm model sets an N × N sliding window on the feature map of the input image, and each sliding window can be mapped to a fixed-dimensional feature. If Anchors set k different aspect ratios, each window will correspond to k candidate regions of different sizes, and then their features will be sent to two fully connected layers for category prediction and border regression. The output dimension of the classification layer is 2k, which represents the probability that the candidate area contains the target or background; the output dimension of the border regression layer is 4k, which represents the location information of the candidate area. The RPN network can be trained separately for effectively reducing the time to generate the feature map of the candidate region. In this way, the training speed of Faster R-CNN is greatly improved over Fast R-CNN.

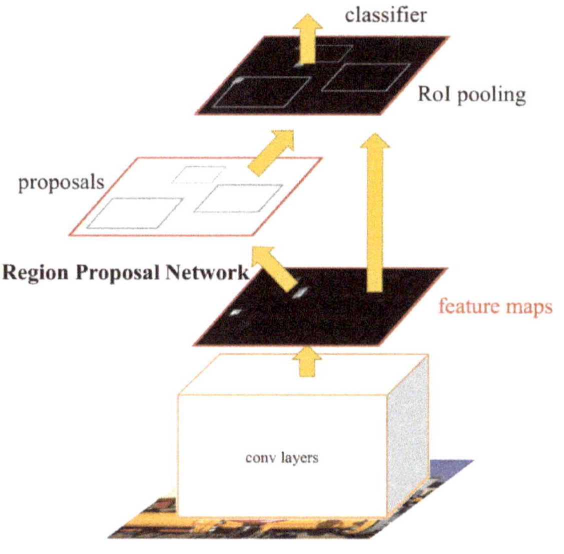

Figure 5. RPN-based Faster-RCNN network structure.

4.2. The Core Feature Extracting Network Commonly Used by the Faster R-CNN Algorithm

4.2.1. ResNet

Residual Network(ResNet) [24], as a residual network, can be understood as a sub-network, which can form a deep network after stacking. As a general rule, the deeper the network, the more information obtained and the richer features. However, the deepening of the network did not improve the optimization effect, and the accuracy of the test data and training data was decreased. To solve the problem of gradient explosion and gradient disappearance caused by network deepening, four Chinese Scholars including Kaiming He of Microsoft Research proposed ResNet. They successfully trained a neural network with a depth of up to 152 layers on the image network data set, and only a 3.57% error rate in the top 5. This result won first place in the ILSVRC 2015 Classification Task. Besides, the parameter quantity is lower than Visual Geometry Group Network(VGGNet), and the effect is outstanding.

4.2.2. VGGNet

The VGG [25] network model, as a deep learning convolutional neural network model, was proposed by the Visual Geometry Group of Oxford. VGGNet explored the relationship between the depth and the performance of a convolutional neural network. In short, VGG successfully builds a convolutional neural network with a depth of 16–19 layers by repeatedly stacking 3 × 3 convolution kernels and 2 × 2 pooling kernels. An improvement of VGG16 compared to Alex Krizhevsky Network(AlexNet) [26] is to use several consecutive 3 × 3 convolution kernels to replace the larger convolution kernels in AlexNet. In this way, it ensures that improving the network depth under the condition of the same receptive field can improve the neural network result to a certain extent. VGGNet won second place in the ILSVRC 2014 competition classification project and first place in the positioning project, with an error rate of 7.5% on the top 5. At present, VGGNet is also widely used in extracting image features.

4.2.3. MobileNet

MobileNet [27] was proposed by Google in 2017. It is a lightweight CNN neural network focusing on mobile devices and embedded devices, and it quickly derived three versions of v1, v2, and v3. The main work of MobileNet is to replace the past standard convolution with depth-level separable convolution to solve the problem of computational efficiency and parameter amount of convolutional network. Deep convolution applies each convolution kernel to each channel, and 1 × 1 convolution is used to combine the output of channel convolution. Compared with the traditional CNN network, it greatly reduces model parameters and computation under the premise of a small reduction in accuracy.

4.2.4. The Classifier Used in this Article

The mainframe of the classifier used in this paper is shown in Figure 6. Combined with the network structure in Table 4, the classifier has only one hidden layer containing 512 dimensions, which is then activated by the ReLu activation function. Afterward, it outputs in two dimensions according to the number of classes in. To prevent over-fitting, convergence is accelerated for increasing the training speed and improving the object detection accuracy. The Batch Normalization (BN) layer is added and the Dropout layer is discarded. and process the final output into a pile to evaluate the accuracy of the model. The classifier uses CrossEntropyLoss [28] as the loss function. In the final output, the loss function has integrated the Log Softmax layer and the NLLLoss layer [29] to reduce computation. This loss is helpful for object detection problems.

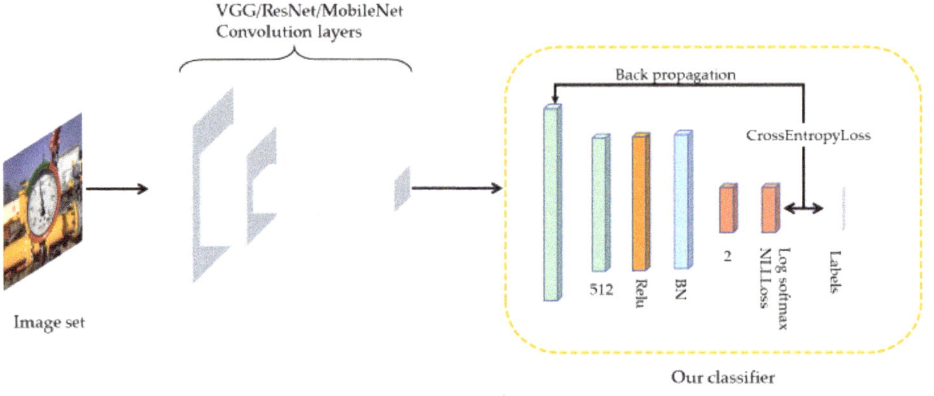

Figure 6. Classification network frame: integrating the convolutional layer of other classification networks after optimizing based on the original classifier.

Table 4. Classification grid structure mentioned in the article. This article uses three classic classification networks. The X in conv X refers to the internal structure of the convolutional layer.

	VGG 16-layer	ResNet 50-layer	MobileNet 16-layer	
Conv1_×	3 × 3.64	7 × 7.64 stride 2	Type/stride	Filter Shape
	3 × 3.64		Conv/s2	3 × 3 × 3 × 32
	3 × 3maxpool	3 × 3maxpool	Conv dw/s1	3 × 3 × 32 dw
	stride 2	stride 2	Conv/s1	1 × 1 × 32 × 64
Conv2_×	3 × 3.128	$\begin{bmatrix} 1 \times 1.64 \\ 3 \times 3.64 \\ 1 \times 1.256 \end{bmatrix} \times 3$	Conv dw/s2	3 × 3 × 64 dw
	3 × 3.128		Conv/s1	1 × 1 × 64 × 128
	3 × 3maxpool,		Conv dw/s1	3 × 3 × 128 dw
	stride 2		Conv/s1	1 × 1 × 128 × 128
Conv3_×	3 × 3.256	$\begin{bmatrix} 1 \times 1.128 \\ 3 \times 3.128 \\ 1 \times 1.512 \end{bmatrix} \times 4$	Conv dw/s2	3 × 3 × 128 dw
	3 × 3.256		Conv/s1	1 × 1 × 128 × 256
	3 × 3.256		Conv dw/s1	3 × 3 × 256 dw
	3 × 3maxpool,		Conv/s1	1 × 1 × 256 × 256
	stride 2		Conv dw/s2	3 × 3 × 256 dw
Conv4_×	3 × 3.512	$\begin{bmatrix} 1 \times 1.256 \\ 3 \times 3.256 \\ 1 \times 1.1024 \end{bmatrix} \times 6$	Conv/s1	1 × 1 × 256 × 512
	3 × 3.512		5 × Conv dw/s1 Con/s1	3 × 3 × 512 dw1 × 1 × 512 × 512
	3 × 3.512		Conv dw/s2	3 × 3 × 512 dw
	3 × 3maxpool,		Conv/s1	1 × 1 × 512 × 1024
	stride 2		Conv dw/s2	3 × 3 × 64 dw
Conv5_×	3 × 3.512	$\begin{bmatrix} 1 \times 1.512 \\ 3 \times 3.512 \\ 1 \times 1.2048 \end{bmatrix} \times 3$	Conv/s1	1 × 1 × 1024 × 1024
	3 × 3.512 3 × 3.512 3 × 3maxpool, stride 2		Classificatin layer	fc 1000 softmax
Classificationlayer	fc 4906 fc 4906 fc 1000 softmax	fc 1000 softmax		

4.3. Faster-RCNN Algorithm Results Based on Different Frames

4.3.1. Experimental Environment

The experimental platform is configured with Intel Core CPU i7-10700k, 3.80 GHz CPU, 10 GB NVIDIA GeForce RTX3080 GPU, 32 GB DDR4 2666 Mhz memory, 1 TB Samsung 970 solid-state drive, and the operating system is Windows 10. The experimental equipment comes from China Lenovo.

4.3.2. Evaluation Parameters

IoU represents the overlapping degree between the generated candidate bound and the ground truth bound, i.e., the ratio of their intersection and union. The larger the value, the higher the correlation. The ideal situation is complete overlapping, i.e., the ratio is 1. The specific formula is as follows:

$$IoU = \frac{area(C) \cap area(G)}{area(C) \cup area(G)} \tag{2}$$

In Formula (2), C represents the candidate boundary, and the candidate bound is generated by the object detection model. G represents the actual labeling truth bound. The most common threshold is 0.5—if $IoU > 0.5$, it will be deemed as correct detection, otherwise it will be deemed as a wrong detection. For those objects that are deemed as correct predictions, they will be used to evaluate the overall accuracy of the final model.

The cross-entropy loss function is used to estimate the gap between the output of the model and the true value, and to guide model optimization. We usually use the Sigmoid function to compress the output of the model to the interval (0,1). $\hat{y}_i \in (0,1)$ is used to represent the probability that the given input $\hat{x}_i$ model is judged to be a positive class. The detailed formula is as follows:

$$p(y_i = 1|xi) = \hat{y}_i \tag{3}$$

$$p(y_i = 0|x_i) = 1 - \hat{y}_i \tag{4}$$

$$p(y_i|x_i) = \hat{y}_i{}^{y_i}(1 - \hat{y}_i)^{1-yi} \tag{5}$$

$$L(x,y) = \prod_{i=1}^{N}(\hat{y}_i)^{y_i}(1 - \hat{y}_i)^{1-y_i} \tag{6}$$

$$NLL(x,y) = -\sum_{i=1}^{N} y_i \log(\hat{y}_i) log(1 - \hat{y}_i) \tag{7}$$

Formula (7) represents the output of the cross-entropy loss function, and loss is used to represent the cross-entropy loss function below. The other evaluation index parameters are defined as the following formulas:

$$P = \frac{N(TurePositives)}{N(TotalObjects)} \tag{8}$$

$$AP = \frac{\sum Precisions}{N(TotalImages)} \tag{9}$$

$$mAP = \frac{\sum AP}{N(AP)} \tag{10}$$

In Formulas (8)–(10), P represents the prediction accuracy of each image; AP represents the overall average accuracy of all data sets; mAP is the mean average precision evaluated from multiple verification sets for use as an index for measuring accuracy in object detection; N represents the number of element types in brackets.

4.3.3. Experimental Results

The detection algorithm is developed based on the deep learning language framework Torch. To achieve the best results, adjust the training hyperparameters for this data set, the training parameters of the Faster-RCNN algorithm are shown in Table 5. The training process is shown in Figure 7.

Table 5. Training hyperparameters.

	Learning Rate	Batch Size	Maximum Number of Epochs	Iterations Per Epoch	Momentum	Decay
Faster-RCNN	0.005	4	25	50	0.9	0.0005

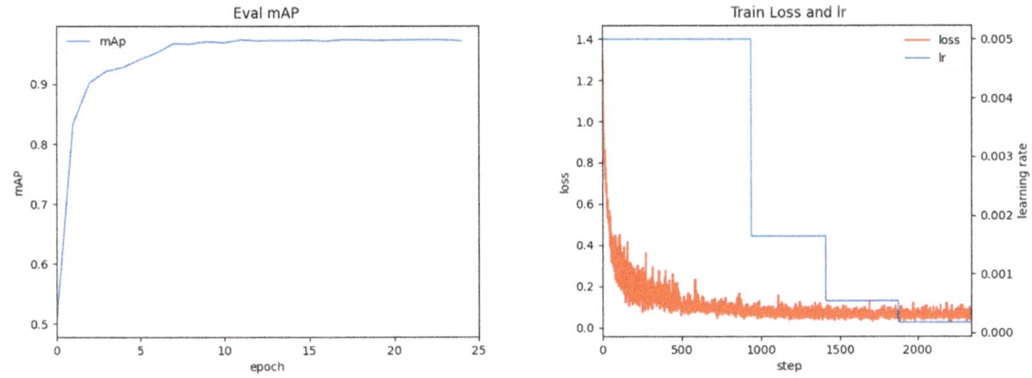

(**a**) VGG-16

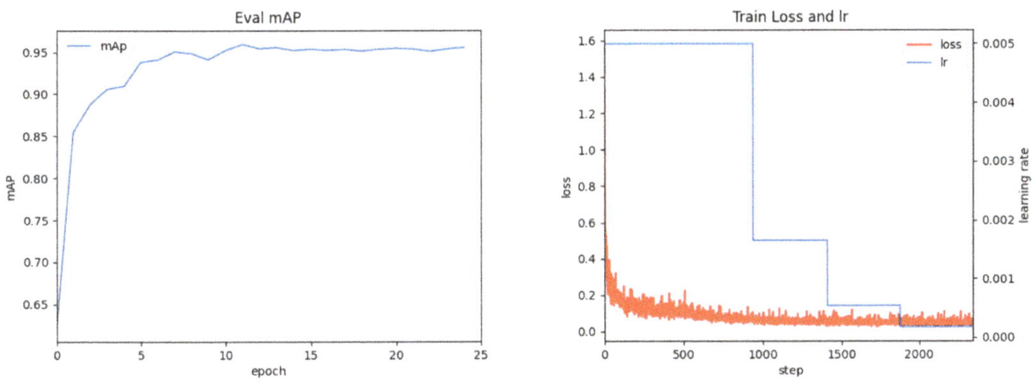

(**b**) Mobilenet-V2

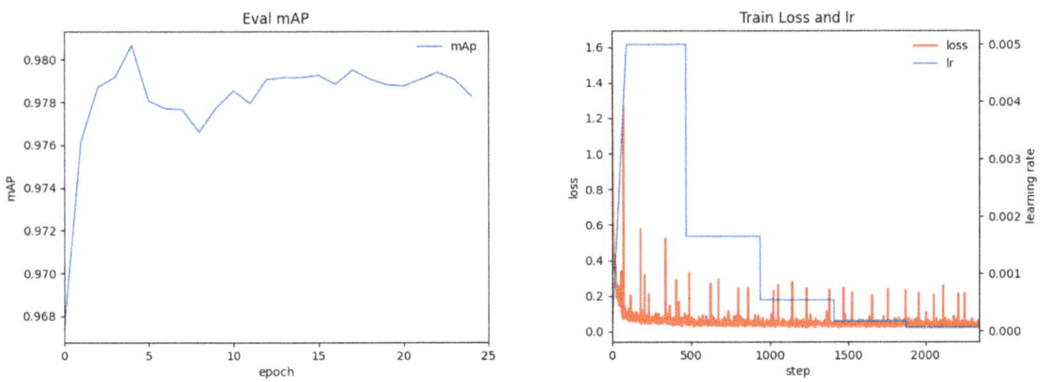

(**c**) Resnet-50

Figure 7. The left figure reflects the change of mAP (mean Average Precision) value in the training process, and the right figure shows the change of Faster-RCNN total loss and learning rate (lr) in the training process. Lr in Figure 7 represents the learning rate (Learning rate), as an important hyperparameter in supervised learning and deep learning, determines whether the objective function can converge to a local minimum and when to converge to the minimum. A proper learning rate can make the objective function converge to a local minimum in a proper time.

Based on the observations in Figure 7, we can draw the following conclusions:

- Through the above experiment, observe the above training process diagram, the best mAP value for the classification network VGG-16 is 97.49%, the best mAP value for the Resnet-50 network is 98.20%, and the best mAP value for the MobileNet-V2 network is 95.46%.
- Resnet-50 performs poorly in global optimization. Its accuracy in the latter period is high, but the final convergence effect is not ideal. The oscillation still exists under the optimized classifier in the later stage, and the experiment still fluctuates greatly.
- MobileNet-V2's early oscillation is more serious, but the later convergence effect is favorable. Given its low mAP value, it is discarded.
- With regard to the data of this sample, the VGG-16 network finally converges well, and its performance in the classifier is also ideal. Experiments under the optimized classifier show good accuracy and convergence. It is proved that the global optimization of VGG-16 is the best classification network, and its mAP value converges to 0.974.

VGG-16 was selected as the model for object detection in this article because the combination of the convolutional layer of VGG-16 and our optimized classifier achieved favorable accuracy and the best convergence.

5. Processing and Analysis

The dial area is extracted from the complex background with the Faster-RCNN algorithm, and then the centerline of the pointer is accurately extracted with the corresponding image processing algorithm. The main content of this part is as follows: First, perform preprocessing such as binarization and image noise reduction on the dial image, then extract the centerline of the indicator pointer with the Hough line detection algorithm, and finally, obtain the pointer reading according to the deflection angle of the pointer.

5.1. Image Preprocessing

To extract pointer information from the picture, it first has to gray the cropped color instrument image to obtain the gray image and get it binarized. Commonly used binarization algorithms include the average grayscale method, maximum between-class variance method, etc. [30]. Comparing the two methods, the average grayscale method has a lower calculation burden, while the maximum between-class variance method can better preserve the image information. According to experiments, for images with uneven brightness, using the maximum between-class variance method cannot balance the details of the brighter and darker parts of the image. Therefore, the local binarization method should be used to preserve the details of the image. Figure 8 shows the effects of different binarization methods. As shown in Figure 9, a local block of 1 square side length is selected from the image, and translated as per step size S to form a series of local blocks. In each local block, the result obtained by the best-performing maximum between-class variance method is taken as the binarization threshold, and each threshold is recorded to form a distribution threshold matrix. When local binarization is carried out with the average gray level method, it is greatly affected by the choice of local block size and the moving step length, and the parameter adaptation range is narrow. However, the effect is stable when local binarization is carried out with the maximum between-class variance. Therefore, this paper adopts the local binarization of maximum between-class variance. In the distribution threshold matrix, the brightness of some local blocks may be exceedingly bright or dark making the threshold deviate far from the average gray value, which causes abnormal spots in the binarized image. To reduce such influence, median filtering is performed on the distribution threshold matrix. The median filtered distribution threshold matrix is smoothly interpolated and boundary expanded to form a complete threshold matrix on the same dimension with the gray image. Binarizing the grayscale image according to the complete threshold matrix, the experimental results show good results.

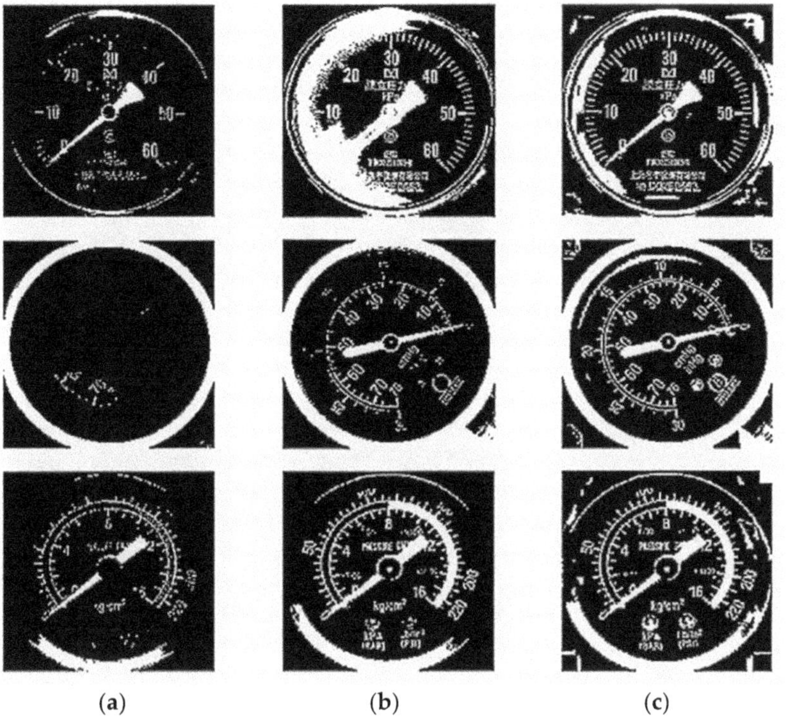

Figure 8. Comparison of binarization methods: (a) average grayscale method; (b) maximum between-class variance method; (c) maximum between-class variance local binarization; The average grayscale method cannot favorably reserve dial information. The maximum between-class variance method performs poorly in poor lighting conditions, so it is not practical in actual scenes. While the maximum between-class variance local binarization balances the integrity of the information and the reduction of the influence of lighting conditions.

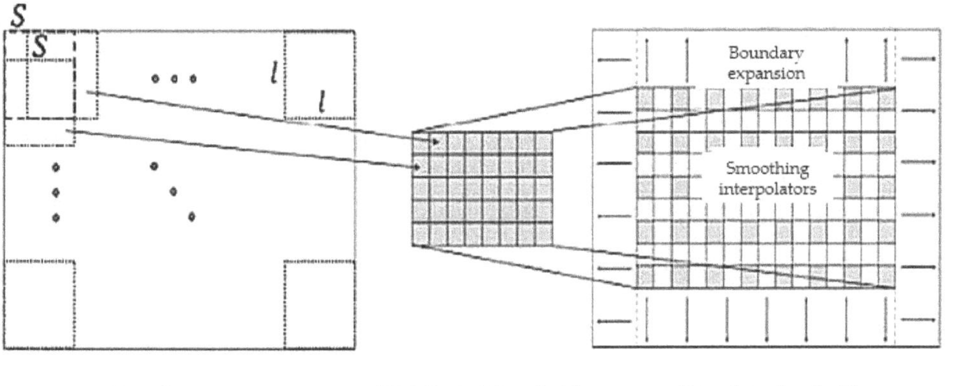

Figure 9. Represents local binarization of local blocks and moving step length.

There may be noise or line breaking on the binarized image due to image quality or interference in the acquisition process, and the image after the meter pointer binarization is thicker but easy to have burrs. To eliminate thinner lines and burrs, smooth the boundary of the thick line, this paper performs image opening and closing processing on the binary image [31] to improve the accuracy of the subsequent Hough transform line detection. Figure 10 shows the effect of image open-close operations.

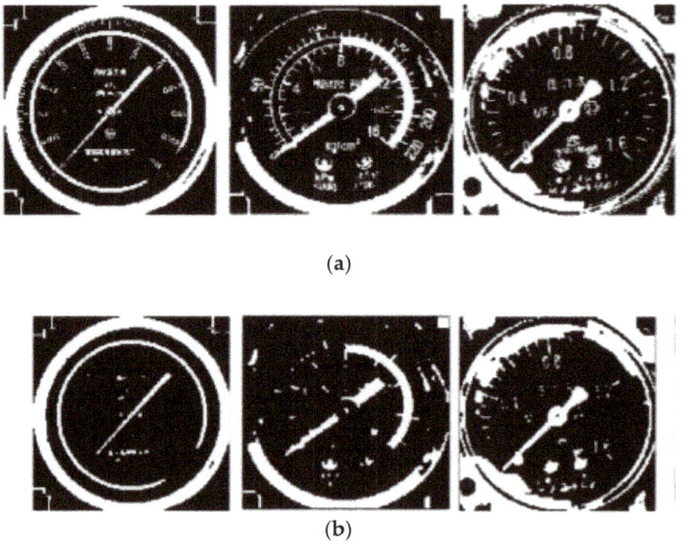

(a)

(b)

Figure 10. The binarization image and the open-close treated image are compared: (**a**) is the original image of the binarization; (**b**) the image after the opening and closing processing.

5.2. Pointer Extraction

The purpose of this article is to detect the pointer. When the meter can be accurately captured in the aforementioned object detection, the pointer will be in the middle of the binarized image. As shown in Figure 11. Divide the image into 5 × 5 divisions, and start from the central grid to find the connected domain [32] and delete the other images.

The central grid may not only retain the pointer image, but also other small spots. We also found in experiments that, as shown in Figure 12, the pointer image is often broken into two segments due to the brightness of the fixed axis is different from that of the pointer, thus the remaining connected domains should be filtered. Sort the connected domains by the area in descending order, keep the part with the largest area, and compare the ratio of the second-largest connected domain to the largest connected domain. If the area ratio is greater than 1/3, keep it, otherwise delete it together with the other connected domains. When there is only the pointer skeleton as shown in Figure 13, straight-line detection is performed on the pointer. As the connected domain of the pointer may contain other graphs, some interference lines may be obtained in the Hough transform line detection. As shown in Figure 14, the lines obtained by the Hough transform should be filtered. In case the object detection is accurate enough, the straight-line boundary will inevitably pass through the central grid of the 5 × 5 grid, so only straight lines that meet this condition are retained.

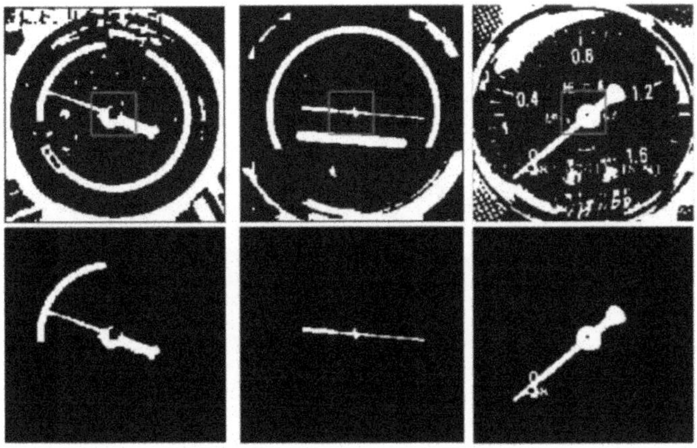

(**a**)

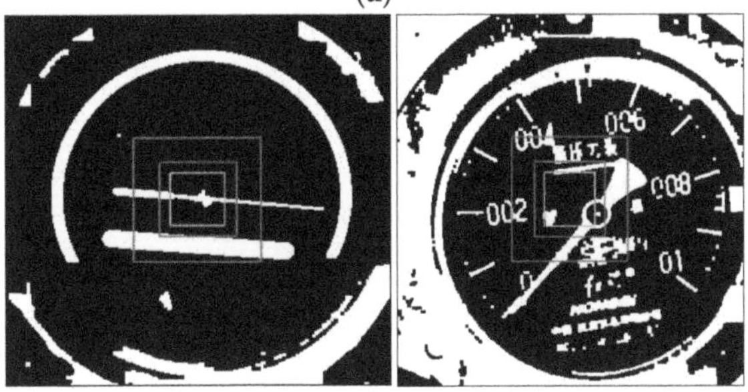

(**b**)

Figure 11. Extract pointer image: (**a**) Search image for the connected domain with the standard of 5 × 5 grid, and the central grid and the connected domain starting from the central grid is shown; (**b**) The comparison of the central grid with different specifications; interference graphics are easily introduced if the central grid is too large (3 × 3), and it cannot overcome the pointer missed detection problem caused by the deviation of the pointer position from the image center caused by the inaccurate object detection if it is too small (7 × 7).

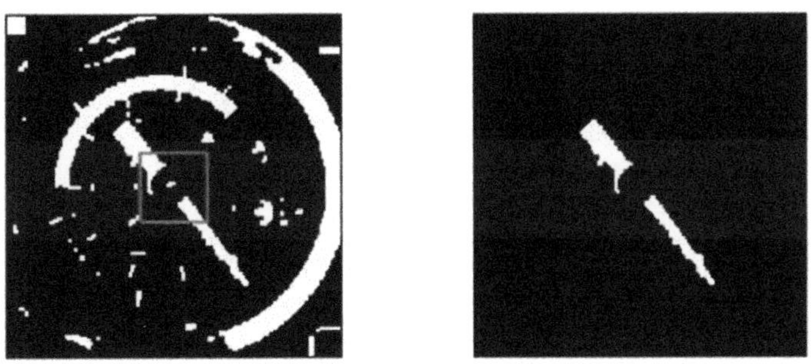

Figure 12. The situation of the broken pointer.

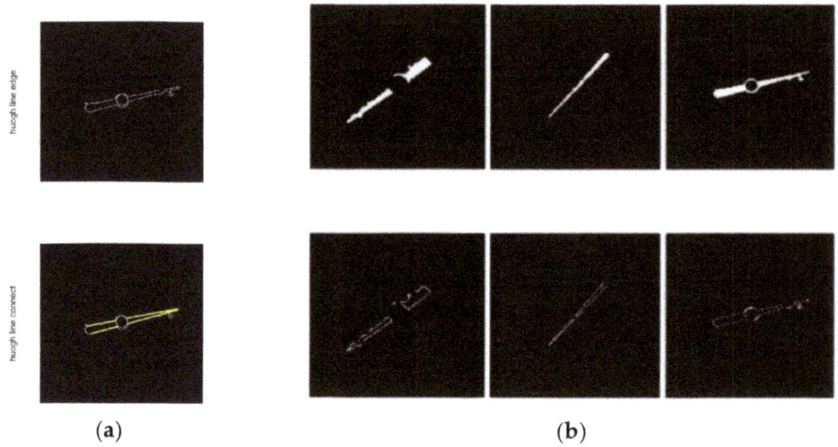

Figure 13. Line detection: (**a**) Perform boundary extraction on the filtered images of the connected domain; (**b**) Use Hough transform for straight line detection.

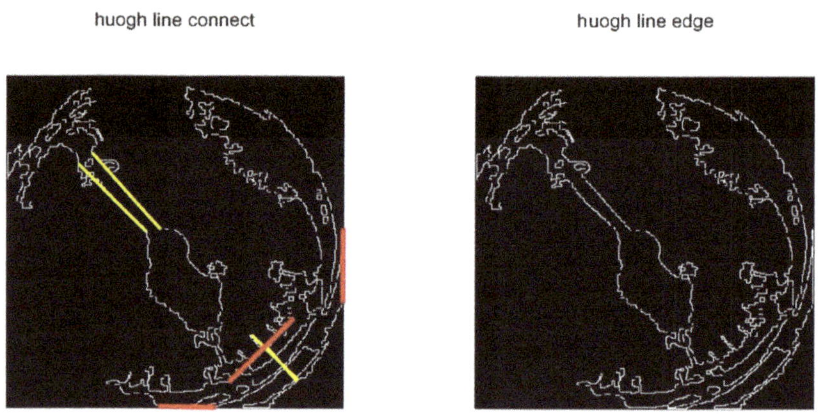

Figure 14. Straight-line filtering diagram: The extension of the yellow straight line passes through the central grid, while the extension of the red straight line does not pass through the central grid.

Put all the endpoints of all conforming straight lines together for linear fitting. Figure 15 shows the linear fitting effect of the endpoints of the straight line. The range limited by the leftmost and rightmost abscissas in the endpoints is used as the defined interval of the fitted straight line for indicating the pointer position. And the pointer center line is mapped to the original dial to observe the extraction effect of the pointer centerline.

The image processing algorithm is tested. The image processing algorithm has a large calculation burden, and the average time from image input to result output is 0.85 s.

5.3. *Image Reading*

After extracting the centerline of the pointer, the reading is obtained according to the deflection angle of the centerline. In this paper, the commonly used thermometers in substations are selected as the test objects, and 1/2 positive and 1/2 negative samples are selected as the test set. Table 6 shows the comparison of manual readings, existing research methods [3,4] readings, and readings obtained by the method used in this article. Since the identification process of different pointer instruments is similar, this conclusion can be applied to other types of instruments.

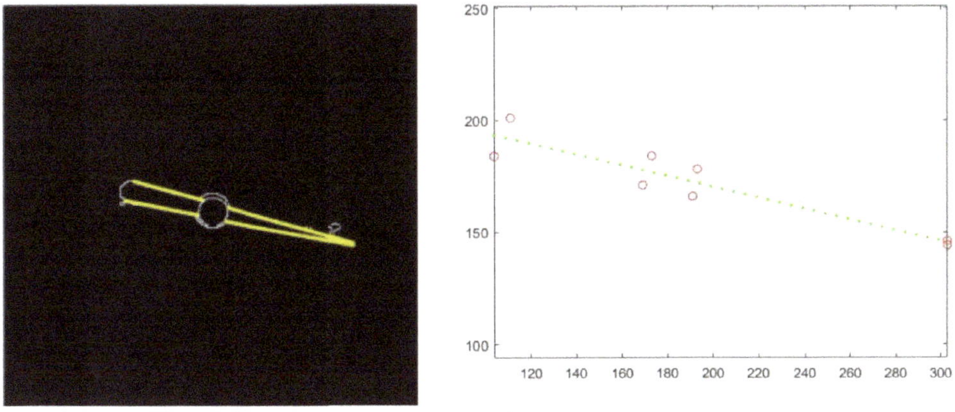

(a)

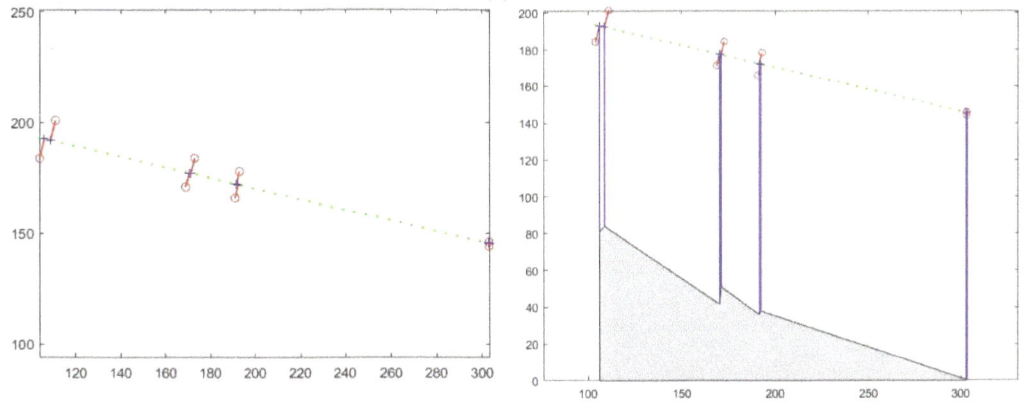

(b)

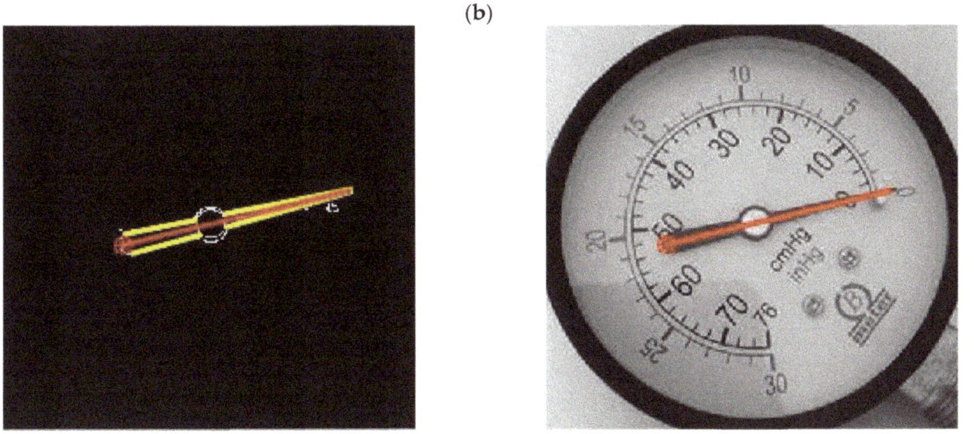

(c)

Figure 15. *Cont.*

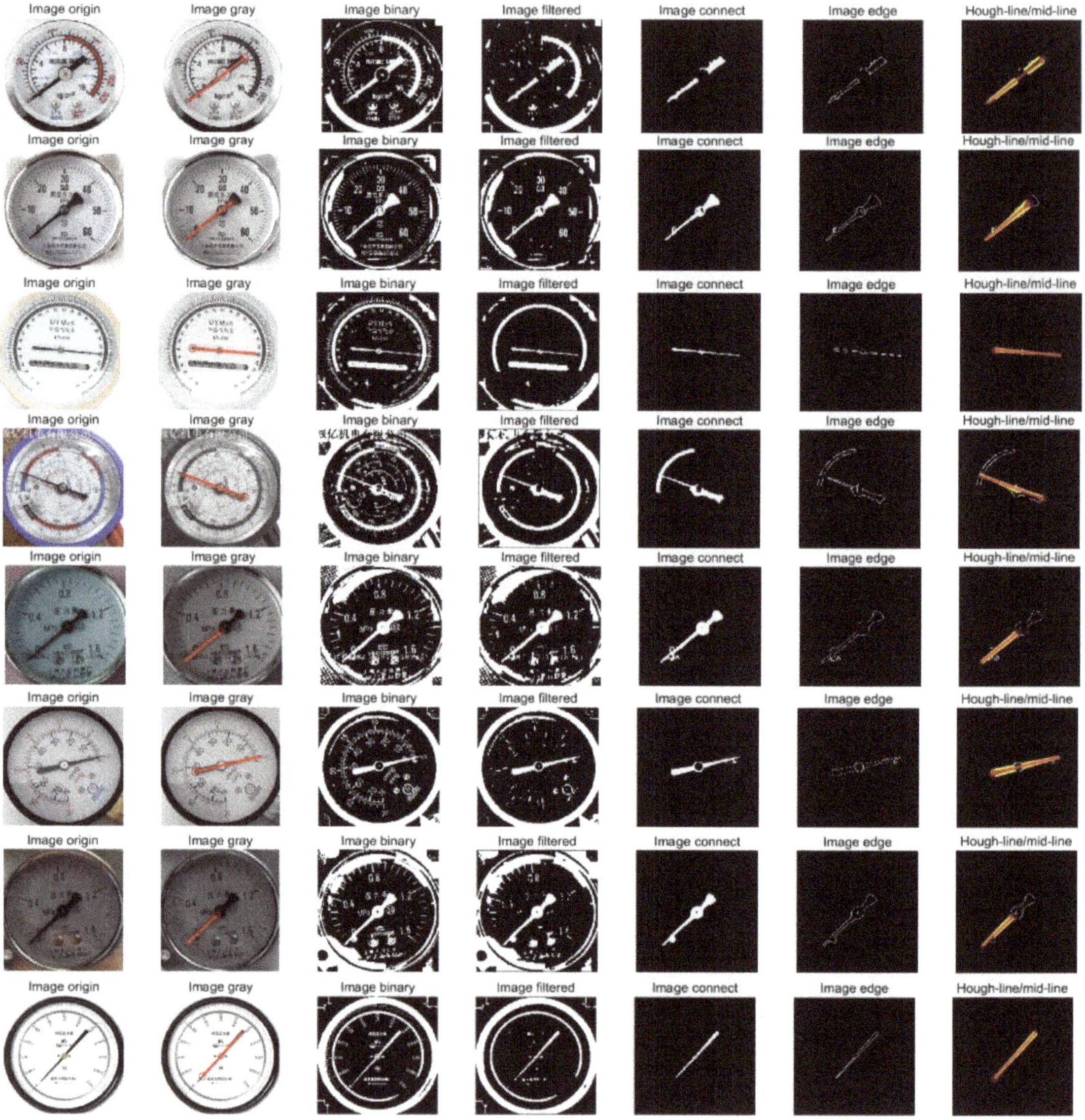

(d)

Figure 15. Pointer centerline extraction effect: (**a**) The linear fitting effect of the boundary line; (**b**) the vertical distance from the point on the boundary line to the fitted line, and the horizontal distribution of its square value; (**c**) the detected pointer center coincides with the pointer in the dial image when it is mapped to the original dial image; (**d**) shows the overall processing flow of extracting the final pointer centerline from the original dial image.

Table 6. Test results of different methods for reading recognition.

Number	Human Vision	Method in [3]	Relative Error (%)	Method in [4]	Relative Error (%)	Our Method	Relative Error (%)
1	41.121	39.140	1.651	39.564	1.297	40.481	0.533
2	38.725	33.548	4.314	34.176	3.790	35.952	2.310
3	30.710	35.954	4.370	25.354	4.463	26.720	3.325
4	47.601	42.364	4.364	42.689	4.093	43.894	3.089
5	39.399	35.021	3.648	35.637	3.135	36.746	2.211
6	43.054	40.364	2.242	40.854	1.833	41.946	0.923
7	36.732	37.962	1.025	35.086	1.372	36.333	0.332
8	34.137	31.541	2.163	31.877	1.883	32.091	1.705
9	57.380	55.100	1.900	59.324	1.620	58.273	0.744
10	46.332	44.326	1.672	44.638	1.411	45.624	0.590
11	54.462	53.920	0.452	53.214	1.040	54.237	0.187
12	86.906	89.785	2.399	90.102	2.663	84.785	1.762
13	41.987	39.542	2.038	39.563	2.020	40.069	1.592
14	33.205	35.248	1.702	36.420	2.679	34.350	0.954
15	36.888	35.147	1.451	34.695	1.827	37.699	0.682
16	74.691	73.210	1.234	72.913	1.481	73.645	0.871
17	40.342	35.124	4.348	36.451	3.242	37.250	2.576
18	58.933	58.322	0.509	58.360	0.477	58.485	0.373
19	40.986	36.853	3.444	37.021	3.304	38.279	2.255
20	41.248	40.985	0.219	40.830	0.348	41.164	0.070
Mean	/	/	2.257	/	2.199	44.905	1.354

To reduce the error of manual readings, we used the average result of readings of five testers. The measuring range of the test instrument is 120 °C. It can be seen from Table 6 that the method proposed in this paper is basically consistent with the manual reading result, with an average error of 1.354% and a maximum relative error of 3.325%. The error of this scheme is obviously smaller than that of the other two methods. It proves that the recognition system proposed in this paper is more accurate.

6. Summary

With regard to recognition of pointer instrument reading, this paper abandons traditional feature point detection algorithms with low robust such as SIFT and ORB but introduces deep learning-based Faster R-CNN model and improved accuracy. Aiming at the number of the data set, we expand the data set with the method of Poisson fusion and use the k-fold cross-validation to preprocess the data set to improve the quality of it to optimize the model parameters. As far as we know, this work is applied in the indicator reading recognition area for the first time. For this data set, we designed a classifier and get it integrated with the convolutional layers of VGG-16, ResNet-50, and MobileNet V2 for experiments. According to the experimental results, the classifier performed well, with VGG-16 achieving the best performance with an average prediction accuracy of 97.49%. Finally, the position of the pointer of the meter is detected by image processing methods such as Hough transform, and finally, the reading of the pointer meter is obtained. The average relative error of the pointer angle obtained by this scheme is no more than 1.354%. The experimental results prove that the accuracy and stability of the detection and recognition system are suitable for practical application.

Author Contributions: Conceptualization, L.W. (Lu Wang) and P.W.; methodology, Z.K.; software, L.W. (Linhai Wu); validation, P.W.; formal analysis, P.W.; investigation, L.W. (Lu Wang); resources, Z.K.; data curation, L.X.; writing—original draft preparation, L.W. (Lu Wang) and P.W.; writing—review and editing, L.X. and Z.K.; visualization, P.W.; supervision, L.W. (Lu Wang); project administration, P.H.; funding acquisition, Z.K. All authors have read and agreed to the published version of the manuscript.

Funding: This work was supported by the National innovation experiment project for College Students (Grant NO. 2021998090).

Data Availability Statement: Publicly available datasets were analyzed in this study. This data can be found here: [https://www.kaggle.com/wl13980027641/automatic-recognition-of-pointer-instruments, accessed on 24 February 2021].

Conflicts of Interest: The authors declare no conflict of interest.

References

1. Huo, F.; Wang, D.; Li, Z. Improved Recognition Method of Industrial Linear Pointer Meter. *J. Jilin Univ. (Inf. Sci. Ed.)* **2018**, *36*, 423–429.
2. Wang, T.; Liu, S. Numerical recognition method of pointer instrument. *J. Zhoukou Norm. Univ.* **2019**, *36*, 57–61.
3. Xu, P.; Zeng, W.; Shi, Y.; Zhang, Y. A Reading Recognition Algorithm of Pointer Type Oil—level Meter. *Comput. Technol. Dev.* **2018**, *28*, 189–193.
4. Fang, H.; Ming, Z.; Zhou, Y. Meter Recognition Algorithm for Equipment Inspection Robot. *Autom. Instrum.* **2013**, *28*, 10–14.
5. Li, W.; Wang, O.; Gang, Y. An automatic reading method for pointer meter. *J. Nanjing Univ. (Nat. Sci.)* **2019**, *55*, 117–124.
6. Xing, H.; Du, Z.; Su, B. Detection and recognition method for pointer-type meter in transformer substation. *Chin. J. Sci. Instrum.* **2017**, *38*, 2813–2821.
7. Duda, R.; Hart, P. Use of the hough transformation to detect lines and curves in pictures. *Commun. ACM* **1972**, *15*, 11–15. [CrossRef]
8. Illingworth, J.; Kittler, J. A survey of the Hough transform. *Comput. Vis. Graph. Image Process.* **1988**, *43*, 280. [CrossRef]
9. Ballard, D.H. Generalizing the hough transform to detect arbitrary shapes. *Read. Comput. Vis.* **1987**, 714–725. [CrossRef]
10. Lowe, D.G. Distinctive image features from scale-invariant keypoints. *Int. J. Comput. Vis.* **2004**, *60*, 91–110. [CrossRef]
11. Cho, M.; Kwak, S.; Schmid, C.; Ponce, J. Unsupervised Object Discovery and Localization in the Wild: Part-based Matching with Bottom-up Region Proposals. In Proceedings of the IEEE Conference on Computer Vision and Pattern Recognition, Boston, MA, USA, 7–12 June 2015.
12. Liu, Z.; Zhang, Y.; Chen, Y.; Fan, X.; Dong, C. Detection of Algorithmically Generated Domain Names Using the Recurrent Convolutional Neural Network with Spatial Pyramid Pooling. *Entropy* **2020**, *22*, 1058. [CrossRef]
13. Wang, W.; Tian, B.; Liu, Y.; Liu, L.; Li, J. Study on the Electrical Devices Detection in UAV Images based on Region Based Convolutional Neural Networks. *J. Geo-Inf. Sci.* **2017**, *19*, 256–263.
14. Girshick, R. Fast R-CNN. In Proceedings of the ICCV 2015: IEEE International Conference on Computer Vision, Santiago, Chile, 11–16 December 2015.
15. Jia, W.; Tian, Y.; Luo, R.; Zhang, Z.; Lian, J.; Zheng, Y. Detection and segmentation of overlapped fruits based on optimized mask R-CNN application in apple harvesting robot. *Comput. Electron. Agric.* **2020**, *172*, 105380. [CrossRef]
16. Joshi, R.C.; Yadav, S.; Dutta, M.K.; Travieso-Gonzalez, C.M. Efficient Multi-Object Detection and Smart Navigation Using Artificial Intelligence for Visually Impaired People. *Entropy* **2020**, *22*, 941. [CrossRef] [PubMed]
17. Liu, W.; Anguelov, D.; Erhan, D.; Szegedy, C.; Reed, S.; Fu, C.Y.; Berg, A.C. SSD: Single shot multibox detector. In Proceedings of the ECCV 2016: European Conference on Computer Vision, Amsterdam, The Netherlands, 7–12 June 2016.
18. Pérez, P.; Gangnet, M.; Blake, A. Poisson image editing. In *ACM SIGGRAPH 2003 Papers*; ACM: San Diego, CA, USA, 2003; Volume 22, pp. 313–318.
19. Sun, H.; Li, Z.; Jiang, Y.; Zhao, H.; Gu, C.; Shi, J.; Sheng, G.; Jiang, X. Approach of Data Augmentation Based on Context Modeling of Foreign Object Detection in Power Inspection. *Power Syst. Technol.* **2020**. [CrossRef]
20. He, S.; Yang, B. Local Structural and Visual Saliency Based Infrared and Visible Image Fusion with Poisson Reconstruction. *J. Univ. South China (Sci. Technol.)* **2020**, *34*, 62–70+76.
21. Shi, J.; Li, Z.; Gu, C.; Sheng, G.; Jiang, X. Research on Foreign Matter Monitoring of Power Grid With Faster R-CNN Based on Sample Expansion. *Power Syst. Technol.* **2020**, *44*, 44–51.
22. Shen, Z. *Research of Image Compositing Technology Based on Deep Learning*; East China Normal University: Shanghai, China, 2018.
23. Bengio, Y.; Grandvalet, Y. No Unbiased Estimator of the Variance of K-Fold Cross-Validation. *J. Mach. Learn. Res.* **2004**, *5*, 1089–1105.
24. He, K.; Zhang, X.; Ren, S.; Sun, J. Deep residual learning for image recognition. In Proceedings of the CVPR 2016: IEEE Conference on Computer Vision and Pattern Recognition, Las Vegas, NV, USA, 26 June–1 July 2016.
25. Simonyan, K.; Zisserman, A. Very deep convolutional networks for large-scale image recognition. *arXiv* **2014**, arXiv:1409.1556.
26. Krizhevsky, A.; Sutskever, I.; Hinton, G.E. ImageNet classification with deep convolutional neural networks. In Proceedings of the NIPS 2012: The Conference and Workshop on Neural Information Processing Systems, Lake Tahoe, CA, USA, 3–8 December 2012.
27. Howard, A.G.; Zhu, M.; Chen, B.; Kalenichenko, D.; Wang, W.; Weyand, T.; Andreetto, M.; Adam, H. MobileNets: Efficient Convolutional Neural Networks for Mobile Vision Applications. *arXiv* **2017**, arXiv:1704.04861.
28. Zhang, Z.; Sabuncu, M.R. Generalized Cross Entropy Loss for Training Deep Neural Networks with Noisy Labels. *arXiv* **2018**, arXiv:1805.07836.

29. Yao, H.; Zhu, D.-L.; Jiang, B.; Yu, P. Negative log likelihood ratio loss for deep neural network classification. *Adv. Intell. Syst. Comput.* **2019**, *1069*, 276–282.
30. Wei, J.; Gao, Z. Prediction Method of Spacecraft On-orbit Environment Based on OTSU Binarization. *Comput. Meas. Control* **2020**, *28*, 75–79.
31. Wang, Y. Workpiece surface image segmentation based on ontrast enhancement and maximum entropy. *Foreign Electron. Meas. Technol.* **2019**, *38*, 37–40.
32. Gao, H.; Wang, W. New connected component labeling algorithm for binary image. *J. Comput. Appl.* **2007**, 2776–2777.

entropy

MDPI

Article

Interpolation with Specified Error of a Point Series Belonging to a Monotone Curve

Yevhen Havrylenko [1], Yuliia Kholodniak [1], Serhii Halko [1], Oleksandr Vershkov [1], Larysa Bondarenko [1], Olena Suprun [1], Oleksandr Miroshnyk [2,*], Taras Shchur [3], Mścisław Śrutek [4] and Marta Gackowska [4]

1 Department of Information Technologies of Design named after V. M. Naidysh, Dmytro Motornyi Tavria State Agrotechnological University, 18 B.Khmelnytsky Ave, 72312 Melitopol, Ukraine; yevhen.havrylenko@tsatu.edu.ua (Y.H.); yuliya.kholodnyak@tsatu.edu.ua (Y.K.); serhii.halko@tsatu.edu.ua (S.H.); oleksandr.vershkov@tsatu.edu.ua (O.V.); larysa.bondarenko@tsatu.edu.ua (L.B.); olena.suprun@tsatu.edu.ua (O.S.)
2 Department of Electricity and Energy Management, Kharkiv Petro Vasylenko National Technical University of Agriculture, 19 Rizdviana Str., 61052 Kharkiv, Ukraine
3 Department of Cars and Tractors, Faculty of Mechanics and Energy, Lviv National Agrarian University, 1 Volodymyr Great Str., 80381 Dubliany, Ukraine; shchurtg@gmail.com
4 Faculty of Telecommunications, Computer Science and Electrical Engineering, UTP University of Science and Technology in Bydgoszcz, 85-796 Bydgoszcz, Poland; mscislaw.srutek@utp.edu.pl (M.Ś.); marta.gackowska@utp.edu.pl (M.G.)
* Correspondence: omiroshnyk@khntusg.info; Tel.: +380-97-950-6033

Citation: Havrylenko, Y.; Kholodniak, Y.; Halko, S.; Vershkov, O.; Bondarenko, L.; Suprun, O.; Miroshnyk, O.; Shchur, T.; Śrutek, M.; Gackowska, M. Interpolation with Specified Error of a Point Series Belonging to a Monotone Curve. *Entropy* **2021**, *23*, 493. https://doi.org/10.3390/e23050493

Academic Editor: Ernestina Menasalvas

Received: 29 March 2021
Accepted: 19 April 2021
Published: 21 April 2021

Abstract: The paper addresses the problem of modeling a smooth contour interpolating a point series belonging to a curve containing no special points, which represents the original curve with specified accuracy. The contour is formed within the area of possible location of the parts of the interpolated curve along which the curvature values are monotonously increased or decreased. The absolute interpolation error of the point series is estimated by the width of the area of possible location of the curve. As a result of assigning each intermediate point, the location of two new sections of the curve that lie within the area of the corresponding output section is obtained. When the interpolation error becomes less than the given value, the area of location of the curve is considered to be formed, and the resulting point series is interpolated by a contour that lies within the area. The possibility to shape the contours with arcs of circles specified by characteristics is investigated.

Keywords: monotone curve; tangent circle; adjacent circle; area of location of the curve; contour

1. Introduction

Modeling is an effective tool for investigating objects, phenomena, and processes. Geometric modeling of an object often determines its functional properties. Such objects are, first of all, items bounded by functional surfaces which ensure the laminar nature of the item's flow around by the environment–gas, liquid, or loose materials [1–3]. Examples of items limited by functional surfaces are automobile and aircraft hulls, working bodies of agricultural machinery, impeller blades of turbines, and compressors.

Complex surfaces are typically modeled based on linear frameworks whose elements are formed by interpolation of point series. The operating performance of the item is ensured by the geometric characteristics of the interpolated curves.

The improved aero- and hydrodynamic properties of the surface ensure the use of lines with regular variations in the values of characteristics and a minimum number of special points according to the statement of the problem as elements of the framework [4,5]. For a plane curve, these are the junction points of the convex and concave parts and the points where the curvature values are extreme. A smooth, plane curve that contains no special points will be referred to as a curve with a monotone curvature change or a monotone curve.

If a surface is modeled with the aim of creating a copy of an existing item (reverse engineering), the task of interpolating a point series becomes even more complex. In this case, the interpolation accuracy requirement is added to the requirement of ensuring the necessary characteristics of the interpolated curve [6].

In order to solve the stated problem, it is necessary to develop an interpolation method which ensures control of the change pattern of its characteristic values along the modeled curve, the possibility of local correction of the resulting solution, and prevention of uncontrolled emergence of special points.

The possibility of local correction of the shape of the interpolated curve while controlling its characteristics at output points is provided by methods of interpolation of the point series with a contour. The contour is formed by sections of analytically defined curves, which are connected at output points with a specified order of smoothness [7]. These are methods of interpolation by sections of second-order curves, Bézier curves, and B-splines [8–12].

Among the methods of interpolation of the point series with a contour, B-spline interpolation provides the greatest possibilities to adjust the sections of the contour.

A spline is defined by a set of control points, each of which has a transition function. It is a composite curve, each segment of which defines a separate equation. The curve approximates the broken line connecting the control points. The configuration of this broken line makes it possible to control the presence of inflection points in the B-spline.

The main disadvantage of B-spline interpolation as well as interpolation with other analytically defined curves is the lack of mechanisms to control the occurrence of points with extreme curvature. This disadvantage reduces the possibilities of using analytically defined curves while modeling shapes with the given characteristics and, above all, contours designed for modeling surfaces with specified functional properties.

Papers [7,9] solve the problem of interpolation of a point series by a contour consisting of smoothly joined arcs of circles. When each section is formed by an arc of one circle, local adjustment of the contour is not possible. The change of the radius of any of the arcs changes the configuration of all the sections of the contour. The problem is solved by forming sections with two or more arcs of circles at fixed positions of tangents to the contour at output points. In this case, adjustment of an individual section does not imply a reconfiguration of other sections of the contour.

The problem of forming a contour along which the radii of the arcs of circles change monotonically as well as the problem of ensuring the specified interpolation accuracy is not considered in [7,9].

The aim of the study is to develop a method for forming a smooth contour interpolating a point series that represents a monotone curve with specified accuracy.

In order to achieve this aim, the following objectives should be pursued:

- develop a method for forming the area of the monotone curve interpolating a given point series whose width does not exceed a specified value;
- to develop a method for forming a smooth contour consisting of arcs of circles, which interpolates the given point series and is located within the area of the monotone curve;
- to investigate the possibilities of the proposed method in solving the problem of interpolation of a given point series.

2. Materials and Methods

Formation of the area of location of the monotone curve is illustrated by the point series, which can be interpolated by a curve with a monotonous increase in the radii of curvature. Let us consider the following tasks:

- analyzing a output point series, which makes it possible to identify its parts that can be interpolated by a monotone curve;
- determining the absolute interpolation error of a point series by a monotone curve.

The analysis of the output point series is based on determination of the radii of adjacent circles (AC), each of which passes through three consecutive points in the series. The part of the point series along which the range of AC increases or decreases may be interpolated by a monotone curve along which the radii of curvature increase or decrease accordingly.

The absolute interpolation error of a section of a point series of a monotone curve is determined by the width of the area bounded by arcs of the corresponding AC (Δ_i^{AC})

For section $(i, i+1)$, this area is bounded by the arc AC_i, which passes through the points $i-1$, i, $i+1$ and the arc AC_{i+1} (Figure 1).

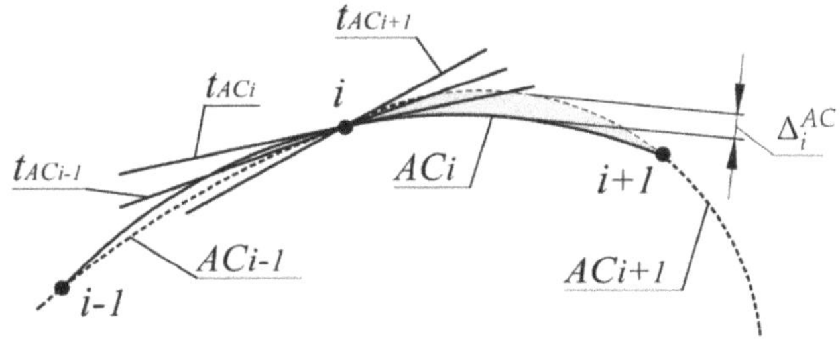

Figure 1. Area of location of the monotone curve.

If the positions of the tangents t_i, t_{i+1} and the radii of curvature R_i, R_{i+1} are known at the points of the monotone curve i and $i+1$, then the absolute error of interpolation of its sections can be determined more precisely.

It is established that, for lines t_i and t_{i+1} tangent to the monotone curve along which the radii of curvature increase monotonously, the distances from the intersection point of the lines $(point\ T)$ to the points of tangency with the curve (Figure 2) correlate as

$$|i, T| < |T, i+1| \tag{1}$$

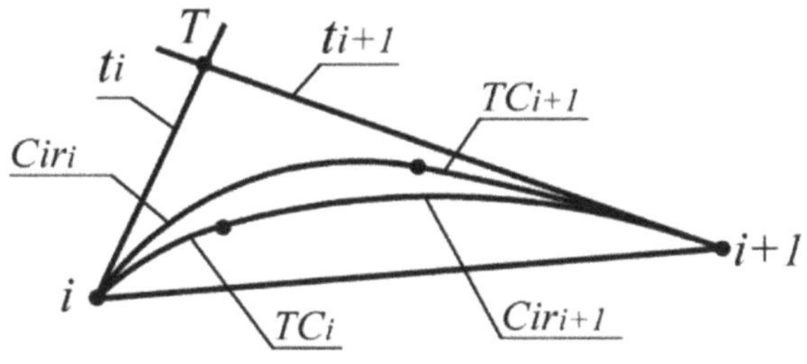

Figure 2. Refinement of the area of location of the monotone curve.

Correlation (1) can be used as a criterion for the correct assignment of tangent lines in the formation of a contour with specified accuracy representing the monotone curve. For section of the curve $(i, i+1)$, the error is estimated by the maximum distance between the two boundaries (Figure 2):

- a boundary consisting of the arc of the tangent circle at point i (TC_i) and the arc of the circle tangent to TC_i and line t_{i+1} at point $i+1(Cir_{i+1})$. For the curve along which the radii of curvature increase, this limit of the area is lower (closest to the chord $[i, i+1]$);
- a boundary consisting of the arc of the tangent circle at point $i+1$ (TC_{i+1}) and the arc of the circle tangent to TC_{i+1} and line t_i at point i (Cir_i).

If the values of absolute interpolation error are greater than the given value, more interpolated curve points specifying the interpolated curve must be assigned to the corresponding sections of the original curve.

If the values of absolute interpolation error Δ_i^{AC} are greater than the given value, more interpolated curve points specifying the interpolated curve must be assigned to the corresponding sections of the original curve.

The final solution can be represented in the form of an accompanying broken line, the distance from which to the curve with the specified geometric characteristics does not exceed a predetermined value. The use of contours consisting of straight line sections in the modeling of functional surfaces reduces their aero- and hydrodynamic properties [7]

Let us define the constraints on the location of the section of the curve located between two consecutive output points.

The most accurate estimate of the area of the monotone curve is possible when the position of the centers of curvature C_i *and* C_{i+1} for points i *and* $i+1$ is known. To define the boundaries of the area, the positions of tangent lines to the monotone curve at output points are predefined. Positions of the tangents must meet condition (1). One option for determining the necessary location of the tangents is to assign them within the ranges limited by AC at output points. For i point of the series, the range is bounded by the tangent to AC_i (t_{ACi}) and the tangent nearest to it AC_{i-1} *and* AC_{i+1} (Figure 1).

Once the tangent to the monotone curve has been assigned at each of the output points, the upper and lower boundaries of the area of possible location of the monotone curve interpolating the point series are determined.

The lower boundary of the area is represented by the arc TC_i and the arc of the circle Cir_{i+1} tangent to the curve at point $i+1$, and with TC_i at some point A (Figure 3).

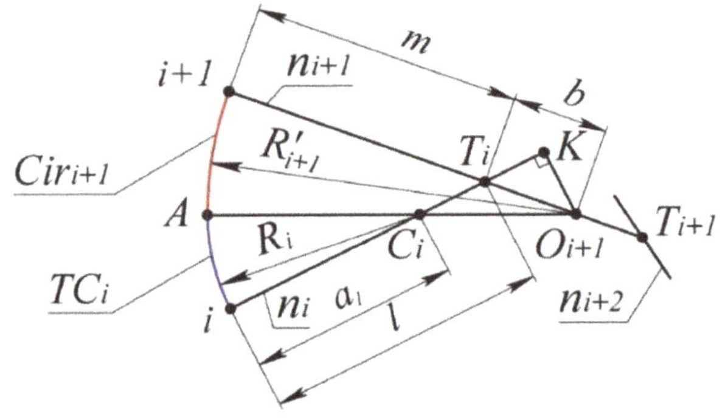

Figure 3. Lower boundary of the area of location of the section of the monotone curve .

The given data for determining the lower boundary of the area are the position of the normal of the curve n_i and n_{i+1} at points i and $i+1$, and the position of the center of curvature C_i.

The problem reduces to determining the position of the center of Cir_{i+1} (point O_{i+1}). Let us introduce the following notations to be used for the problem: T_i—the point where normals n_i and n_{i+1} intersect (Figure 3); T_{i+1}—the point where normals n_{i+1} and n_{i+2} intersect; $b = |T_i, T_{i+1}|$; $a_1 = |C_i, T_i|$; $m = |i+1, T_i|$, S_i - the area of the triangle C_i, T_i, T_{i+1}.

The position of the center of Cir_i within the segment $[T_i, T_{i+1}]$ is unambiguously determined by the correlation:

$$M_{i+1} = \frac{|T_i, O_{i+1}|}{b} \tag{2}$$

Based on the tangency of the arcs $i - A$ and $A - i + 1$:

$$|C_i, O_{i+1}| = m - R_i + M_{i+1} \cdot b \tag{3}$$

Based on the triangle C_i, O_{i+1} K:

$$|C_i, O_{i+1}|^2 = |K, C_i|^2 + |K, O_{i+1}|^2 = a_1^2 + 2a_1|T_i, K| + M_{i+1}^2 \cdot b^2 \tag{4}$$

The length of the segment $[T_i, K]$ shall be determined as a leg of the triangle T_i, K, O_{i+1} : $|T_i, K|^2 = M_{i+1}^2 \cdot b^2 - |K, O_{i+1}|^2$.

The length of the segment$[K, O_{i+1}]$ shall be expressed through the area (S_1) of the triangle C_1, O_{i+1}, T_i as $|K, O_{i+1}| = \frac{2S_1}{a_1}$.

By expressing $S_1 = MS_i$, let us bring expression (4) to:

$$|C_i, O_{i+1}|^2 = a_1^2 + 2M_{i+1}\sqrt{a_1^2 b^2 - 4S_i^2} + M_{i+1}^2 \cdot b^2 \tag{5}$$

Having plugged (3) into (5) after transformations, we obtain:

$$M_{i+1} = \frac{a_1^2 - (m - R_i)^2}{2(b(m - R_i) - \sqrt{a_1^2 b^2 - 4S_i^2})} \tag{6}$$

The upper boundary is formed by the arc TC_{i+1} and the arc of the circle Cir_i, which is tangent to the curve at point i, and with TC_i at some point B.

The problem of determining the upper boundary of the area of the monotone curve is reduced to calculating the coordinates of the center of Cir_i (point O_i). The given data for the problem are the position of the normals n_i, n_{i+1}, n_{i+2} and the center of curvature C_{i+1}.

The position of O_i within the segment $[T_{i+1}, T_i]$ (Figure 4) is determined by the correlation:

$$M_i = \frac{|O_i, T_i|}{a} \tag{7}$$

where a = $|T_{i+1}, T_i|$.

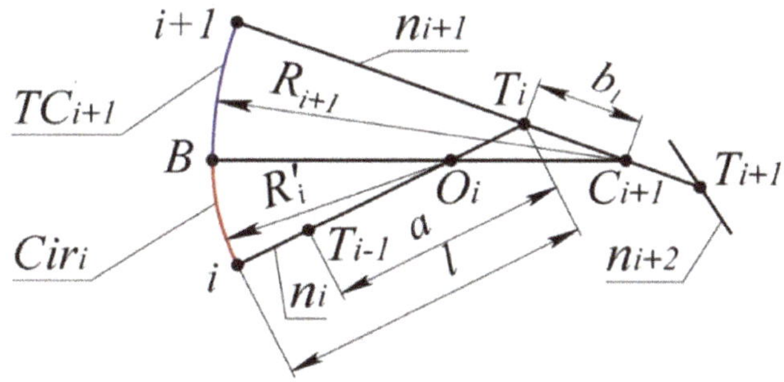

Figure 4. Upper boundary of the area of location of the section of the monotone curve .

As a result of manipulations similar to those done in the derivation of expression (6), we obtain:

$$M_i = \frac{b_1^2 - (R_{i+1}^2 - l)^2}{2(a(R_{i+1} - l) - \sqrt{a^2 b_1^2 - 4S_{i+1}^2})} \tag{8}$$

where $l = |i, T_i|$, $a = |T_{i+1}, T_i|$, $b_1 = |T_i, C_{i+1}|$, and S_{i+1} is the area of the triangle T_{i+1}, T_i, C_{i+1}.

At the specified position of normals n_i, n_{i+1} and the center of curvature C_{i+1}, the radius of Cir_i is the maximum possible radius of curvature of the monotone curve at point i. Similarly, the position of C_i determines the minimum possible radius of curvature at point $i + 1 - R_{i+1}' = |O_{i+1}, i + 1|$.

Based on this criterion, let us determine the area of the monotone curve, driven by a random point series.

Provided that the radii of curvature along the curve increase monotonously, the minimum radius of curvature at point i R_i^{min} equals zero. Here, point i is considered to be the tangent circle of zero radius. In this case, the lower boundary of the interpolated curve is the arc of circle Cir_{i+1}, which is tangent to line t_{i+1} at point $i+1$ and passes through point i (Figure 5). The minimum possible radius of curvature that can be assigned at point $i+1$, (R_{i+1}^{min}) is equal to the radius of Cir_{i+1}.

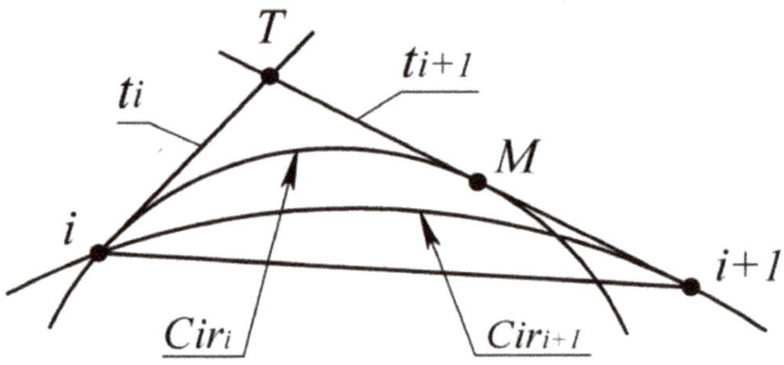

Figure 5. Finding possible values of the radii of curvature.

The maximum radius of curvature that can be assigned at point $i+1$ (R_{i+1}^{max}) is equal to infinity. In this case, the upper boundary of the area of location of the formed curve is the curve consisting of the arc of the circle tangent to T_i at point i and to t_{i+1} at point M Cir_i and the segment $[M; i+1]$ (the tangent circle of infinite radius). At point i, the maximum possible radius of curvature (R_i^{max}) equals the radius of Cir_i.

All curves with monotonous increase of the radii of curvature, having at points i and $i+1$ tangents t_i and t_{i+1} respectively, pass within the area bounded by the resulting composite curves. The radii of curvature of the curve satisfying the conditions of the problem at points i and $i+1$ must belong to the following ranges:

$$0 \le R_i \le R_i^{max} \; and \; R_{i+1}^{max} \le R_{i+1} \le \infty \tag{9}$$

If a monotone curve interpolates a sequence of points 1...n, then the minimum radius of curvature can be equal to zero only at the first point, and the maximum radius can be equal to infinity only at the last point.

In this case, the lower boundary of the area in which the curve is located shall be determined in the following way.

1. In sections 1 and 2, the area is bounded by the arc of the circle passing through point 1 and having a common tangent to the curve at point 2. The specified circle is taken as the tangent circle at point 2, whose radius is the minimum possible. Let us mark this circle as TC_2^{min}
2. Taking TC_2^{min} as the tangent circle at point 2, we determine the circle tangent to it and to the monotone curve at point 3. Let us denote this circle TC_3^{min}. The position of the center shall be determined by the procedure presented in Figure 3 for point O_{i+1}. The lower boundary of the area of location of the curve in sections 2 and 3 is formed by smoothly conjoined arcs of circles TC_2^{min} и TC_3^{min}
3. Based on the location and the size of TC_3^{min}, the lower boundary of the area in sections 3 and 4 is similarly determined, as well as the boundaries of the remaining sections.

The upper boundary of the area of location of the monotone curve is determined starting with the last section by the following scheme.

1. In section $(n-1, n)$, the boundary of the area consists of the tangent line segment at point n (t_n) and the arc of circle TC_{n-1}^{max}, which is tangent to t_n and has a common

tangent to the curve at point $n-1$. The specified circle is taken as the tangent circle at $n-1$, whose radius is the maximum possible.

2. In section $(n-2, n-1)$, the upper boundary of the area consists of the arcs of circles TC_{n-1}^{max} and TC_{n-2}^{max} tangent to it which has a common tangent to the monotone curve at point $n-2$. The center of TC_{n-2}^{max} shall be defined by the procedure presented in Figure 4 for point O_i.
3. The upper boundary of the area of location of the curve is determined in sequence from section to section, similarly to the boundary of section $(n-2, n-1)$.

The absolute error of interpolation (δ_i) of a point series by a monotone curve is estimated by the width of its possible location. In section $i \ldots i+1$, the width of the area shall be defined as the distance between lines t_i^u and t_i^d, which are parallel to segment $[i, i+1]$ and are tangent to the lower and upper boundaries of the area respectively (Figure 6).

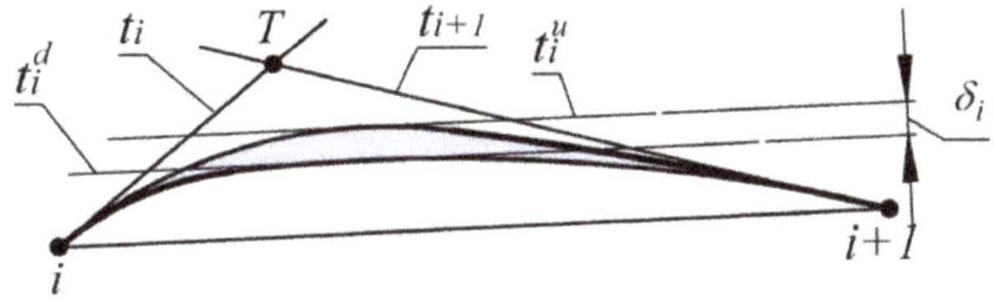

Figure 6. Finding the absolute interpolation error.

If the interpolation error of the output point series is greater than the assigned value, the width of the area of location of the curve is reduced by assigning intermediate points belonging to it. The intermediate point shall be assigned to the line passing through the middle of segment $[i, i+1]$, at right angles to this segment within the area of location of the curve. By assigning each intermediate point, we obtain the location of two new sections of the curve which lie within the area of the corresponding output section.

After the interpolation error becomes less than the specified value, the area of the monotone curve is considered to be formed, and the resulting point series is interpolated by a contour which lies within the area of location of the curve. It is appropriate to take the minimum possible processing error on a numerical control machine, which is 10^{-3} mm, as the value that cannot exceed the specified interpolation error [6].

To solve the posed problem, it is necessary to ensure the following:

- the presence of a common tangent to the contour and the monotone line curve at output points;
- the increase of the radii of curvature along the contour in the same direction as the monotone curve.

Let us consider the possibility of forming contours with specified characteristics of the arcs of circles.

The formation of a contour consisting of two arcs of circles which replace section $[i, i+1]$ with a monotone curve along which the radii of curvature monotonously in-crease is represented in Figure 7.

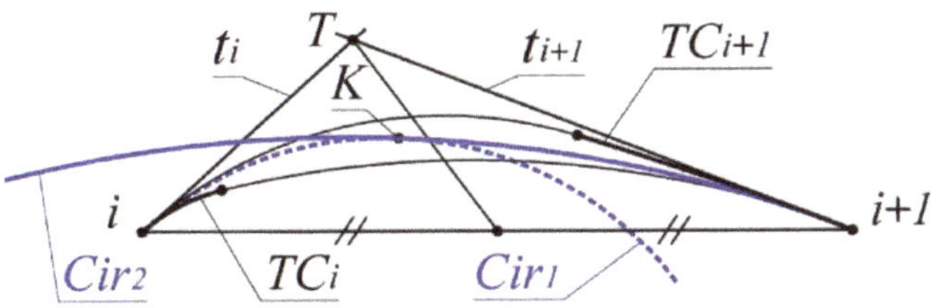

Figure 7. Replacement of the monotone curve with arcs of circles.

The circle defining the first arc (Cri_1) is driven by a tangent with a monotone curve at point i and passing through some point assigned within the area of location of the monotone curve. For instance, this point may be the middle of a segment that is bounded by the intersection points of the median of triangle $i, T, i+1$ with the boundaries of the area of location of the monotone curve.

The circle to which the second arc (Cri_2) belongs is defined by touching the monotone curve at point i+1 and by touching (Cri_1) at some point K. As a result, we obtain a circle with the correlation of the radii:

$$R_i^{min} < R_1 < R_i^{max} \text{ and } R_{i+1}^{min} < R_2 < R_{i+1}^{max} \text{ and } R_1 < R_2 \tag{10}$$

where R_1 and R_2-radii of (Cri_1) and (Cri_2), respectively; R_i^{min}, R_i^{max}, R_{i+1}^{min}, R_{i+1}^{max}-radii of TC_i^{min}, TC_i^{max}, TC_{i+1}^{min}, TC_{i+1}^{max}.

By similarly forming the arcs replacing the remaining sections of the monotone curve, we get a contour interpolating the whole point series along which the radii of the arcs of circles increase monotonously.

The monotonic increase of the radii of the circles along the contour in the same direction as the increase of the radii of curvature along the monotone curve, as well as the common tangents of the contour and the curve at output points, ensure the position of the contour within the area of location of the monotone curve.

The main drawback of forming the contour by arcs of circles is the irregular variation of the values of curvature at the points at which they are conjoined. Reducing the effect of the mentioned drawback on the functional characteristics of the surface while using the contour as an element of the framework is possible by increasing the number of arcs constituting the contour while reducing the difference between the values of the radii of the circles that determine them.

3. Results and Discussion

The possibilities of the method for shaping contours proposed in the paper are investigated by the example of interpolation of a sequence of ten points. The position of the specified points was determined based on the condition of a monotonous increase in the AC radii along the point series. The characteristics of the point series: coordinates of output points—$i(x_i, y_i)$; distance between adjacent output points—$|i, i+1|$; radii of AC defined by the point series—R_{Aci}; absolute error of interpolation of a point series by a monotone curve, specified by the coordinates of output points—Δ_i^{AC} are given in Table 1.

Table 1. The characteristics of the point series.

i	x_1	y_1	$\|i, i+1\|$	R_{Aci}	Δ_i^{AC}
1	0	0	5.28	-	-
2	0.21	−5.28	7.20	7.41	0.4966
3	−5.11	−10.13	15.57	13.30	0.9431
4	−20.48	−7.65	25.24	20.04	1.8869
5	−28.68	16.23	28.82	32.08	1.4296
6	−13.87	40.95	34.90	53.22	0.8614
7	17.95	55.28	36.29	74.21	1.1654
8	54.24	53.04	49.65	151.68	1.3118
9	100.94	36.19	155.96	420.54	–
10	230.38	−50.82	–	–	–

The monotonous increase of RAC_i along the point series allows for interpolating it by the monotone curve along which the values of the radii of curvature increase. The absolute interpolation error is estimated as the maximum distance between the arcs AC_i and AC_{i+1}, bounded by points i and $i+1$ (Figure 1). In the first and the last sections, the value Δ_i^{AC} is not defined as AC_1 and AC_{10} do not exist.

The next step in formation of the contour at output points is denoting the position of the tangents to the monotone curve interpolating the point series and the refined area of its possible location at each of the sections. The positions of the tangents are defined within the ranges bounded by the corresponding AC (Figure 2). The length of the sides of triangles obtained from assigning the tangent lines–$|i, i+1|$, $|i, T|$, $|i+1, T|$; the absolute error of interpolation of a point series by a monotone curve driven by the coordinates of output points and the position of tangents at these points at each of the sections–Δ_i; the exceedance of the lower boundary of the area of location of the interpolated curve over the segments, connecting the corresponding output points h_i, is given in Table 2.

Table 2. The result in formation of the contour.

	BT				
i	$\|i, i+1\|$	$\|i, T\|$	$\|i+1, T\|$	h_i	Δ_i
1	5.28	3.07	3.12	0.7327	0.0140
2	7.20	3.78	3.83	0.5906	0.0078
3	15.57	7.80	9.97	1.8639	0.0218
4	25.24	13.53	16.52	3.9238	0.0595
5	28.82	13.67	16.71	2.2022	0.0629
6	34.90	16.95	19.53	2.6298	0.0922
7	36.29	15.63	21.32	1.5500	0.2184
8	49.65	20.41	29.53	1.2163	0.2257
9	155.96	69.13	88.31	4.8396	1.1281

The boundaries of the area of location of the interpolated curve are determined by the procedure presented in Figure 6. The absolute interpolation error in section $i \ldots i+1$ is defined as the distance between the lines tangent to the upper and lower boundaries of the area of location of the curve and parallel to section $[i, i+1]$.

The maximum absolute interpolation error was 1.1281 mm in section 9...10.

In order to reduce the maximum absolute interpolation error in reference section 9...10, an intermediate point and a tangent line to the monotone curve are assigned. The intermediate point is assigned within the area of location of the interpolated curve in section 9...10 on a line which passes through the middle of section [9,10] at right angles to segment B. As a result, we obtain a point series consisting of 11 points where the intermediate point is assigned No. 10. The position of the point is determined by the coordinates x_i= 168.62, y_i = −2.92.

The characteristics of the area of possible location of the monotone curve interpolating the point series are given in Table 3.

Table 3. The characteristics of the area of possible location.

	BT				
i	$\|i, i+1\|$	$\|i, T\|$	$\|i+1, T\|$	h_i	Δ_i
1	5.28	3.07	3.12	0.7327	0.0140
2	7.20	3.78	3.83	0.5906	0.0078
3	15.57	7.80	9.97	1.8639	0.0218
4	25.24	13.53	16.52	3.9238	0.0595
5	28.82	13.67	16.71	2.2022	0.0629
6	34.90	16.95	19.53	2.6298	0.0922
7	36.29	15.63	21.32	1.5500	0.2184
8	49.65	20.41	29.53	1.2163	0.2857
9	78.16	37.03	41.39	1.5356	0.1753
10	78.16	37.63	40.65	1.0480	0.0975

For the obtained point series, the maximum absolute interpolation error was 0.2257 mm in section 8...9. By assigning intermediate points, it is possible to form a point series consisting of any number of points, which determines the arbitrarily small interpolation error of the monotone curve.

The characteristics of the contour consisting of smoothly joined arcs of circles interpolating a sequence of eleven points are given in Table 4.

Table 4. The characteristics of the contour.

i	R_1^i	R_2^i	H_i	δ_i	R_1^i
1	4.9902	5.1701	0.7461	0.0134	4.9902
2	11.0191	11.3487	0.5962	0.0056	11.0191
3	11.4630	19.9296	1.8851	0.0212	11.4630
4	19.9359	34.1886	3.9828	0.059	19.9359
5	34.5156	53.4423	2.2646	0.0624	34.5156
6	53.5381	73.6735	2.7214	0.0916	53.5381
7	76.7344	145.9776	1.7641	0.2141	76.7344
8	163.8066	352.0208	1.4783	0.2620	163.8066
9	428.4307	537.2404	1.6269	0.0913	428.4307
10	660.3963	775.5817	1.0971	0.0483	660.3963

Each of the sections of the contour consists of two arcs of circles of radius R_1^i and R_2^i, accordingly, which possess a common tangent to the monotone curve at output points. The maximum exceedance of the section of the contour over the segment connecting the respective output points is indicated in the table as H_i. The absolute error with which the contour replaces the monotone curve δ_i is estimated by the maximum distance from the arcs of circles comprising the contour to the boundaries of the area of location of the corresponding sections of the monotone curve.

Formation of the sections of the contour by the procedure presented in Figure 7 ensured its location within the area of possible location of the monotone curve. The maximum absolute error with which the contour replaces the monotone curve was 0.2620 mm in section 8...9.

Figure 8 represents the contour formed in the CAD system Solidworks, the characteristics of which are presented in Table 4. With the aid of the CAD system, the graph of change in curvature along the contour is formed. The resulting graph shows its monotonous change and irregular variation at joining points of the arcs of circles. The possibility of increasing the number of arcs forming the contour makes it possible to reduce arbitrarily the difference of the curvature values at their joining points.

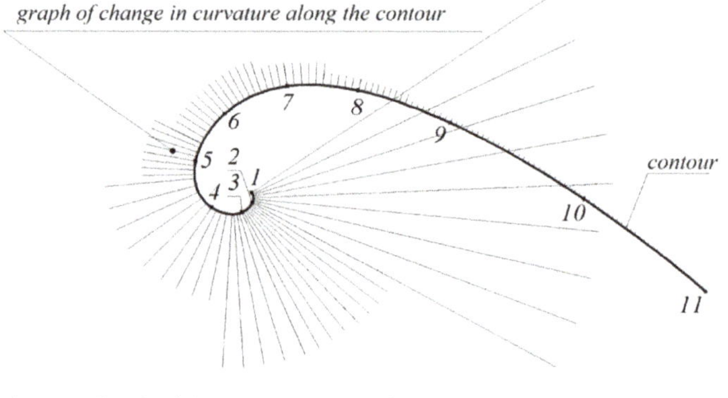

Figure 8. Graph of change in curvature along the contour.

The method proposed in the article was tested when creating a model of an impeller of an axial and radial flow turbocharger. The initial data for the design of the geometric model of the impeller were taken from a drawing designed for quality assurance of the finished product. The drawing contains tabulated data specifying an ordered array of

77 points belonging to the blade surface. The hub is driven by the axis of rotation and the generating line.

The model of the blade surface was designed on the basis of a framework, consisting of eleven plane sections. Each plane section was initially represented by seven points. Analysis of the original point series showed a monotone dynamics of changes in the values of the radii of the AC along each of them.

For example, the original point series representing the fourth section of the blade surface specifies the following radii of successive AC—23.65, 29.18, 40.37, 65.74 and 160.08 mm. The maximum error with which the original point series represents a monotone curve is 3.017×10^{-2} mm. The specified error is determined by the width of the band limited by a sequence of adjacent circles specified by a point series. As a result of assigning the positions of the intermediate points, a point series is obtained consisting of twenty-five points representing the horizontal section of the blade surface. At the points obtained, the positions of the adjacent circles of the monotone curve are assigned and the area of its possible location is determined. The error values with which the point series represents the sections of the monotone curve are within the range from 2×10^{-4} mm to 5×10^{-4} mm. The points are interpolated by the obtained compound curve of the circles.

The eleven formed contours, each of which interpolates a point series consisting of twenty-five points, make up a family of generating lines of the framework of the blade surface model. The contours were imported into the CAD system SolidWorks and a model of the working surface of the blade was obtained using the "Loft Surface" function of the CAD system (Figure 9).

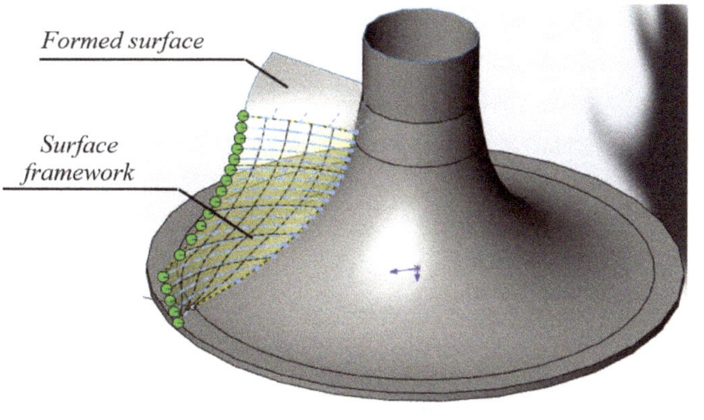

Figure 9. Model of the blade surface.

Based on the obtained frameworks of the hub and the blades, a CAD model of the compressor impeller was created. The resulting model was used as the given data for creating a program for processing the impeller surfaces on a CNC machine in the CAM system PowerMill. The model of the product was imported from the CAD system SolidWorks to the CAM system PowerMill is using direct data translators. The program control for NC manufacturing was created using the standard functions of the CAM system.

4. Conclusions

The problem of forming a smooth contour which interpolates a point series and represents a monotone curve with specified accuracy is solved in the paper. The following methods have been developed in order to achieve the objective:

- forming the area of possible location of the monotone curve interpolating a given point series the width of which does not exceed the given value;
- forming a smooth contour consisting of the arcs of circles which is located within the area of location of the monotone curve.

The area of possible location of the monotone curve is formed as closed contours which are joined at output points. The width of the area is determined by the maximum possible distance between the lines with the specified characteristics. All the monotone lines interpolating the point series are located within the area. The developed method provides an arbitrarily wide area by assigning intermediate points for the output point series.

The developed method for forming a contour within the area of the monotone curve ensures its replacement with a contour with an error which does not exceed the width of the area.

The possibilities of the proposed method have been investigated in solving the problem of interpolation of a point series consisting of 10 points using the CAD system SolidWorks. By solving the test case, it has been established that the minimum possible interpolation error is provided by a contour along which curvature values change monotonously and which has common tangents with the monotone curve at output points. The developed method for forming a contour by the arcs of circles meets the stated requirements.

The main field of use of the developed method is modeling of linear elements of frameworks of surfaces with improved aero- and hydrodynamic properties, including the use of reverse engineering.

The main disadvantage of the proposed method is that the curvature values at junction points of the sections of the contour are irregular. While solving applied problems, the effect of this disadvantage can be eliminated by increasing the number of sections.

Author Contributions: Conceptualization, Y.H. and O.M.; methodology, Y.H. and Y.K.; Project administration, T.S.; validation, O.V., S.H., and M.Ś.; formal analysis, Y.H. and L.B.; data curation, Y.H.; writing—original draft preparation, Y.K. and M.Ś.; writing—review and editing, O.S., S.H., M.Ś., and M.G.; supervision, O.V. All authors have read and agreed to the published version of the manuscript.

Funding: This research received no external funding.

Institutional Review Board Statement: Not applicable.

Informed Consent Statement: Not applicable.

Data Availability Statement: The data presented in this study are available on request from the corresponding author.

Conflicts of Interest: The authors declare no conflict of interest.

References

1. Chekalin, A.A.; Reshetnikov, M.K.; Shpilev, V.V.; Borodulina, S.V. Design of Engineering Surfaces Using Quartic Parabolas. *IOP Conf. Ser. Sci. Eng.* **2017**, *221*, 012015. [CrossRef]
2. Chovnyuk, Y.V.; Shutovskiy, O.M. Analysis of the dynamic behavior of the elastic elements of machines and designs: Effects of wave generation caused by the motion of the source. *Bul. Kherson Natl. Tech. Univ.* **2016**, *3*, 164–170.
3. Czerech, L.; Kaczyski, R.; Werner, A. Machining error compensation for objects bounded by curvilinear surfaces. *Acta Mech. Autom.* **2012**, *6*, 26–30.
4. Fooladi, M.; Foroud, A.A. Recognition and assessment of different factors which affect flicker in wind turbine. *IET Renew. Power Gener.* **2015**, *10*, 250–259. [CrossRef]
5. Gavrilenko, E.A.; Kholodnyak, Y.V. Discretely geometrical modeling of one-dimensional contours with a regular change of differential-geometric characteristics. In Proceedings of the 2014 Dynamics of Systems, Mechanisms and Machines (Dynamics), Omsk, Russia, 11–13 November 2014.
6. Havrylenko, Y.; Kholodniak, Y.; Vershkov, O.; Naidysh, A. Development of the method for the formation of one-dimensional contours by the assigned interpolation accuracy. *East.-Eur. J. Enterp. Technol.* **2018**, *1*, 76–82. [CrossRef]
7. Hoschek, J.; Müller, R. Turbine blade design by lofted B-spline surfaces. *J. Comput. Appl. Math.* **2000**, *119*, 235–248. [CrossRef]
8. Kholodniak, Y.; Havrylenko, E.; Pykhtieieva, I.; Shcherbyna, V. Design of Functional Surfaces in CAD System of SolidWorks via Specialized Software. In *Modern Development Paths of Agricultural Production*; Springer: Cham, Switzerland, 2019; pp. 63–74. [CrossRef]
9. Kotov, I.; Utishev, E. Trace Curvatures with Shared Curvatures. *Themat. Collect. Sci. Pap.* **1972**, *414*, 20–21.
10. Light, R.; Gossard, D. Modification of Geometric-Models Through Variational Geometry. *Comput.-Aided Des.* **1982**, *14*, 209–214. [CrossRef]

11. Okaniwa, S.; Nasri, A.; Lin, H.; Abbas, A.; Kineri, Y.; Maekawa, T. Uniform B-Spline Curve Interpolation with Prescribed Tangent and Curvature Vectors. *IEEE Trans. Vis. Comput. Graph.* **2016**, *18*, 1474–1487. [CrossRef] [PubMed]
12. Osipov, B. *Machine Methods for Designing Continuous Wireframe Surfaces*; Mechanical Engineering, USSR: Moscow, Russia, 1979; 248p.

entropy

Article

Representation of a Monotone Curve by a Contour with Regular Change in Curvature

Yevhen Havrylenko [1], Yuliia Kholodniak [1], Serhii Halko [2], Oleksandr Vershkov [1], Oleksandr Miroshnyk [3,*], Olena Suprun [4], Olena Dereza [1], Taras Shchur [5] and Mścisław Śrutek [6]

1. Department of Technical Mechanics and Computer Design Named after Prof. V.M. Naidysh, Dmytro Motornyi Tavria State Agrotechnological University, 18 B.Khmelnytsky Ave, 72312 Melitopol, Ukraine; yevhen.havrylenko@tsatu.edu.ua (Y.H.); yuliya.kholodnyak@tsatu.edu.ua (Y.K.); oleksandr.vershkov@tsatu.edu.ua (O.V.); olena.dereza@tsatu.edu.ua (O.D.)
2. Department of Electrical Engineering and Electromechanics Named after Prof. V.V. Ovharov, Dmytro Mo-tornyi Tavria State Agrotechnological University, 18 B.Khmelnytsky Ave, 72312 Melitopol, Ukraine; serhii.halko@tsatu.edu.ua
3. Department of Electricity and Energy Management, Kharkiv Petro Vasylenko National Technical University of Agriculture, 19 Rizdviana Street, 61052 Kharkiv, Ukraine
4. Department of Foreign Languages, Dmytro Motornyi Tavria State Agrotechnological University, 18 B.Khmelnytsky Ave, 72312 Melitopol, Ukraine; olena.suprun@tsatu.edu.ua
5. Department of Cars and Tractors, Faculty of Mechanics and Energy, Lviv National Agrarian University, Volodymyr Great 1 Street, 80381 Dubliany, Ukraine; shchurtg@gmail.com
6. Faculty of Telecommunications, Computer Science and Electrical Engineering, UTP University of Science and Technology in Bydgoszcz, 85-796 Bydgoszcz, Poland; mscislaw.srutek@utp.edu.pl

* Correspondence: omiroshnyk@khntusg.info

Citation: Havrylenko, Y.; Kholodniak, Y.; Halko, S.; Vershkov, O.; Miroshnyk, O.; Suprun, O.; Dereza, O.; Shchur T.; Śrutek, M. Representation of a Monotone Curve by a Contour with Regular Change in Curvature. *Entropy* **2021**, *23*, 923. https://doi.org/10.3390/e23070923

Academic Editor: Ernestina Menasalvas

Received: 11 June 2021
Accepted: 19 July 2021
Published: 20 July 2021

Abstract: The problem of modelling a smooth contour with a regular change in curvature representing a monotone curve with specified accuracy is solved in this article. The contour was formed within the area of the possible location of a convex curve, which can interpolate a point series. The assumption that if a sequence of points can be interpolated by a monotone curve, then the reference curve on which these points have been assigned is monotone, provides the opportunity to implement the proposed approach to estimate the interpolation error of a point series of arbitrary configuration. The proposed methods for forming a convex regular contour by arcs of ellipses and B-spline ensure the interpolation of any point series in parts that can be interpolated by a monotone curve. At the same time, the deflection of the contour from the boundaries of the area of the possible location of the monotone curve can be controlled. The possibilities of the developed methods are tested while solving problems of the interpolation of a point series belonging to monotone curves. The problems are solved in the CAD system of SolidWorks with the use of software application created based on the methods developed in the research work.

Keywords: interpolation; tangent line; curvature; area of location of the curve; contour; error; ellipse; B-spline

1. Introduction

The technology of manufacturing products on numerical control machines (CNC) allows machining surfaces of arbitrary form. The program control for the CNC machine is created in automatic mode in the CAM system. The source data for the CAM system are represented by a three-dimensional computer model of the product formed in the CAD system (SolidWorks, AutoCAD, NXCAD etc.).

The accuracy of machining is determined by the accuracy of forming the three-dimensional model, the accuracy of determining the cutting tool trajectory by the CAM system and the accuracy of performing the programmed operations by the machine. The correct formation of the three-dimensional model is a necessary condition for manufacturing a high-quality product [1,2].

One of the most challenging problems in geometric modelling is creating models of surfaces bounding the items whose functional purpose is interaction with the environment (gas, liquid or loose materials). Examples of such products are automobile and aircraft hulls, working units of agricultural machinery and turbine blades. The main functional characteristic of the surfaces bounding such items is the laminar nature of their flow around by the environment [3–5].

The modelling technology in the CAD system is based on forming the surfaces on the basis of linear frameworks. Furthermore, the functional properties of the modelled surfaces were determined by the geometric characteristics of the curves comprising the framework. The laminar nature of the interaction of the surface with the environment was ensured by representing the elements of the framework by lines with regular changes in values of curvature and a minimum possible number of special points (according to the conditions of the problem). For a plane curve, these are the junction points of the convex and concave parts and the points with extreme curvature values [6].

In solving reverse engineering problems, when the given data for surface formation are represented as an array of points, the linear elements of the framework are formed by interpolating a point series. Thus, developing methods of interpolation for point series allowing the representation of the original curve with specified accuracy ensuring the given characteristics is a topical problem.

A smooth plane curve containing no special points will be called a curve with monotone curvature, or a monotone curve.

The problem of interpolating a point series specifying a monotone curve is solved by forming a contour—a curve consisting of sections of analytically determined curves joined at the reference points. At present, the most developed methods of forming contours are second-order curves, Bézier curves and B-splines [7–11].

Forming a smooth curve consisting of arcs of second-order curves [7] guarantees no uncontrolled emergence of inflection points along the contour; however, preventing the emergence of points with extreme values of curvature is impossible. Adjusting the shape of the contour is possible by changing the positions of the tangents to the contour at the reference points. With the positions of tangents fixed to the contour at the reference points, there is a possibility of locally adjusting the form of the curve while forming it by arcs of ellipses. The purpose of adjustment may be to equalize the radii of the curvature at junction points of the sections of the contour.

The use of Bézier curves [8,9] and B-splines [10,11] of third order and above when forming the contour provides greater control of its shape. B-spline is defined by the reference points, each of which has a corresponding conjugate function. It is represented by a composite curve, each segment of which determines a separate equation, the degree of which is equal to the degree of conjugate functions (k). Furthermore, at the junction points of the segments, the continuity of the derivative values to the $k-1-th$ order inclusively is automatically ensured. The curve approximates the control polygon—a polygonal line connecting the reference points. The shape of the B-spline is adjusted by changing the position of the reference points. Control of curvature dynamics is carried out using the graph available in the CAD system. Increasing the quality of joining the sections of the contour requires increasing the degree of the spline equation resulting in an increase in the number of the reference points. As a result, local adjustments of the spline's configuration and characteristics become more complex. It is difficult to achieve a monotone curvature along a B-spline interpolating a large number of reference points.

Modern CAD systems make it possible to form contours by a cubic B-spline, which automatically ensures the regularity of curvature values at the points of the contour. Control of the occurrence of special points within the sections of the contour can be manually performed by moving the phantoms of the control points specifying the spline with a cursor of a mouse. The need for manual adjustments depresses the possibility of using B-spline while forming contours with monotone change in curvature.

The work [12] proposes a method for forming a contour by sections representing a monotone curve with specified accuracy. The problem is solved in the following stages:

- Analysis of the reference point series is carried out, which results in identifying its sections that can be interpolated by a monotone curve;
- At the reference points, the positions of the tangent and the radii of the curvature are assigned, at which the problem of forming a monotone curve has a solution;
- The area of the possible location of the monotone curve interpolating sections of the point series is determined;
- Within the obtained area, a contour is formed, representing a monotone curve with specified accuracy.

The analysis of the reference point series is based on defining the radii of the adjacent circles (AC), each of which passes through three consecutive point series. Part of the point series along which the radii of the AC increase or decrease may be interpolated by a monotone curve along which the radii of curvature increase or decrease, respectively.

A criterion for the correct assignment of tangent lines of the monotone curve interpolating the point series along which the radii of curvature increase is the fulfilment of the ratio $|i, T| < |T, i+1|$ (Figure 1), where T is the point of intersection of the tangents to the curve at nearly reference points (t_i and t_{i+1}).

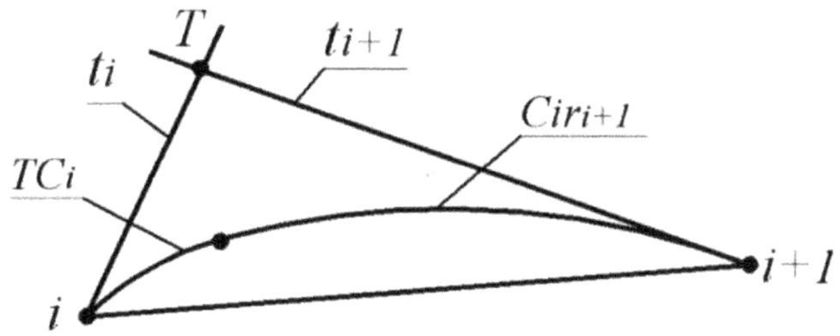

Figure 1. The lower boundary of the area of possible location of the monotone curve.

Based on this criterion, ranges of the positions of tangents are determined, at which a curve with monotone curvature can be formed. The final positions of tangents at the reference points are assigned to the centre of the obtained ranges.

Subsequently, the radii of curvature for which the problem of forming a monotone curve has a solution are assigned. A method for defining the ranges of possible values of curvature radii of a monotone curve interpolating a point series at fixed positions of tangents at the reference points is proposed in work [13]. For a curve along which the radii of the curvature increase, the minimum possible values of the radii of curvature at each point are defined proceeding from the condition that the radius of curvature equal to zero is assigned at the first point. The maximum radii of curvature are similarly defined by assigning the radius of curvature equal to infinity at the first point. The final values of the radii of the curvature at the reference points are assigned to the centre of the obtained ranges.

The positions of tangent lines and tangent circles assigned to the reference points define the area of the possible location of the monotone curve, interpolating a point series and possessing the specified characteristics at its points. The lower boundary of the area consists of the arc of the tangent circle at point i (TC_i) and the arc of the circle which is tangent to (TC_i) and line t_{i+1} at point $i+1$ (Cir_{i+1}) (Figure 1). The upper boundary of the area is similarly defined based on the radius of curvature assigned at point $i+1$ and the position of line t_i.

Similar areas defined along the whole point series determine the possible location of the monotone curve interpolating the reference point series.

Work [12] proposes a method for forming arcs of circles whose radii increase in the same direction as the radii of the curvature of the monotone curve interpolating a point series. The disadvantage of this method is that the regularity of curvature values at the points at which the arcs of circles are conjoined is disturbed.

The aim of the present work was to develop a method for forming a contour with regular curvature change, representing a monotone curve with specified accuracy.

In order to achieve the aim of the research, the following objectives should be pursued:

- To develop a method for evaluating the accuracy with which the convex contour represents a curve with monotone curvature that can be used to interpolate the reference point series;
- To develop methods for forming a smooth convex curve with regular curvature interpolating with specified accuracy a point series by arcs of ellipses and B-spline;
- To compare the potential of the proposed methods in solving the problem of interpolating a point series.

2. Materials and Methods

To ensure the regularity of change in the curvature values along the contour at the joining points of its sections, it is necessary to provide common tangents and curvature values. Such a problem can be solved by using the sections of curves whose parametric number (the number of fixed parameters that unambiguously determine the geometric image) is not less than four. The problem of forming a regular contour representing a monotone curve with specified accuracy is investigated using curves with a corresponding parametric number, which is used in SolidWorks. These are parabola, ellipse and B-spline.

The accuracy with which the contour represents a monotone curve will be evaluated by the maximum absolute interpolation error, a value which cannot exceed the distance between the contour and the monotone curve.

The use of the curves listed above does not ensure a monotone change in curvature values along the contour, but makes it possible to control its convexity. Since the monotone curve is a convex curve, we will use the larger of the distances between the opposite boundaries (upper and lower) as the absolute interpolation error at each of the sections of the areas of the possible location of the monotone curve and convex curves, respectively.

If the positions of the tangents to the contour have been assigned at the reference points, the area of possible location of its section is triangle $i, i+1, T$ bounded by segment $[i, i+1]$ connecting the successive reference points to tangents t_i and t_{i+1} (Figure 2).

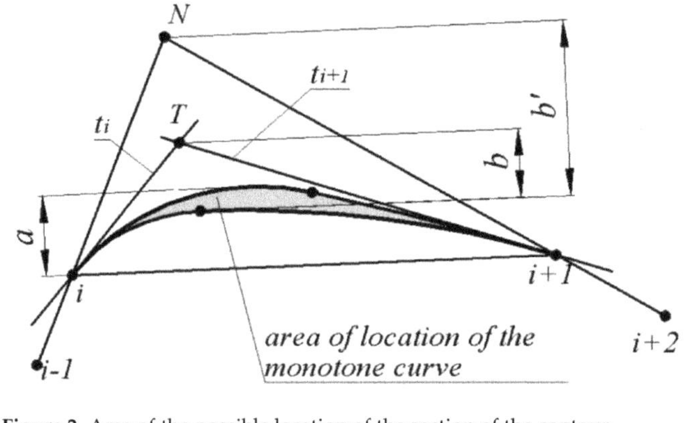

Figure 2. Area of the possible location of the section of the contour.

In this case, the maximum interpolation error of section $i \ldots i+1$ is estimated to be greater than a or b, where:

a—distance from the side of triangle $[i, i+1]$ to the tangent to the upper boundary of the area of location of the section of the monotone curve which is parallel to segment $[i, i+1]$;

b—distance from the apex of triangle T to the lower boundary of the area of location of the monotone curve.

If the positions of the tangents to the convex contour are unknown, the area of the possible location of the section of the contour $i \ldots i+1$ can be defined as the triangle $i, i+1, N$, bounded by lines $[i-1, i], [i, i+1], [i+1, i+2]$ (Figure 2). In this case, the error of interpolation of the sections is estimated to be greater than a or b', where b' is the distance from the lower boundary of the area of location of the monotone curve to the intersection point of lines $[i-1, i], [i+1, i+2]$ (point N). The value of b' exceeds the value of b, and it will usually determine the error of interpolation by the contour at the reference points of which the positions of the tangents are not defined.

When forming a contour representing a monotone curve, it is appropriate to assign positions of the tangents at the reference points where a point series can be interpolated by a monotone curve [12].

Smooth joining of the sections can be achieved by forming them with a curved line, whose parametric number is not less than four. For section $i \ldots i+1$, the two conditions applied to the curve are a passage through the fixed reference points i and $i+1$, and contingency with lines t_i and t_{i+1} produces the third and fourth conditions, respectively. In order to ensure the regularity of change in curvature along the contour, the parametric number of curves forming the contour must be larger than four.

The parametric number of curves specified by an algebraic equation can be defined by the number of coefficients of its equation. For a parabola whose general equation is $Ay^2 + Bxy + Cx + Dy + 1 = 0$ [14], the parametric number is equal to four. Therefore, the conditions of passage through two fixed points and contingency with two fixed lines completely specify the curve. Imposing additional conditions on the section of the parabola ensuring equal curvature values at the junction points of the sections of the contour is not possible.

The general equation of an ellipse has the following form:

$$Ax^2 + Bxy + Cy^2 + Dx + Ey + 1 = 0 \tag{1}$$

The number of coefficients of the equation determines the parametric number of the curve equal to five. Let us demonstrate how the possibility of imposing five conditions on the section of the curve ensures its regularity.

Once the tangents to the monotone curve are assigned at the reference points, four of the five conditions specifying each of the sections of the contour are defined.

There is a set of ellipses meeting the four indicated conditions. To extract an unambiguous solution from the set, it is necessary to impose a fifth condition on the arc of the ellipse. As a fifth condition, let us define that the ellipse passes through another fixed point M. In order to reduce the number of calculations required, let us assign point M on the median of triangle $i, i+1, T$. At the same time, to ensure the minimum error with which the arc of the ellipse replaces the section of the monotone curve, point M is assigned within the area of the possible location of the monotone curve.

The conditions of the passage of the arc of the ellipse through points $1(x_1,y_1)$ and $2(x_2,y_2)$ bounding the area and point $M(x_M, y_M)$ define the following correlations:

$$Ax_1^2 + Bx_1y_1 + Cy_1^2 + Dx_1 + Ey_1 + 1 = 0 \tag{2}$$

$$Ax_2^2 + Bx_2y_2 + Cy_2^2 + Dx_2 + Ey_2 + 1 = 0 \tag{3}$$

$$Ax_M^2 + Bx_My_M + Cy_M^2 + Dx_M + Ey_M + 1 = 0 \tag{4}$$

By differentiating Equations (2) and (3), we obtain:

$$2Ax_1 + B(y_1 + x_1y_1') + 2Cy_1y_1' + D + Ey_1' = 0 \quad (5)$$

$$2Ax_2 + B(y_2 + x_2y_2') + 2Cy_2y_2' + D + Ey_2' = 0 \quad (6)$$

If the tangent lines at points 1 and 2 are determined by the equations $y = xa_1 + b_1$ and $y = xa_2 + b_2$, respectively, the values of the derivatives of the function in Equations (5) and (6) are defined as $y_1' = a_1, y_2' = a_2$.

By solving the system of Equations (2)–(6) we define the values of coefficients A, B, C, D, E and obtain an equation that defines the first section of the curve of form (1).

In order to determine the value of the second derivative of the function defining the ellipse at point 2, it is necessary to differentiate Equation (1) twice and substitute the coordinates of point $2(_{x2,y2})$. By extracting the second derivative of the function from the equation, we obtain:

$$y'' = -\frac{2Bx_2 + 2Cy_2 + E}{2A + 4B_y' + 2C(y')^{2'}} \quad (7)$$

where:

$$y' = -\frac{2Bx_2 + 2Cy_2 + E}{2Ax_2 + 2By_2 + D}$$

For the regular joining of the first and second sections of the curve, in addition to the common tangent at point 2, it is sufficient to ensure the equality of the second derivatives of the functions specifying the arcs of ellipses at that point.

To determine the coefficients of the equation specifying the ellipse that forms the second section of the curve, it is necessary to solve a system of five equations:

- Equations (3) and (6);
- Equation (7) with substituted values of y'' defined by coefficients of the equation specifying the arc of the ellipse for the first section of the contour;
- Equations defining characteristics of the contour at point 3:

$$Ax_3^2 + Bx_3y_3 + Cy_3^2 + Dx_3 + Ey_3 + 1 = 0$$

$$2Ax_3 + B(y_3 + x_3y_3') + 2Cy_3y_3' + D + Ey_3' = 0$$

The first and second sections of the contour, the characteristics of which are calculated according to the method described above, are shown in Figure 3.

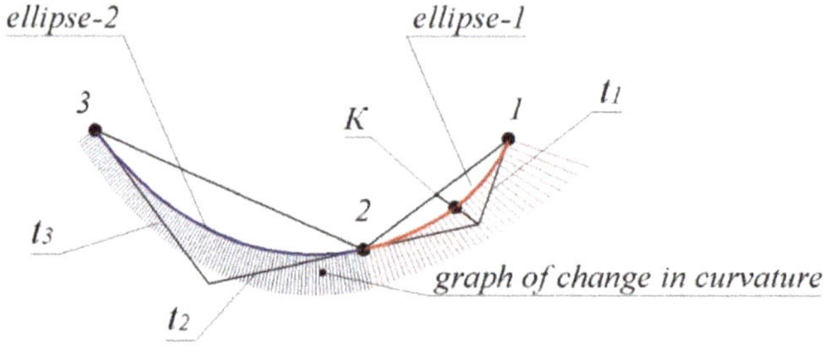

Figure 3. Forming a contour by arcs of ellipses.

The contour was formed in the CAD system SolidWorks with the use of a specially developed software module. The given data are represented by a point series belonging to a monotone curve interpolation which was studied in the work [12]. The graph of curvature values along the contour formed with the CAD system demonstrates the regularity of joining of the sections. The maximum absolute interpolation error was 0.8735 mm in the first section and 0.6328 mm in the second section.

The maximum deflection of the contour from the boundaries of the areas of the possible location of the monotone curve in each of the sections was 0.0121 and 0.0053 mm, respectively. The monotone change in curvature values along the arcs of ellipses forming the sections of the curve determined their location within the area of the location of the monotone curve. However, the ways to predict the absence of points of extreme curvature within the arc of the ellipse specified by the characteristics at boundary points are unknown.

After determining the coefficients defining the equations of the ellipse, which represent the second section of the contour, the equations of the third and subsequent sections were similarly determined. Furthermore, the fifth condition defining the configuration of each of the arcs of ellipses is the equality of second derivatives with the previous section of the contour at the point of junction.

The disadvantage of the proposed method for forming the contour is the lack of possibility to locally adjust its form. The reconfiguration of the first section will change the configuration of all other sections. The reason for this deficiency is the imposition of as many conditions as possible on each of the sections of the contour.

The possibility to locally adjust the form of the sections is a major advantage of forming the contours by B-spline. The degree of the equation specifying the B-spline determines the order of the derivative functions whose values coincide at the junction points of the sections of the spline [15]. The third-degree equation specifying the sections of B-spline that is used in the CAD system SolidWorks automatically ensures the regularity of change in curvature values along the contour of arbitrary configuration.

The interpolation of a point series in the CAD system is performed interactively—the reference points are sequentially indicated with the cursor. If the reference point series can be interpolated by a convex curve, then the B-spline has the form of a convex contour, the number of sections of which is minimal—as each section of the contour is formed by one section of the B-spline.

The specified sequence of commands is easily implemented by the software application, providing the automatic formation of a convex contour in the CAD system SolidWorks.

The imposition of the tangency condition with fixed straight lines on the B-spline requires an increase in the parametric number of curves by increasing the number of spline sections. This could lead to a disturbance of the convexity of the contour. In CAD systems, the convexity of B-spline can be achieved interactively by adjusting the position of the vertices of the specifying polygon. With a large number of reference points and tangents assigned to them, the task becomes excessively laborious.

Creating an application that automatically generates a convex regular B-spline specified by a sequence of reference points and positions of tangents at those points is a complex task. Examples of solving this problem are unknown. When forming a B-spline representing a monotone line with specified accuracy, the simplest and most effective solution is to increase the number of reference points that can be interpolated by a monotone curve.

3. Results and Discussion

Figure 4 shows a cubic non-periodic B-spline interpolating a point series consisting of 10 points created in the CAD system SolidWorks. The sequence of points that was used in [12] was taken as the reference point series to determine the area of location of the monotone curve.

The spline consists of nine segments whose configuration is determined by the 12 specifying points. The graph of change in the curvature values along the B-spline formed in the CAD system allows determining whether it has special points. The sections of the spline contain seven points with extreme values of curvature in total.

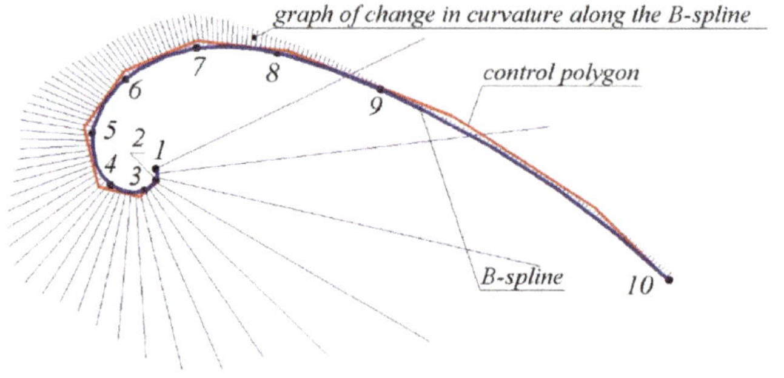

Figure 4. B-spline specified by the coordinates of the reference points.

Table 1 presents the characteristics of the reference point series and error values of its interpolation by B-spline, where:

1. x_i, y_i—coordinates of the reference points;
2. $h_i = |i, i+1|$—lengths of segments connecting the reference points;
3. δ_i—maximum absolute error of interpolation of a point series by a monotone curve;
4. The width of the area of the possible location of the segments of the convex contour, which is equal to:
   - Distance H_i from section $[i, i+1]$ to the intersection point of lines $(i-1, i)$ and $(i+1, i+2)$;
   - Distance H_i^t from section $[i, i+1]$ to the intersection point of tangents t_i and t_{i+1};
5. The absolute error with which a monotone curve is represented by sections of the contour is:
   - Δ_i, when the curve is specified by a point series;
   - Δ_i^t when the curve is specified by a point series and by the location of the tangents at the reference points;
6. f_i—the maximum distance from the formed B-spline to the furthest boundary of the area of possible location of the monotone curve.

Table 1. Characteristics of the reference point series and error values of its interpolation by B-spline.

i	x_i	y_i	h_i	δ_i	H_i	H_i^t	Δ_i	Δ_i^t	f_i
1	0	0	5.28	0.0140	3.1408	1.6068	2.4075	0.8735	0.1500
2	0.21	−5.28	7.20	0.0078	4.4045	1.2234	3.8139	0.6328	0.0824
3	−5.11	−10.13	15.57	0.0218	11.7259	4.2358	9.8524	2.3623	0.1440
4	−20.48	−7.65	25.24	0.0595	18.3244	8.1017	14.3855	4.1628	0.6637
5	−28.68	16.23	28.82	0.0629	12.6391	4.7848	10.4325	2.5782	0.4586
6	−13.87	40.95	34.90	0.0922	10.4670	5.2981	7.8280	2.6591	0.3054
7	17.95	55.28	36.29	0.2184	6.8370	3.2593	5.2685	1.6908	0.1534
8	54.24	53.04	49.65	0.2257	6.7010	2.6676	5.4764	1.4320	0.1712
9	100.94	36.19	155.96	1.1281	19.7329	10.6813	14.8909	5.8393	1.5414
10	230.38	−50.82	-	-	-	-	-	-	-

Meeting the condition of tangency with ten fixed tangents to the monotone curve assigned at the reference points required increasing the parametric number of the spline by increasing the number of its segments. The resulting B-spline consists of 14 segments whose configuration is determined by 16 specifying points. By adjusting the configuration of the B-spline using the tools of the CAD system SolidWorks "Add Tangency Control", "Add Curvature Control" and "Spline Handles", it was possible to achieve the convexity of the entire contour with no more than one special point on sections 6…7, 7…8, 8…9 and 9…10 of the curve.

The resulting B-spline was a convex curve that contains thirteen points with extreme values of curvature. On sections containing special points, the spline was located outside the area of a possible location of the monotone curve with a maximum deflection of 0.1182 mm on sections 3…4. On sections 6…7, 7…8, 8…9, 9…10, where the curvature values change monotonously, the spline is located within the area of the location of the corresponding sections of the monotone curve.

In Table 2, the error values of the interpolation of a point series by the arcs of ellipses are presented. For the correctness of the comparison of the characteristics of the contours formed by arcs of various curves, the same point series that was used in interpolation by B-spline (Table 1) was taken as source data.

Table 2. Characteristics of the reference point series and error values of its interpolation by a contour formed by arcs of ellipses.

i	x_i	y_i	h_i	δ_i	H_i^t	Δ_i^t	f_i
1	0	0	5.28	0.0140	1.6068	0.8735	0.0121
2	0.21	−5.28	7.20	0.0078	1.2234	0.6328	0.0053
3	−5.11	−10.13	15.57	0.0218	4.2358	2.3623	0.0714
4	−20.48	−7.65	25.24	0.0595	8.1017	4.1628	0.1642
5	−28.68	16.23	28.82	0.0629	4.7848	2.5782	0.0415
6	−13.87	40.95	34.90	0.0922	5.2981	2.6591	0.0661
7	17.95	55.28	36.29	0.2184	3.2593	1.6908	0.2948
8	54.24	53.04	49.65	0.2257	2.6676	1.4320	0.3041
9	100.94	36.19	155.96	1.1281	10.6813	5.8393	1.3420
10	230.38	−50.82	-	-	-	-	-

The error of interpolation by a contour consisting of arcs of ellipses was analogous to the interpolation error of the B-spline which has common tangents with the monotone curve at the reference points. The maximum interpolation error of the entire point series was determined by the section of the contour where the values of Δ_i or Δ_i^t are the greatest. These will be the longest sections or the ones where the curvature of the contour is extreme. For the contours, represented in Tables 1 and 2, it is section 9 for which $\Delta_i = 14.8909$, $\Delta_i^t = 5.8393$. The values of Δ_i and Δ_i^t are determined by the configuration of the reference point series and do not depend on the curve sections of which the contour is formed.

The actual deflection of the convex curve from the boundaries of the area of the possible location of the monotone curve (f_i) depends on the characteristics of the contour on each section. In the case of the monotonous change in curvature along the contour and common tangents with the monotone curve on its boundaries, the deflection of the contour from the boundaries of the area of location of the monotone curve does not exceed the width of that region. If the section of the contour contains points with extreme values of curvature, ways to determine in advance the position of the contour within the area of the convex curve are unknown. The value of f_i is determined by the characteristics of the curve forming the sections of the contour and by the location of tangents to the contour at the reference points.

It can be guaranteed that the error in the representation of a monotone curve by a convex curve does not exceed the maximum distance between the boundaries of their possible location. This error can be arbitrarily reduced by increasing the number of reference points.

Table 3 presents the characteristics of a point series consisting of 19 points and error values of its interpolation by a contour formed by the arcs of ellipses. The point series consists of 10 reference points whose coordinates are presented in Tables 1 and 2, and nine intermediate points. The indicated 19 points can be interpolated by a monotone curve. The position of intermediate points and the position of lines tangent to the monotone curve at 19 points were determined by a method proposed in [12].

Table 3. Characteristics of the point series and error values of its interpolation by a contour formed by arcs of ellipses.

i	x_i	y_i	h_i	δ_i	H_i^t	Δ_i^t	f_i
1	0	0	2.74	0.0021	0.3770	0.1955	0.0091
2	0.85	−2.61	2.74	0.0014	0.3909	0.1972	0.0085
3	0.21	−5.28	3.65	0.0018	0.2660	0.1373	0.0031
4	−2.05	−8.14	3.65	0.0011	0.3289	0.1695	0.0039
5	−5.11	−10.13	8.01	0.0026	0.9969	0.5291	0.0523
6	−13.09	-10.74	8.01	0.0028	0.9623	0.4704	0.0861
7	−20.48	−7.65	13.23	0.0071	2.2027	1.0666	0.1481
8	−28.32	3.00	13.23	0.0075	1.7039	0.8308	0.0721
9	−28.68	16.23	14.58	0.0077	1.1985	0.5771	0.0014
10	−23.19	29.74	14.58	0.0084	1.1187	0.5399	0.0011
11	−13.87	40.95	17.65	0.0115	1.2179	0.5808	0.0102
12	0.94	50.54	17.65	0.0121	1.3186	0.6311	0.0092
13	17.95	55.28	18.26	0.0268	0.8529	0.4051	0.2185
14	36.20	55.82	18.26	0.0271	0.7170	0.3416	0.2651
15	54.24	53.04	24.86	0.0281	0.8677	0.4574	0.2861
16	78.05	45.89	24.86	0.0296	0.4971	0.2600	0.7652
17	100.94	36.19	78.17	0.1516	2.9496	1.4995	1.2075
18	168.69	−2.81	78.17	0.1682	2.3530	1.1534	1.4265
19	230.38	−50.82	-	-	-	-	-

By reducing the distances between the reference points by approximately two-fold, the guaranteed interpolation error Δ_i^t was reduced by more than four times.

Reducing the interpolation error Δ_i by increasing the number of reference points can be demonstrated by the example of replacing the involute of the circle with a convex B-spline.

The involute of the circle is a monotone curve along which the curvature values are reduced. Its parametric equation has the following form [16]:

$$x = rcos\varphi + r\varphi sin\varphi;$$

$$y = rcos\varphi - r\varphi sin\varphi,$$

where, r—radius of the circle (evolute) defining the involute; φ—the angle corresponding to the arc of the involute.

The coordinates of the points belonging to the involute of the circle and making up the original point series for the interpolation by B-spline were obtained for the step of changing the values of the parameter $\Delta\varphi = \varphi_{i+1} - \varphi_i = 1^\circ$ starting from the point determined by the value of the parameter $\varphi_i = 0$.

We assigned 270 points on the involute of the circle, which were interpolated by B-spline. The characteristics of the point series are given in Table 4, and the B-spline interpolating the point series is shown in Figure 5.

The values of the interpolation of the first 10 points and the last of the point series by a convex contour at points of which the position of the tangents is not defined (Δ_i) are presented in Table 4.

Table 4. Values of the interpolation of a point series by a convex contour.

i	x_i	y_i	h_i	Δ_i
1	35	0	0.5×10^{-4}	-
2	35	0.6×10^{-4}	1.60×10^{-2}	1.30×10^{-4}
3	35.02	0.5×10^{-3}	2.67×10^{-2}	2.289×10^{-4}
4	35.05	1.7×10^{-3}	3.73×10^{-2}	3.233×10^{-4}
5	35.09	3.97×10^{-3}	4.80×10^{-2}	4.169×10^{-4}
6	35.13	7.74×10^{-3}	5.86×10^{-2}	5.103×10^{-4}
7	35.19	1.34×10^{-2}	6.93×10^{-2}	6.036×10^{-4}
8	35.26	2.12×10^{-2}	7.97×10^{-2}	6.968×10^{-4}
9	35.34	3.16×10^{-2}	9.06×10^{-2}	7.90×10^{-4}
10	35.43	4.51×10^{-2}	10.13×10^{-2}	8.83×10^{-4}
...	...	...	...	...
261	−162.49	−6.89	2.777	2.423×10^{-2}
262	−162.95	−9.63	2.787	2.433×10^{-2}
263	−163.36	−12.39	2.798	2.442×10^{-2}
264	−163.73	−15.16	2.809	2.452×10^{-2}
265	−164.73	−17.95	2.819	2.461×10^{-2}
266	−164.32	−20.76	2.830	2.470×10^{-2}
267	−164.54	−23.58	2.841	2.479×10^{-2}
268	−164.71	−26.42	2.851	2.489×10^{-2}
269	−164.83	−29.27	2.862	2.498×10^{-2}
270	−167.91	−32.13	2.873	-

With the use of a software developed by us, the obtained point series consisting of 270 points is interpolated by a cubic B-spline which is imported into the CAD system SolidWorks. Figure 5 shows a section of the involute and a graph of change in the curvature along it.

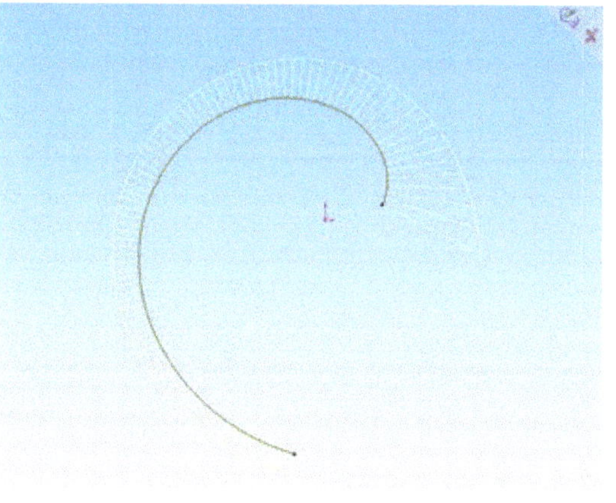

Figure 5. Representation of the involute of the circle by B-spline.

The accuracy with which the B-spline represents the evolution of of the circle is provided by a graph of change in curvature values along the contour, which are defined by the CAD system tools as monotone.

4. Conclusions

This work investigated the possibility of forming a smooth regular contour, which with specified accuracy, represents a monotone curve. The curves whose sections can form a regular contour do not provide a monotonous change in curvature values along the contour, but ensure the control of its convexity. Therefore, it is proposed that the largest distance between the areas of the possible location of the monotone and the convex curves is used, specified by the same point series as an absolute error on each of the sections.

The main criteria for evaluating the computational efficiency of interpolation methods are error, stability and convergence. The interpolation method is stable if negligible errors of the source data lead to minor changes in the result. The use of stable interpolation methods reduces the accuracy requirements of the source data and make it possible to adjust them. The interpolation convergence consists in reducing the error by increasing the number of reference points. When interpolating a point series of arbitrary configuration consisting of a large number of points, most of the known methods do not ensure the convergence and stability of the solution. The reason for this deficiency of interpolation methods is the uncontrolled emergence of inflection points of convexity/concavity of the interpolating curve.

The maximum absolute interpolation error is estimated by the value that cannot be exceeded by the distance between the contour and any monotone curve interpolating the reference point series. Such an error is determined by the width of the area bounded by a sequence of triangles. The sides of the triangles belong to lines connecting the reference points or lines tangent to the monotone curve. By assigning intermediate points, the width of the area of location of the curve is reduced.

The main advantage of the proposed method over known interpolation methods is the possibility to form the contour within a bounded spatial domain. The domain was defined by the expected characteristics of the reference curve. The most important of these characteristics is the minimum (based on the configuration of the reference point series) number of special points of the curve.

The methodology for estimating the error of existing interpolation methods is as follows. A point series is assigned on the reference curve. After interpolating the point series, the deflection of the interpolating curve from the reference curve is determined. The error with which the interpolating curve represents the reference curve is estimated as the maximum deflection between the specified curves. When solving other tasks, the engineer operates on the premise that the order of error of interpolation will be identical with the test examples. Generally, when the reference curve is not known, such an assumption might be wrong.

Defining the area of the possible location of the reference curve not only allowed to determine the absolute interpolation error, but also to arbitrarily decrease this error by increasing the number of reference points or assigning intermediate points by the method proposed in [12].

The convergence and stability of the proposed method are ensured by the preliminary analysis of the configuration of the reference point series and formation of the contour based on sections of the point series which can be interpolated by a monotone curve.

The stability of the developed method to changes of reference conditions is ensured by forming a contour within the area of the possible location of the convex parts of the reference curve. With any adjustment of the positions of the reference points, the distance from the segments connecting the reference points to the boundaries of the possible location of the corresponding sections of the reference curve remains limited.

The consistent localization of the area of the location of the curve within its original boundaries guarantees the convergence of the interpolation process. Adding each additional point brings the boundary of the area closer to the accurate solution, namely to a monotone curve with a given set of characteristics.

The number of sections of point series that can be interpolated by a monotone curve, as well as the number of reference points that make up each of these sections, does not affect the convergence or stability of interpolation.

Any curve can be considered to consist of monotone parts. Thus, the assumption that if a sequence of points can be interpolated by a monotone curve, then the reference curve on which these points were assigned is monotone, allows implementing the proposed approach for estimating the error of interpolation of a point series of arbitrary configuration.

Defining the area of the possible location of the reference curve enables developing methods for forming contours by arcs of various curves, the convexity of which can be controlled. Based on the proposed criteria for estimating the interpolation error, methods for forming convex regular contours by arcs of the ellipses and B-spline were developed. The methods were implemented in a software application that automatically creates contours in the CAD system SolidWorks. The method for forming a contour by arcs of an ellipse is based on the derivation of equations specifying ellipses and the determination of parameters by these equations, which make it possible to form sections of the curve in the SolidWorks system. Such parameters are the coordinates of the points bounding the contour, the positions of the tangent lines to the monotone curve at these points and the coordinates of another point belonging to the ellipse.

The application of forming contours by B-spline ensures its passage through a sequence of points specified by their coordinates.

The method for forming contours by arcs of ellipses takes into account the positions of tangents to the contour and provides for greater interpolation accuracy. The method for forming contours by B-spline requires fewer calculations and is more easily executed in the form of software.

The possibilities of the method for forming curves by B-spline were investigated while solving the following problems:

- Interpolation of a point series consisting of ten points specifying a monotone curve;
- Interpolation of a point series belonging to the involute of the circle.

This study proved that an arbitrarily small interpolation error can be achieved by reducing the distances between the reference points belonging to the monotone curve by increasing their number. The sequence of interpolated points can be increased either by refining the given data or by assigning intermediate points according to the methodology proposed in [12].

The developed methods are appropriate for modelling linear elements of the frame of functional surfaces with improved aerodynamic and hydrodynamic properties, for which the minimum number of special points is the main criterion of the quality of the obtained solution.

The advantages of the methods proposed in this work over the method developed in [12] for presenting a monotone curve by a contour formed by arcs of circles is the regular change in curvature values. The advantage of the method proposed in [12] is the monotonous change in curvature values along the contour. Both of these characteristics are important for modelling functional surfaces. While solving particular problems, it does not seem possible to prioritise one of the characteristics in advance as it will depend on the conditions of the problem. Such conditions are the speed of the flow of the environment along the surface and the physical characteristics of the environment (density, viscosity etc.).

The influence degree of the quality of joining the sections of the curve and the conformity of the change in curvature values along the curve can be determined by experiment.

Increasing the number of interpolated points is a universal method of increasing the accuracy with which a regular contour represents a monotone curve and reducing the difference in curvature values at points where arcs of circles are joined.

Author Contributions: Conceptualization, Y.H.; methodology, Y.H. and Y.K.; project administration, O.S., T.S.; validation, O.V., S.H. and O.M.; formal analysis, Y.H., O.M. and O.D.; data curation, Y.H.; writing—original draft preparation, Y.K., O.M. and M.Ś.; writing—review and editing, O.S., S.H., O.M. and M.Ś.; supervision, O.V. and S.H. All authors have read and agreed to the published version of the manuscript.

Funding: This research received no external funding.

Institutional Review Board Statement: Not applicable.

Informed Consent Statement: Not applicable.

Data Availability Statement: The data presented in this study are available on request from the corresponding author.

Conflicts of Interest: The authors declare no conflict of interest.

References

1. Czerech, L.; Kaczyski, R.; Werner, A. Machining error compensation for objects bounded by curvilinear surfaces. *Acta Mech. Auto.* **2012**, *6*, 26–30.
2. Jadhav, Y.P.; Chougule, V.N.; Mulay, A.V. Free-form surface models generation using reverse engineering techniques—An investigation. *J. Mech. Civ. Eng.* **2006**, 379–385. [CrossRef]
3. Pérez-Arribas, F.; Pérez-Fernández, R. A B-spline design model for propeller blades. *Adv. Eng. Softw.* **2018**, *118*, 35–44. [CrossRef]
4. Fooladi, M.; Foroud, A.A. Recognition and assessment of different factors which affect flicker in wind turbine. *J. Abbr.* **2015**, *10*, 250–259. [CrossRef]
5. Kholodniak, Y.; Havrylenko, E.; Pykhtieieva, I.; Shcherbyna, V. Design of Functional Surfaces in CAD System of SolidWorks via Specialized Software. *Mod. Dev. Paths Agric. Prod.* **2019**, 63–74._7. [CrossRef]
6. Osipov, B. *Machine Methods for Designing Continuous Wireframe Surfaces*; Mechanical Engineering; USSR: Moscow, Russia, 1979; 248p.
7. Chekalin, A.A.; Reshetnikov, M.K.; Shpilev, V.V.; Borodulina, S.V. Design of Engineering Surfaces Using Quartic Parabolas. *IOP Conf. Ser. Sci. Eng.* **2017**, *221*, 012015. [CrossRef]
8. Shen, W.; Wang, G.; Huang, F. Direction monotonicity of a rational Bézier curve. *Appl. Math.* **2016**, *31*, 1–20. [CrossRef]
9. Cai, H.; Wang, G. Udging or setting weight steady-state of rational Bézier curves and surfaces. *Appl. Math.* **2014**, *29*, 391–398. [CrossRef]
10. Kvasov, B. Monotone and convex interpolation by weighted cubic splines. *Adv. Comput. Math.* **2014**, *40*, 91–116. [CrossRef]
11. Okaniwa, S.; Nasri, A.; Lin, H.; Abbas, A.; Kineri, Y.; Maekawa, T. Uniform B-Spline Curve Interpolation with Prescribed Tangent and Curvature Vectors. *IEEE Trans. Vis. Comput. Graph.* **2016**, *18*, 1474–1487. [CrossRef] [PubMed]
12. Havrylenko, Y.; Kholodniak, Y.; Halko, S.; Vershkov, O.; Bondarenko, L.; Suprun, O.; Miroshnyk, O.; Shchur, T.; Śrutek, M.; Gackowska, M. Interpolation with Specified Error of a Point Series Belonging to a Monotone Curve. *Entropy* **2021**, *23*, 493. [CrossRef] [PubMed]
13. Havrylenko, Y.; Cortez, J.I.; Kholodniak, Y.; Alieksieieva, H.; Garcia, G.T. Modelling of surfaces of engineering products on the basis of array of points. *Teh. Vjesn.* **2020**, *27*, 2034–2043. [CrossRef]
14. Ashyralyev, A.; Sobolevskii, P. *New Difference Schemes for Partial Differential Equations*; Birkhäuser: Basel, Switzerland, 2004; 446p.
15. Fritsch, F.N.; Carlson, R.E. Monotone Piecewise Cubic Interpolation. *Siam. J. Numer. Anal.* **1980**, *17*, 238–246. [CrossRef]
16. Hilbert, D.; Cohn-Vossen, S. *Geometry and the Imagination*; Chelsea: New York, NY, USA, 1999; 357p.

entropy

MDPI

Article

An Improved Encoder-Decoder Network Based on Strip Pool Method Applied to Segmentation of Farmland Vacancy Field

Xixin Zhang [1], Yuhang Yang [1], Zhiyong Li [1,2,*], Xin Ning [3], Yilang Qin [4] and Weiwei Cai [5]

1 College of Information Engineering, Sichuan Agricultural University, Ya'an 625000, Sichuan, China; zhangxixin@stu.sicau.edu.cn (X.Z.); yangyuhang@stu.sicau.edu.cn (Y.Y.)
2 Sichuan Key Laboratory of Agricultural Information Engineering, Ya'an 625000, Sichuan, China
3 Institute of Semiconductors, Chinese Academy of Sciences, Beijing 100083, China; ningxin@semi.ac.cn
4 Institute of Agricultural Economy and Information, Henan Academy of Agricultural Sciences, Zhengzhou 450002, Henan, China; indexlang@outlook.com
5 College of Logistics and Transportation, Central South University of Forestry and Technology, Changsha 410004, Hunan, China; vivitsai@ieee.org
* Correspondence: lzy@sicau.edu.cn; Tel.: +86-138-822-13811

Abstract: In the research of green vegetation coverage in the field of remote sensing image segmentation, crop planting area is often obtained by semantic segmentation of images taken from high altitude. This method can be used to obtain the rate of cultivated land in a region (such as a country), but it does not reflect the real situation of a particular farmland. Therefore, this paper takes low-altitude images of farmland to build a dataset. After comparing several mainstream semantic segmentation algorithms, a new method that is more suitable for farmland vacancy segmentation is proposed. Additionally, the Strip Pooling module (SPM) and the Mixed Pooling module (MPM), with strip pooling as their core, are designed and fused into the semantic segmentation network structure to better extract the vacancy features. Considering the high cost of manual data annotation, this paper uses an improved ResNet network as the backbone of signal transmission, and meanwhile uses data augmentation to improve the performance and robustness of the model. As a result, the accuracy of the proposed method in the test set is 95.6%, mIoU is 77.6%, and the error rate is 7%. Compared to the existing model, the mIoU value is improved by nearly 4%, reaching the level of practical application.

Keywords: semantic segmentation; farmland vacancy segmentation; strip pooling; crop growth assessment; encoder–decoder

Citation: Zhang, X.; Yang, Y.; Li, Z.; Ning, X.; Qin, Y.; Cai, W. An Improved Encoder-Decoder Network Based on Strip Pool Method Applied to Segmentation of Farmland Vacancy Field. *Entropy* **2021**, *23*, 435. https://doi.org/10.3390/e23040435

Academic Editor: Michal Choras

Received: 11 March 2021
Accepted: 1 April 2021
Published: 8 April 2021

1. Introduction

Since a few years ago, the segmentation of farmland based on remote sensing images has been studied extensively owing to the development of deep learning technology, making intelligent farmland a hot trend. Based on their shooting distance, satellites and unmanned aerial vehicles are two common photographing tools. Satellite-based remote sensing images are mainly used in the segmentation of farmland forest areas such as Han Lin Aung, Burak Uzkent, etc., using space–time convolution networks to circle the farmland plots [1]. Guang Wang, Fan Yu, and others used the segmentation models of farmland and woodland with the help of remote sensing images to segment the farmland and woodland [2]. Ganchenko et al. used convolutional neural networks based on agricultural vegetation monitoring image semantic segmentation [3]. Drone-based remote sensing images are mainly used in specific scene analysis of a field, such as Yang's [4] deep semantic segmentation techniques, Fully Convolutional Networks (FCN), SegNet, etc., which segment the farmland covered by plastic mulch to study environmental and soil contamination. There are crop and weed segmentations studied by Lei F. Tian and Heping Zhu et al. [5] using drones to capture low-altitude images of crops for crop morphological segmentation [6].

Wheat is the world's largest ration crop; it has high economic, medicinal, and nutritional value, according to the Journal of Experimental Botany [7]. Therefore, we take wheat crops as an example to study.

The satellite image monitoring used previously is a very macro means of observation. Many problems are often faced in the early stages of wheat cultivation. Many factors can make wheat crops grow poorly in the early stages. The main factors leading to their initial dysplasia can be classified as environmental factors and their own factors. Environmental factors mainly include weather, pests and diseases, soil, and other such external factors. In the previous application, the use of semantic segmentation techniques for early development monitoring of wheat included wheat insect identification [8] and wheat field weed segmentation [5], combined with sensor equipment to measure soil acidity and alkalinity, detect nutrients, monitor weather [9], and so on. During the study, it was found that a small proportion of wheat planted in the field could be stunted or lodged, or even directly died due to these influencing factors. This leads to a common phenomenon: the creation of vacancies in farmland. This performance is a direct reflection and manifestation of the poor growth of wheat. Therefore, we propose a method of segmenting wheat vacancies in the early stages, as it is an important factor in the growth of wheat, making it easy to consider in estimating yield.

This paper mainly studies the problems prevailing in the application of existing semantic segmentation technologies in farmland vacancy segmentation, analyzes the advantages and disadvantages of each method according to the comparison of training process and experimental results, and proposes a new semantic segmentation network based on encoder–decoder architecture. At the same time, according to the characteristics of the dataset, the algorithm is improved adaptively. This paper is divided into six parts. The background section introduces several mainstream semantic segmentation techniques and summarizes the contribution of this paper, followed by the dataset section, and the methods used in this paper and the training process are described later. The validity of the new model is then proved by experimental results.

2. Related Work

The introduction mentions that semantic segmentation technology has developed rapidly in recent years, which makes its application scope gradually expand and gradually cover all areas of life. However, in the field of agricultural application, the research on farmland vacancy segmentation is rarely involved. Therefore, we first study the development process and basic principles of the semantic segmentation algorithm, explore the application of these technologies in similar scenes, and analyze their advantages and disadvantages. According to these algorithms, we redesign and improve the algorithms in order to get the optimal algorithm for the scene.

As we all know, the FCN model [10] opens a new path for semantic segmentation research and solves the semantic-level image segmentation, and the U-Net [11] has achieved significant results in the field of biomedical segmentation, which mainly solves the problem of small sample training to extract detailed texture features, such as using it for yellow embroidery surveillance in wheat fields [12] and retinal vessel segmentation [13]. Thus, the semantic segmentation algorithm has made a leap forward. Both algorithms originate from the Convolutional Neural Network.

The Convolutional Neural Network (CNN) gained popularity because AlexNet [14] had achieved a higher score than traditional methods in an ImageNet image recognition competition. However, as it defines every pixel, the relationship between pixels cannot be fully considered, and the lack of spatial unity leads to poor performance in semantic level segmentation. Thus, the Full Convolutional Network (FCN) [10] was born, which changed the entire connection layer in the last layer of the earlier classification network. It thus changed the previous result of the output one-dimensional probability vector and realized output characteristic graph. At the same time, a combined deconvolution, upsampling, and skip structure method was used to achieve end-to-end training results. Similarly, combined with deconvolution, the method of upper sampling and jump structure was used, and it achieved pioneering results of end-to-end training.

In order to reduce the cost of computing in the process of network training, researchers propose a dense network architecture called DenseNet [15]. DenseNets are built from dense block sand pooling operations, where each dense block is an iterative concatenation of previous feature maps. This structure extension is also applied to FCN to form FC-DenseNet [16]. It exploits the feature reuse by extending the more complex DenseNet architecture by skipping connections in the original FCNs, while avoiding feature overflow on the sampling path on the network. Later, in order to better apply the semantic segmentation algorithm to the segmentation of complex scenes, PSPNet [17] was born. PSPNet, with an atrous convolution of FCN as a baseline, focusing on the three aspects of mismatched relations, confusion categories, and inconspicuous classes, designed a hierarchical global priority containing information on different scales between different sub-regions, called the pyramid pooling module. The module incorporates four different pyramid-scale features using a feature chart, which is pooled with four sizes of pooled cores, resulting in four different scales of the feature map. To maintain the weight of the global feature, we use a 1×1 convolution layer to reduce the dimension. Then a bilinear interpolation method is used to restore the size of the feature map and concatenate four different levels of features, and finally, the global features of the pyramid pool are obtained.

The idea of dense connection is applied to atrous spatial pyramid pooling (ASPP) to increase the receptive field. DenseASPP [18] has revealed that atrous convolution becomes less and less effective and gradually loses the modeling ability as the dilation rate of ASPP increases. Therefore, it is very important to find a network structure that can encode multi-scale information while obtaining a large enough receiving domain. DenseASPP combines the advantages of parallel and cascading using atrous convolution to map multi-scale features on a larger scale. Through a series of feature connections, neurons on each intermediate feature map encode semantic information from multiple scales, and different intermediate feature maps encode multi-scale information from different scales. Through a series of atrous convolutions, the neurons at later levels can obtain larger receptive fields, as there is no degeneration of the convolution kernel of ASPP. Thus, the final output of the DenseASPP model is a feature map that covers a wide range of semantic information in a very dense way.

DeepLabv3+ is an improved version of the DeepLab series [19–21]. After a series of developments, it is a mature algorithm for scene segmentation at present. DeepLabv3+ [22] added a decoder considering the problem of reduced feature map resolution and reduced prediction accuracy caused by the network layer with stride=2 in the ResNet network structure. The original DeepLabv3 as an encoder of the network structure, unlike the direct bilinear downsampling recovery feature map, has features that are first bilinearly upsampled by a factor of 4 and then concatenated with the corresponding low-level features from the network backbone that have the same spatial resolution. In addition, because of this improvement, the parameter volume increases. Here, Xception is used instead of ResNet as the backbone for training.

Through the development of these algorithms, we find that most researchers are improving their accuracy in some specific data sets, so that they can be applied to large scene analysis, automatic driving, and other fields. However, for some unique scenarios, these algorithms may not be able to show their advantages well. Of course, the development of these technologies also drives the research on the application of semantic segmentation algorithm in some specific scenes. Among them, the most similar to the research field of this paper is crack segmentation.

The rapid emergence of this technology has also led to the development of semantic segmentation in the field of crack segmentation. Mark David Jenkins, Thomas Arthur Carr [23], and others proposed a deep convolutional neural network for the semantic pixel-wise segmentation of road and pavement surface cracks, which also considered the cost of data acquisition and chose to conduct training in a small sample dataset. It finally achieved good results in a dataset with only 80 images. Similarly, Henrique Oliveira, Paulo Lobato Correia et al. [24] applied improved segmentation methods based on pixel refinement to pave the way for crack detection. Additionally, Martin Mayr, Mathis Hoffmann [25] used the improved ResNet50 network to segment the cracks in the EL image of the solar cell.

It is worth noting that farmland vacancy segmentation is more difficult as compared to cracks segmentation, because the vacancies are comprised of various sizes and shapes. Therefore, in most of the above-mentioned work, there are some drawbacks in the application of vacancy segmentation. Our work is mainly to improve the existing algorithm in this field to be suitable for the precise segmentation of crops and vacancies. Inspired by the DeepLabv3+ model [22], we used the encoder–decoder's network architecture to better recover image information. At the same time, we leveraged a combination of strip pooling and spatial pyramid pooling to achieve a better segmentation of vacancies and crops. SPINet, the end-to-end method proposed by us, uses an improved ResNet network as an encoder and combines SPM and MPM modules to establish a self-attention mechanism. The decoder concatenates the output of the last layer of encoder and the output of the middle layer, and obtains the prediction graph by sampling up and recovering pixels.

Our contributions can be summarized as follows:

(1) We propose a semantic segmentation network, based on encoder–decoder architecture, which takes IResNet network as the backbone and integrates SPM and MPM modules to build the model self-attention mechanism.
(2) We take advantage of strip pooling to capture vacancies with greater precision.
(3) We provide a farmland dataset taken by Unmanned Aerial Vehicle, containing 320 pieces of training data, 80 pieces of test data, and 40 pieces of verification data.
(4) The model that we designed can adapt to the training of small samples and reach a good level.

3. Materials and Methods

3.1. Materials

The image data was collected in January 2019 at the Modern Agricultural Research and Development Base in Henan, China, covering an area of 948 acres. The filming site was a winter wheat test field of the Provincial Academy of Agricultural Sciences, and the experimental data were provided by the Agricultural Economy and Information Research of the Henan Academy of Agricultural Sciences. The shooting equipment used was DJI Phantom 4 Pro V2.0 Professional Intelligent 4K Ultra Clear Aerial Shooting Drone with shooting altitude of 20 meters. A total of 42 images were taken in the field, each one with more than 4000 × 4000 pixels. The pixels of each picture are too large, which is not conducive to training. Thus, we first cut out the excess areas of the image, and then cropped the image into four equal parts, by removing the edges of each image, ultimately obtaining 110 images, each of size 1024 × 1024. We randomly divided the data set into a training set, verification set, and test set according to the ratio of 8:1:2. In order to expand the training set, we cut the image size of 1024 × 1024 into four equal parts, and obtained 320 training images and 40 verification images. However, we used the size of 1024 × 1024 images for testing to better show the segmentation effect. Perez, Luis, and Jason Wang et al. [26] demonstrated the effectiveness of data enhancement in improving network performance through experimental comparison. A.Mikołajczyk and M.Grochowski et al. [27] describe some of the available methods of data augmentation and propose a better one. Therefore, we also use the data augmentation methods to improve the performance and robustness of the model.

The following Figure 1 shows the various actions we performed on the image.

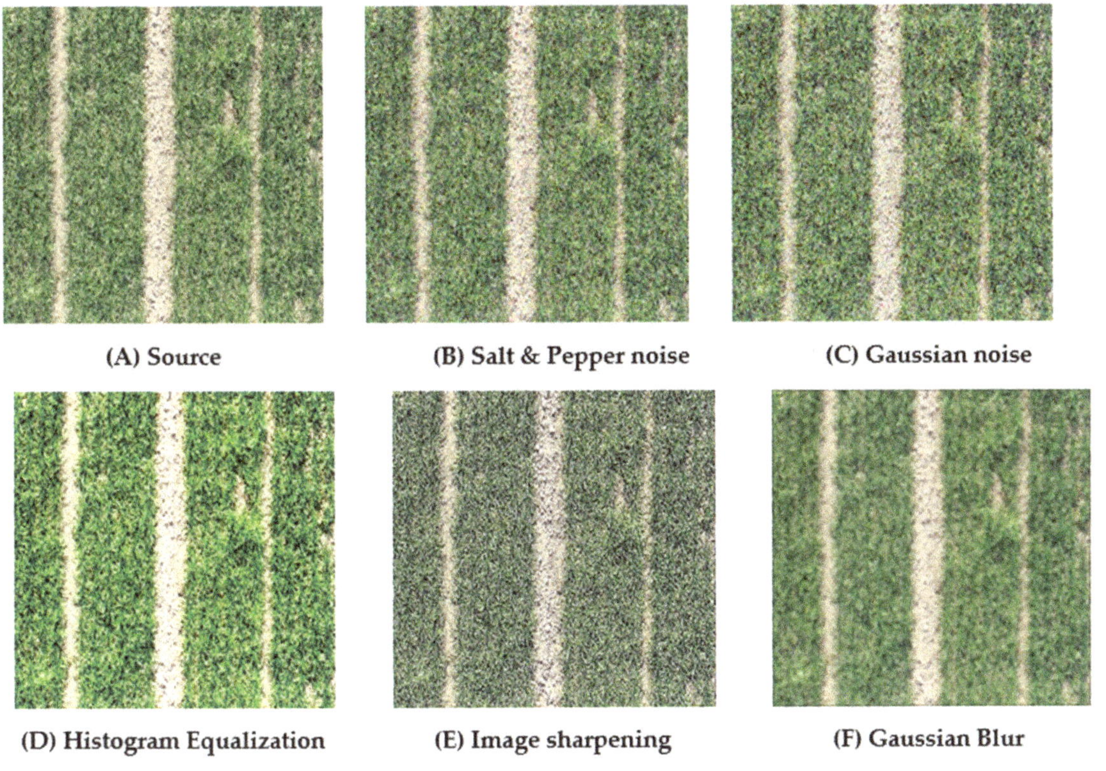

Figure 1. Five kinds of data augmentation operations for images. Explanation for subfigure (**A**) represents the original image, (**B**) represents the image after adding salt and pepper noise, (**C**) represents the image after adding Gaussian noise, (**D**) represents the image after histogram equalization, (**E**) represents the image after sharpening, and (**F**) represents the image after Gaussian blur.

It is worth noting that in this application field, crop category and crop vacancy category in the image need to be segmented. The vacancies in the wheat crop are many and very messy, as seen in the fields filmed, making the marking difficult and increasing the labeling time. The image acquisition cost is very high, and its marking requires the identification and judgment of experts, just like in the field of medical image segmentation. This, to some extent, aggravates the cost of using the model and requires researchers to design a practical model that can achieve good results in a small sample training set. Therefore, in this paper, considering the cost of data annotation, we design the small sample learning method. The algorithm improvement of small sample training will be explained later in detail. In this model, labelme was used for data labeling, which marked crops as red, vacancy as green, and backgrounds as black.

3.2. Methods

Based on the method proposed above (including the FC-DenseNet56, PSPNet, DenseA-SPP, and DeepLabv3+), we carried out a detailed comparative experiment and found that there exist certain disadvantages of the segmentation effect that need to be improved. By further analyzing the data characteristics and the training effect of each model on the dataset, a strip pool was re-selected to capture the characteristics of crop and vacancy, and the effect was much higher. In view of this, we redesigned a semantic segmentation method based on encoder–decoder architecture. In order to avoid the vacancy of other shapes that are difficult to catch in the strip pool, the MPM module, or mixed pool module, is built by combining PPM, which sets up different types of context information through various pool operations. Xin Ning et al. proposed a model combining weak saliency and attentional perception to obtain a more complete global feature by weakening the saliency feature [28]. This is another reason why we designed MPM.

Simultaneously, to improve the model's predictive pixel accuracy for crops and the edge of the vacancy, a decoder module was added to recover more detailed information of the image, avoiding the loss of information by shrinking directly from 512-size channels to three channels. Finally, based on the particularity of little sample data training, inspired by Improve Residual Network, we have improved the signal transmission mode of the original ResNet to accommodate negative signal transmission during the initial training of low-sample training. This avoids eliminating too much information during transmission. Moreover, this paper also adopts IResNet as the backbone of architecture for the transmission characteristics of a few samples of training information, which changes the original means of signal transmission as compared to ResNet to preserve more useful information.

Below, we elaborate on our proposed methods.

3.2.1. Encoder–Decoder Architecture

We designed a semantic split network with encoder–decoder architecture using IResNet, which combines SPM and MPM modules as a part of the encoder portion of the architecture and a decoder layer that combines the encoder layer output and the encoder middle layer output to sample and restore pixels. In most of the known segmentation methods, after sampling the features by a more granular convolution, it is often used directly to reduce the channels to three, i.e., revert to a visual 3D feature map. In case of such an operation, which can be understood as a minimalist decoder this is not conducive to the recovery of image features. Especially in our vacancy segmentation scene, there is a greater disadvantage in the segmentation of small vacancies as well as the segmentation of the edge of the vacancy. This arouses the need to design a more sophisticated decoder.

Figure 2 shows that the image input network structure begins with a series of convolutional normalization operations with IResNet backbone, and that in the process, there is a pool operation to extract feature of vacancy. It is important to note that the arrows in the figure do not refer to sequential operations but the directionality. Thus, it does not mean that the image enters MPM module before entering the SPM module. Additionally, the decoder can combine information of different scales. Therefore, we choose to combine the information of Layer2 and Layer4 in IResNet. Firstly, the channel of characteristic graph output by Layer4 in the encoder layer is converted into 512 by a 1×1 convolutional layer, and then sampled up by a factor of four connected to the output of Layer2 in the IResNet network (the output of Layer2 is also reduced by a 1×1 convolutional layer). Finally, the result is obtained by four-fold bilinear up-sampling recovery, after three 3×3 convolutional layers. To take into account the over-fitting of prevention model training, the output results of last layer of the network were not directly upsampled and recovered, and the contents of previous layers of the network were also combined. The comparison of model effects based on this architecture will be described in detail later, in the discussion section.

3.2.2. SPM and MPM Modules

Strip pooling is a long, narrow pool kernel represented by $1 \times N$ or $N \times 1$. This pooling method helps to capture remote context information and establish a remote dependency, which can be seen as a self-attention mechanism. Compared to other mechanisms such as dilated convolutions and global/pyramidal pooling, the differences lie in the shape of the pooling nucleus and more suitable application scenes. A pyramid-like/pooling-like method is primarily a square window to detect input information to capture useful features. In this case, large receptive fields of the pooling kernel in the backbone network contribute to the scene analysis. The object characteristics of segmentation, in this case, are obvious and special, and it is more necessary to capture the features by using strip pooling. It is also known that a part of the vacancy tends to have long and narrow characteristics, consistent with the design thinking of strip pooling.

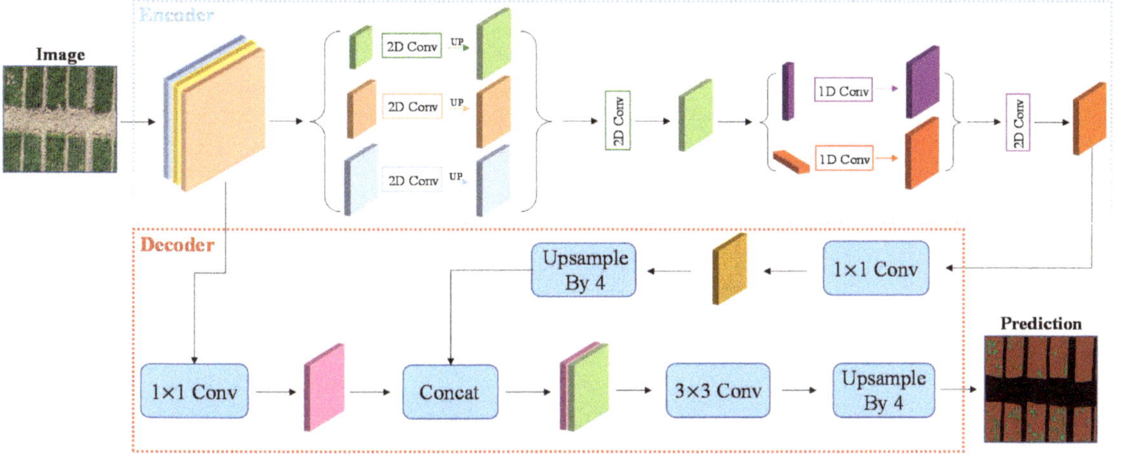

Figure 2. Encoder–decoder network architecture: The upper part of Figure 2 shows the encoder, taking IResNet as the main backbone, combining SPM and MPM modules. The lower part shows the decoder. It combines output of the last layer of encoder and the output connection of the middle layer to carry out up-sampling and restore feature map pixels. The first three-tier network in encoder represents IResNet + Strip Pooling Module (SPM) + Mixed Pooling Module (MPM). The arrows represent operations.

The difference between this pooling method and two-dimensional pooling is that strip pooling averages all feature values of rows or columns. As mentioned above, the spatial extent of pooling is 1 × N or N × 1. If given a two-dimensional vector $x \in R^{H \times W}$, the output $y^h \in R^H$ after horizontal strip pooling can be written as:

$$y_i^h = \frac{1}{W} \sum_{0 \le j \le W} x_{i,j}. \quad (1)$$

Similarly, the output $y^v \in R^w$ after vertical strip pooling can be written as:

$$y_j^v = \frac{1}{H} \sum_{0 \le i \le W} x_{i,j}. \quad (2)$$

where H and W are the spatial height and width, respectively. For this kind of pooling in the horizontal and vertical directions, it can build long-range dependencies to the remote information distributed in the horizontal and vertical directions, thus playing a role similar to the self-attention mechanism. It is also worth noting that because of its narrow kernel in a single direction, it is also more suitable for capturing local details. In Figure 3, we can see that the square-shape kernels are different from the previous one. At the same time, in the practical application process, it extends from horizontal and vertical directions, respectively, from H × 1 to H × W (or from 1 × W to H × W), and then through the fusion of the two (that is, to do the multiplication of elements) to do a further series of operations.

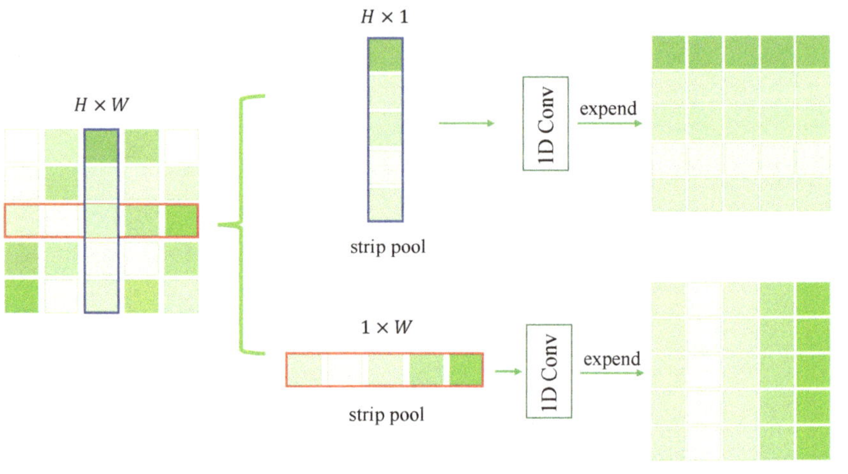

Figure 3. Schematic illustration of the strip pooling.

Based on this idea, two modules are constructed in this article, namely SPM and MPM. SPM contains 1 × N and N × 1 pool operations, followed by a 1D convolutional layer for modulating the current location and its neighboring features. Then, the two results are combined together. We then use a 1×1 convolution to reduce the number of channels of the feature graph and to increase the nonlinearity of the network. Later, ReLU layer normalization is used. Finally, the element-based multiplication is performed with the previous eigenmatrix. This module is primarily used to capture information about the long vacancy, and due to the characteristics of its kernel. As compared with the global average pooling operation, the number of parameters is less. The mixed pooling module (MPM) is designed primarily to address the vacancy of capturing various other shapes. As a result, the mixed pooling module is formed with a combination of standard pooling methods in PPM. As shown in Figure 2, after IResNet, three different sizes are formed through a standard pooling operation with three pooling kernel sizes, which are then fused. Another convolution layer is gone through to reduce the aliasing effect of the down-sampling operation. A new SPM module is entered, which traverses the input image with two strip pooling, 1 × N and N × 1, to capture features and scale horizontally and vertically to form a feature map, respectively. Additionally, another convolution layer should be gone through. In this case, the channel output of IResNet is reduced to 1024 from 2048 with a 1 × 1 convolutional layer before each submodule.

3.2.3. Improved ResNet

In the actual experimental procedure, the information in the data is limited because of the specificity of the small sample dataset. After the training period, without learning new features, it is easy to overfit. It was found that, as shown in Figure 4A, ResNet first reduces the number of channels by 1 × 1 convolution layer, then captures features using a 3 × 3 convolution layer, and finally restores the number of channels through a 1 × 1 convolution layer. ReLU operation directly shields all negative signals on the straight connected road. In the small sample training period, we produce more negative signals and directly remove those negative signals, leading to the huge loss of information. In the case of small samples, data training is not worthwhile. Therefore, we have designed an effective way to communicate information, drawing on the idea of improving the residual network [29], so that the network cannot lose too much useful information in vain during the training of small sample data, and to avoid over-fitting the network.

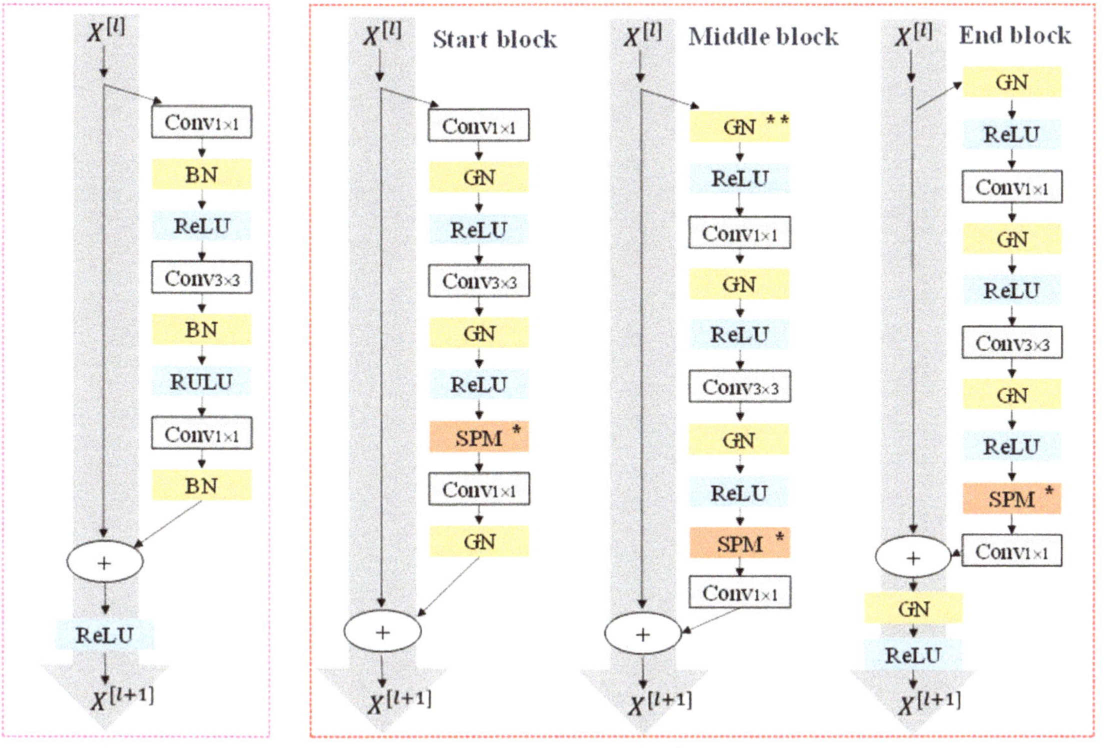

Figure 4. Improved ResNet block structure (**A**) original, (**B**) the structure of IResNet (** means the first GN in the first Middle ResBlock is eliminated in each stage, * means adding the module when channels of the feature map are 512).

We divide signal transfer in the ResBlock into three stages, where different stages mean that convolutional normalization will be performed in a different order, and the end of each stage would ensure nonlinearity, in every layer. These stages of change are determined by the size of space and the number of channels of the network output. As shown in Figure 4B, the ReLU operation on the straight connected channel is eliminated in the Start block phase. At the start of the next phase, ReLU is used for signal activation before the signal transmission continues. In the final stage, GN and ReLU are added to the straight connected channel to ensure the overall normalization and activation of the signal. Such a phased signal transmission, in essence, does not increase the complexity of the network but rather transmits the signal better. Later in the discussion section, we demonstrate this experimentally.

The SPM module is added to establish a self-attention mechanism for the feature target capture (when the number of network output channels is 512 and used). We used the group norm [30] to normalize the feature graph, because the normalization effect of batch normalization with a small batch size is poor, and the small sample dataset we trained is not suitable for using a large batch size, which may exacerbate the possibility of over-fitting. The group norm is a way to avoid this by naturalizing in the form of channel grouping. The effect of normalization depends on the number of groups and not the size of the batch. For the selection of the number of groups, we refer to the original paper [30] and choose the group with the best experimental effect in the paper.

4. Experiments and Results

4.1. Experimental Details

4.1.1. Accuracy Assessment

We mainly used two methods to evaluate the accuracy of our model. However, in order to more vividly highlight the advantages and disadvantages of different algorithms in the application field of this paper (vacancy segmentation), we propose the better error calculation method, which will be mentioned later. The first is the widely used pixel accuracy (PA, which is calculated by dividing the number of truly predicted pixels by the total number of pixels), followed by the mean cross over union (mIoU). The mIoU formula used is:

$$mIOU = \frac{1}{m}\sum_{i=1}^{m}|A_i \cap B_i|/|A_i \cup B_i| \tag{3}$$

where A_i and B_i indicate the area of class i in label and the area of class i in result mapping, respectively, and m indicates the number of classes.

These two methods are most commonly used for evaluating semantic segmentation models and naturally, both evaluation models are also used during our training. Given the particularity of our application scene, we also propose a more intuitive way to evaluate the effects of the actual application of the model, and we define the formula as follows:

$$VR = \frac{V}{C+V} \tag{4}$$

where V represents vacant area, C represents crop area, and VR stands for vacancy ratio. Essentially, we use the vacancy ratio to measure the vacancy between the result of model's segmentation of farmland and the actual manual segmentation result.

Based on this vacancy ratio, we further constructed the error calculation formula for the objective evaluation of the model effect as follows:

$$error = \frac{|VR_f - VR_p|}{VR_f} \tag{5}$$

where VR_f represents the actual vacancy rate, VR_p represents the predicted vacancy rate, and the error value is obtained by dividing the absolute value of the difference by the actual vacancy rate. To distinguish the predictive effect when farmland is in a different state of vacancy, direct subtraction method is not chosen. For example, when there are fewer farm vacancies, such as between 1% and 2%, the fault tolerance of the predicted vacancy will become lower, because it is directly related to the visual effect. Fault tolerance becomes higher when the actual vacancy rate is already high, such as between 5% and 10%.

4.1.2. Training Parameter Setting

Due to the large size of the training images, we use the NVIDIA Tesla V100 32GB to ensure normal training and experimental comparison of data. We selected cross-entropy loss as a loss function to measure the error between prediction result and manual standard. Due to particularity of the dataset, there are only three classes, and there is a huge vacancy between categories. Hence, we specially designed weights for each category. By analyzing the quantitative relationship between categories, we designed three weights, and finally, the experiment proved that weight = [0.8, 1, 1.2] has the best effect. The Loss formula is defined as follows:

$$loss(x, class) = weight[class]\left(-x[class] + \log\left(\sum_j exp(x[j])\right)\right) \tag{6}$$

where class represents three categories of different labels in the experiment. 0 represents the background class, 1 represents the crop category, and 2 represents the vacancy category, which is not directly involved in the calculation but acts as an index. Further, x represents the input vector, and each value in the vector represents the probability prediction value corresponding to each category.

For the optimizer settings, we chose Adam, the adaptive moment estimation, which is essentially the drive of RMSprop [31], to ensure that it learns when to adjust the parameters during training process, while achieving faster speeds and smaller parameters. The base learning rate is set at 0.01. In the training process, 10% random rotation, random horizontal and vertical rotation, and Gaussian noise are adopted. Details on basic tasks can be found in reference [32].

4.2. Results

In the methods described above, we have a comparative analysis of each method, with the main training results as Table 1. From the table, it can be easily seen that the SPINet network based on encoder–decoder architecture in the test set achieved the best results in both accuracy and mIoU indicators, with an increase of nearly 4% points on the original basis and the lowest error rate of 7%. We then analyzed the results of the various indicators of the training corresponding to the different models in detail. From Figure 5, we can see that SPINet trains the fastest and quickly reaches higher accuracy. As can be seen from the training process of mIoU in Figure 5, compared to SPNet, the mIoU value in the training process of our proposed method becomes less volatile, smoother, and reaches a higher value. We mention our loss calculation in Part III. It can be found from the loss training curve, shown in Figure 5, that the loss value of SPINet is higher than that of other models, which indicates the addition of double loss so that the loss value does not fall rapidly to a lower value. The model can continue to learn more new features.

Table 1. Results of several methods on the test set, where error is the result calculated from the formula above.

Methods	Accuracy (%)	mIoU (%)	Error
FC-Densenet56 [16]	94.2%	73.3%	32%
PSPNet [17]	94.3%	73.0%	23%
DenseASPP [18]	93.3%	69.3%	46%
DeepLabv3+ [22]	95.0%	73.8%	24%
SPNet [32]	95.1%	75.5%	19%
SPINet (Ours)	**95.6%**	**77.6%**	**7%**

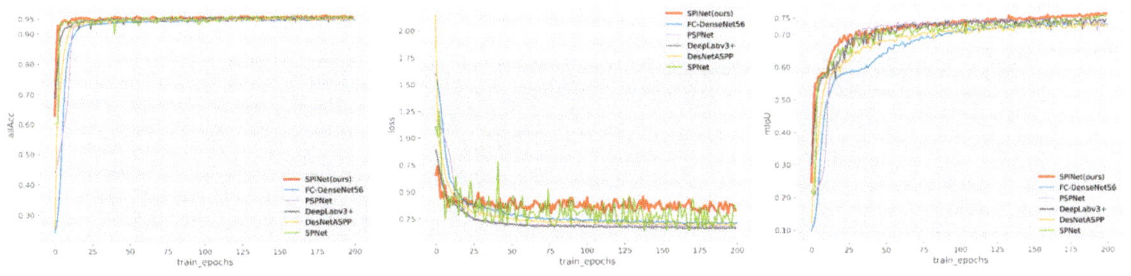

Figure 5. Accuracy, loss value, and mIoU of training process.

In order to better display the comparative results of the models, we took out two pictures with obvious characteristics for analysis. Figure 6 shows the segmentation results of these two images by different models, which will be analyzed one by one later.

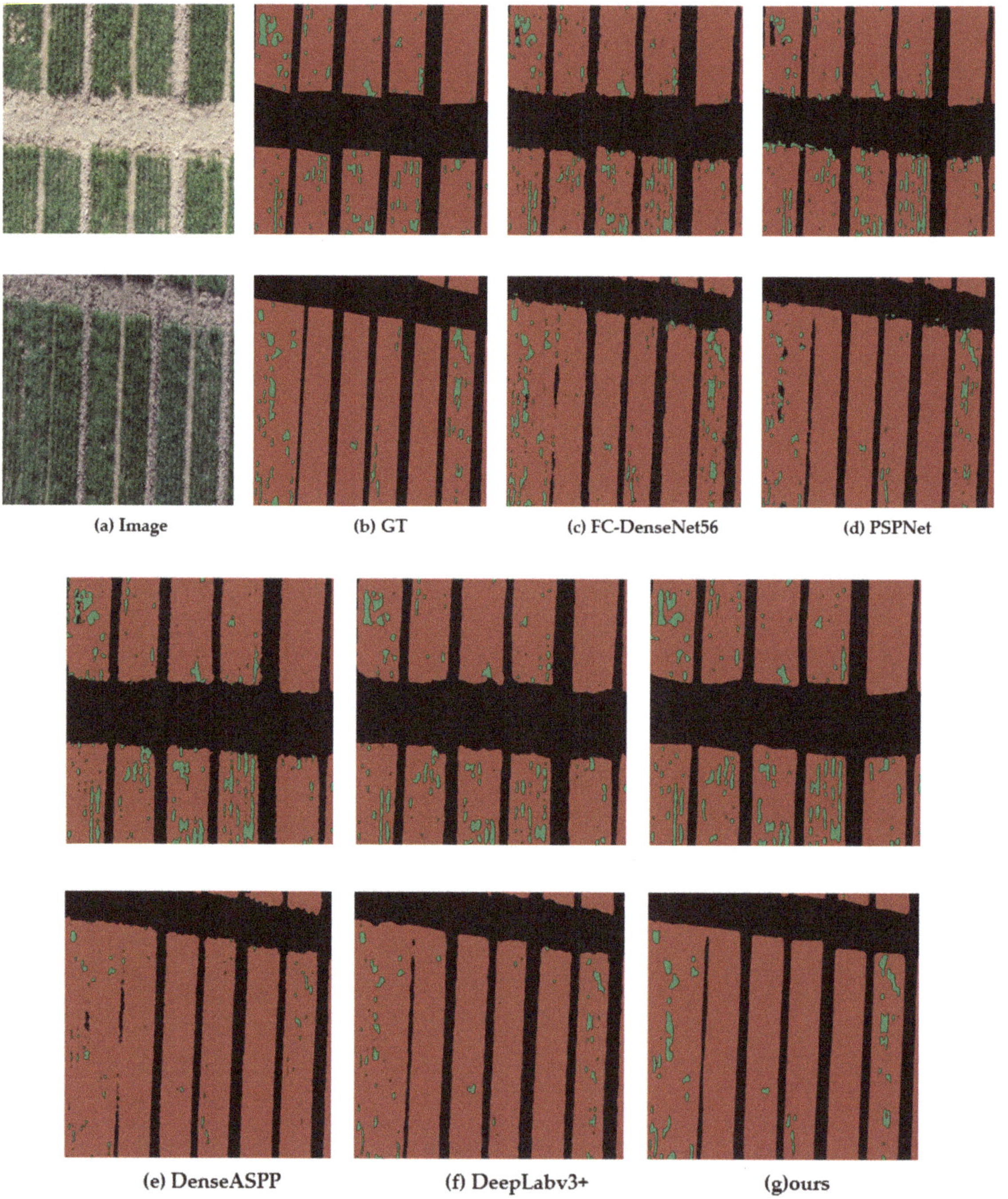

Figure 6. Segmentation results of different models. (**a**) Original image, (**b**) ground truth, and (**c**) visual results of FC-DenseNet56, and (**d**) for the visual results of PSPNet, (**e**) visual results of DenseASPP, (**f**) visual results of DeepLabv3 plus, and (**g**) visual results of our proposed.

In Figure 6, one of the two images have a large vacancy, and the other one has many continuous strip vacancies. From the results of various models, it can be seen that the prediction of FC-DenseNet56 on the vacancy is incomplete, and there is a principal error; that is, the predicted vacancy position is outside the crop region and appears very messy. The prediction results of PSPNet are similar to those of FC-DenseNet56, precisely the rough edges. DenseASPP has largely improved the first two effects, but it still has flaws in its prediction of a big vacancy. DeepLabv3+ has had the most dramatic improvement because of a decoder, which is better at predicting edges and large vacancies. The comparison with SPNet model will be discussed in detail in the discussion section. Finally, the predicted effect of SPINet model is very close to the ground truth. Due to the use of strip pooling, which facilitates the capture of characteristics of vacancies, the segmentation of vacancies is very accurate, and the visualization effect is also very good.

5. Discussion

Vacancy segmentation is a branch of semantic segmentation and has a huge potential for its application in image crack detection (e.g., road crack detection). Farmland vacancies are handled in a different way than in previous datasets because of their relatively large randomness and irregularity. In this paper, a method of semantic segmentation of farmland vacancies is proposed, which is more accurate than several existing typical semantic segmentation methods. Based on this, our discussion is as follows:

5.1. Contribution to Cracks Segmentation

The difference between crack segmentation and general semantic segmentation application is that the image of crack segmentation has subtle features and small targets, making the segmentation method unsuitable for all kinds of traffic and character scenes. The segmentation model, described in this paper, based on farmland vacancy, can better solve this problem. Firstly, we analyze the differences and relation between vacancy segmentation and cracks segmentation. Although farmland vacancy and crack both belong to vacancy segmentation, it is obvious that farmland vacancy has more randomness in shape and size and is denser in distribution than ordinary cracks, which can also be confirmed from the actual images. Therefore, algorithms suitable for simple crack segmentation cannot be well applied in this field. Compared with other methods, the method proposed by us has proved by experiments that all indices have been significantly improved, especially by nearly 4% on mIoU. This method provides an idea for complex cracks segmentation and can be directly transferred to the field for use.

5.2. Contribution to Estimate of Crop Growth and Yield

As mentioned in the Introduction section, the growth process of crops is accompanied by considerable randomness, and all stages from germination to seedling to maturity may be affected by various environmental factors. Therefore, in order to analyze and understand the crop growth situation in time, scientific researchers have put forward many methods. Huilin Tao, Liangji Xu et al. [33], using a hyperspectral sensor installed on an Unmanned Aerial Vehicle (UAV), obtained vegetation index and red edge parameters and their combination. Using stepwise regression (SWR) the partial least squares regression (PLSR) method, above ground biomass (AGB), and the leaf area index (LAI), two kinds of plant growth parameters were estimated. In addition, Thomas Moeckel, Supriya Dayananda et al. [34] used UAV to collect images for 3D point cloud analysis of crop phenotypes. Yi Ma, Shenghui Fang et al. [35] used hyperspectral data to collect dry AGB, because it is an important parameter in assessing crop growth and predicting the yield. Vacancy ratios are not only meaningful in long-term estimation, but also an important indicator in crop growth analysis. At the same time, the segmentation of crops can be derived from the actual planting area of crops in the area, so that it can be directly applied to the crop yield estimate and also combined with the long-potential estimate and other data to carry out the yield estimate.

From the macro point of view, dividing crops and vacancies directly to obtain crop planting vacancy ratio can obtain not only the theoretical crop cover area, but also the actual crop area, which is a direct, effective, and simple way to estimate crop growth. It can be seen from our research that vacancy rate is important in growth estimation and is an important indicator in crop growth analysis. At the same time, the segmentation of crops in this paper can obtain the actual planting area of crops in this area. It can thus be directly used in crop yield estimation or combined with growth prediction and other data for yield estimation.

5.3. *Different from Existing Methods*

Based on the thin, long, and narrow characteristics of most cracks and vacancies, our model adopts the strip pooling idea of deploying a narrow kernel shape along a spatial dimension, to capture long-distance distributions of isolated areas. This is also, in fact, conducive to capturing the characteristics of cracks. At the same time, integration of the spatial pyramid module allows the model to detect effectively and split when other shapes appear in the crack. However, because the crack edge information is too low, in the normal upper sampling process, it is easy to lose its original information.

5.3.1. Effects of Strip Pooling and Decoder

In order to verify the actual effect of the improved method proposed by us, we conducted detailed comparative experiments, and the results are presented in Table 2. SPNet101 is kept constant here. With the improved ResNet as backbone, mIoU increased by about 1% in the validation set. If only the decoder layer is added, the original SPNet101 as the encoder layer increases in accuracy by approximately 0.6% and mIoU by about 2%. When both are added, the accuracy improvement is small compared to just adding the decoder, but mIoU increases by nearly 1%. Hence, we believe that our improvements are effective, especially in the pixel recovery of feature maps, which can accurately restore the information on the vacancy edge of the crop and correctly split out the larger vacancies in the crop.

Table 2. Impact of decoder addition and use of IResNet on results of validation set.

Setting	IResNet	Decoder	Pixel Acc. (%)	mIoU (%)
SPNet-101			94.7%	75.8%
SPNet-101	√		94.5%	76.4%
SPNet-101		√	95.3%	78.8%
SPNet-101	√	√	**95.6%**	**79.8%**

In order to better illustrate this point, we chose an actual picture. From the original picture in Figure 7, it can be found that there are many vacancy distributions in crops, and part of the vacancy is long, in the shape of strips. Therefore, in the segmentation results of SPNet101 model, these vacancies can be well segmented. However, compared with ground truth, its excessive ability to predict vacancy leads to more wrong categories that should not be predicted. In comparison, the model proposed by us achieves a balance in this process, with neither excessive prediction nor little prediction, the prediction result is very similar to the ground truth.

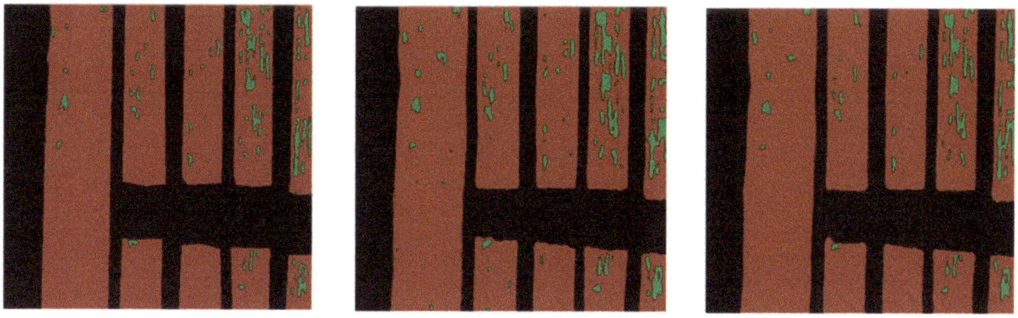

Figure 7. The effect of the decoder: From left to right- the ground truth, the segmentation result of SPNet101, and the segmentation result of the method we proposed.

5.3.2. Comparison with Other Methods

During the experiment, we compared the application of FC-DenseNet56, PSPNet, and other methods in the dataset and obtained the experimental results of each method. In this section, we discuss each method along with the experimental results and the experimental process.

Here, we will macroscopically analyze the parameters of different methods used in this paper. As we all know, DenseNet structure benefits from the design of a dense block, which makes the network narrower (that is, fewer channels) and reduces parameters to a certain extent. The parameters of the two models (FC-DenseNet56 and DenseASPP) derived from this are relatively small. However, as we will discuss later, the DenseNet network is not suitable for this application field, especially as there are large defects in the visualization results. In contrast, PSPNet has increased parameters on the basis of FCN, but the application effect is not good. DeepLabv3+ increases the number of parameters in a series of improvements, but it designs the depthwise separable convolution, which drastically reduces computation complexity, and its application effect is also worthy of praise. However, it is worth noting that the number of additional parameters added in SPM and MPM in this paper is only about half of that of the pyramid pooling module (PSPNet). Moreover, the increase of parameters brought by a simple and effective decoder is almost negligible. At the same time, the information transmission mode is improved for ResNet, but the number of parameters is not increased. Therefore, by comprehensive comparison, the model achieves good results under the condition of limited parameter increase.

First of all, with DenseNet as the backbone, we found that its features of dense jump connections are not suitable for applications in this field. In this field, only two categories need to be separated, crops and vacancies, where crops are the big targets and vacancies are the small ones. Multi-scale fusion and feature reuse in DenseNet grasps the characteristics of crop categories but does not distinguish between the vacant part of crops during subsampling, which leads to negligence of vacant part in the process of upsampling and image recovery. FC-DenseNet56's fully connected approach mitigates this shortcoming, but makes the prediction of vacancies very messy and learning less effective. As DenseNet, the backbone of DenseASPP, can generate a very large range of receiver field features by using atrous spatial pyramid pooling (ASPP), due to the particularity of the vacancy features, can lose more information in the process of final upsampling and recovery. Similarly, for PSPNet, in the training process, we found that the spatial pyramid pooling network is not good at vacancy segmentation, and the prediction of a big vacancy is often wrong in the background. Instead, DeepLabV3 constructed with an encoder–decoder structure shows better adaptability in this dataset and obtains a better model with faster training effect. It also uses multiple sampling rates to expand convolution. The encoder–decoder structure captures clear target boundaries by gradually restoring spatial information. In these training processes, we have summarized the following points, which are the direct sources of our improvement:

- The limited amount of information learned in the early stages of the backbone model, as a model of ResNet, leads us to wonder if much information was lost during the original transmission of ResNet.
- The encoder–decoder structure recovers the target boundary information satisfactorily, especially the prediction accuracy of the vacant edge. This prompts us to think about building a network model of an overall encoder–decoder architecture.
- As a single loss equation is used for all the previous training processes, training in small samples of those methods will quickly drop to a relatively low value, and the ability to learn new features of pictures will significantly reduce. In this regard, we use the method of loss reuse to improve this effect. The value of loss in Layer1 is added to the final calculation of loss with a certain proportion so that the value of loss will not be reduced to the point that it cannot be learned, but will be repeatedly learned in the process of oscillation. This is also determined by the characteristics of small-sample training.

5.4. Limitations and Future Work

The manual labeling of the vacancy segmentation of farmland images is very tedious and requires a lot of effort, which is also the reason why we chose the small sample training network. Moreover, it is difficult for the naked eye to distinguish between the vacant categories that are not obvious, making the results obtained by training and manually marked results erroneous. Such errors are not due to the model itself, but due to the fact that the two measures of vacancy are not exactly the same (mainly for a small vacancy). Therefore, we also propose an evaluation method that can better reflect the effect of the model to alleviate this problem. Additionally, this study is aimed at the training of a small sample of farmland in a certain region. Although the training speed is fast and the training effect is good for particular farmland, to be applied to other farmlands, their information needs to be collected for further training to achieve generalization. Therefore, in the future, farmland data of different regions and different terrains will be collected, and further enhancement in the adaptability and application range of the model through data augmentation training will be made.

Secondly, this paper considers the improvement of prediction accuracy and analyzes the problem of increasing parameters macroscopically, but does not calculate it in detail. In the future, more detailed studies will be made on this point to optimize the algorithm to reduce the number of parameters as much as possible.

6. Conclusions

Earlier, satellite image or aerial image labeling was directly used for farmland segmentation. However, it was too complicated for farmland segmentation in small areas and could not directly highlight the crop growth situation of farmland in such areas. Direct and rapid observation of crop growth in a certain region plays a significant role in rational planting density, crop growth prediction, crop growth analysis, and other aspects. Hence, we establish a semantic segmentation module based on encoder–decoder architecture, where SPINet as an encoder can automatically extract in-depth crop and vacancy information, set-up the decoder to recover extracted feature information, realize end-to-end training and prediction, and achieve good segmentation effect. Thus, from our experiments and comparative analysis, we draw the following conclusions:

Firstly, we construct a dataset of field crop and vacancy segmentation, which provides a new idea for studying field crop segmentation.

Secondly, the application effect of strip pooling in vacancy segmentation is very significant, which can adaptively capture the useful information of vacancy and then conduct accurate segmentation. Meanwhile, the encoder–decoder network structure is very effective for boundary pixel recovery of segmented targets.

Finally, the farmland segmentation method proposed in this paper reached 95.6%-pixel accuracy, and the mIoU value reached 77.6%, reaching the level of practical application. This method can be applied to crop growth analysis, assessment of crop growth, yield estimation, and other practical fields.

Author Contributions: Conceptualization: X.Z. and Y.Y.; methodology: X.N. and X.Z.; software: X.Z.; validation: X.Z., Y.Y. and Z.L.; formal analysis: Z.L.; investigation: Y.Y.; resources: X.N.; data curation: Y.Q.; writing—original draft preparation: X.Z.; writing—review and editing: X.N. and Z.L.; visualization: X.Z.; supervision: Y.Y.; project administration: Y.Q. and W.C.; funding acquisition: Z.L. and X.N. All authors have read and agreed to the published version of the manuscript.

Funding: This work was supported by the National Nature Science Foundation of China under Grant 61901436. General project of Education Department of Sichuan Province, Research on Intelligent Control System of New Agricultural Internet of Things Based on ZigBee Technology, Project No: 17ZB0336.

Informed Consent Statement: "Not applicable" for studies not involving humans.

Data Availability Statement: The following are available online at: https://drive.google.com/drive/folders/1tMsFK1swDTPZt_0Np14B_PAPFM7GsXJB?usp=sharing (accessed on 10 March 2021). The dataset of farmland used in the experiment can be obtained here. As we are still doing more research on the dataset, we will upload our dataset to the same sharing link later.

Acknowledgments: Thanks to the experimental field data shot by Henan Academy of Agricultural Sciences. Additionally, thanks to Jinpu Li and Yuhao Wu for their help and suggestions on data annotation.

Conflicts of Interest: The authors declare no conflict of interest.

References

1. Aung, H.L.; Uzkent, B.; Burke, M.; Lobell, D.; Ermon, S. Farmland Parcel Delineation Using Spatio-temporal Convolutional Networks. *arXiv* **2020**, arXiv:2004.05471. Available online: https://arxiv.org/abs/2004.05471 (accessed on 20 July 2020).
2. Zhang, C.; Wan, S.; Gao, S.; Yu, F.; Wei, Q.; Wang, G.; Cheng, Q.; Song, D. A Segmentation Model for Extracting Farmland and Woodland from Remote Sensing Image. *Preprints* **2017**. [CrossRef]
3. Ganchenko, V.; Doudkin, A. Image Semantic Segmentation Based on Convolutional Neural Networks for Monitoring Agricultural Vegetation. In *International Conference on Pattern Recognition and Information Processing*; Springer: Berlin/Heidelberg, Germany, 2019.
4. Yang, Q.; Liu, M.; Zhang, Z.; Yang, S.; Ning, J.; Han, W. Mapping Plastic Mulched Farmland for High Resolution Images of Unmanned Aerial Vehicle Using Deep Semantic Segmentation. *Remote Sens.* **2019**, *11*, 2008. [CrossRef]
5. Jeon, H.Y.; Tian, L.F.; Zhu, H. Robust crop and weed segmentation under uncontrolled outdoor illumination. *Sensors* **2011**, *11*, 6270–6283. [CrossRef] [PubMed]
6. Zheng, H.; Zhou, X.; He, J.; Yao, X.; Cheng, T.; Zhu, Y.; Cao, W.; Tian, Y. Early season detection of rice plants using RGB, NIR-GB and multispectral images from unmanned aerial vehicle (UAV). *Comput. Electron. Agric.* **2020**, *169*, 105223. [CrossRef]
7. Shewry, P.R. Wheat. *J. Exp. Bot.* **2009**, *60*, 1537–1553. [CrossRef] [PubMed]
8. Deisy, C.; Francis, M. Image segmentation for feature extraction: A study on disease diagnosis in agricultural plants. In *Feature Dimension Reduction for Content-Cased Image Identification*; IGI Global: Hershey, PA, USA, 2018; pp. 232–257.
9. Hui, L.; Mao-hua, W.; Yue-xuan, W.; Dao-kun, M.; Hai-xia, L. Development of farmland soil moisture and temperature monitoring system based on wireless sensor network. *J. Jilin Univ. (Eng. Technol. Ed.)* **2008**, *3*, 604–608.
10. Long, J.; Shelhamer, E.; Darrell, T. Fully convolutional networks for semantic segmentation. In Proceedings of the IEEE Conference on Computer Vision and Pattern Recognition, Boston, MA, USA, 7–12 June 2015.
11. Ronneberger, O.; Fischer, P.; Brox, T. U-net: Convolutional networks for biomedical image segmentation. In *International Conference on Medical Image Computing and Computer-Assisted Intervention*; Springer: Berlin/Heidelberg, Germany, 2015.
12. Su, J.; Yi, D.; Su, B.; Mi, Z.; Liu, C.; Hu, X.; Xu, X.; Guo, L.; Chen, W.H. Aerial Visual Perception in Smart Farming: Field Study of Wheat Yellow Rust Monitoring. *IEEE Trans. Ind. Inform.* **2020**, *17*, 2242–2249. [CrossRef]
13. Wang, C.; Zhao, Z.; Ren, Q.; Xu, Y.; Yu, Y. Dense U-net Based on Patch-Based Learning for Retinal Vessel Segmentation. *Entropy* **2019**, *21*, 168. [CrossRef] [PubMed]
14. Krizhevsky, A.; Sutskever, I.; Hinton, G.E. Imagenet classification with deep convolutional neural networks. *Adv. Neural Inf. Process. Syst.* **2012**, *25*, 1097–1105. [CrossRef]
15. Huang, G.; Liu, Z.; Van Der Maaten, L.; Weinberger, K.Q. Densely connected convolutional networks. In Proceedings of the IEEE Conference on Computer Vision and Pattern Recognition, Honolulu, HI, USA, 21–26 July 2017.

16. Jégou, S.; Drozdzal, M.; Vazquez, D.; Romero, A.; Bengio, Y. The one hundred layers tiramisu: Fully convolutional densenets for semantic segmentation. In Proceedings of the IEEE Conference on Computer Vision and Pattern Recognition Workshops, Honolulu, HI, USA, 21–26 July 2017.
17. Zhao, H.; Shi, J.; Qi, X.; Wang, X.; Jia, J. Pyramid scene parsing network. In Proceedings of the IEEE Conference on Computer Vision and Pattern Recognition, Honolulu, HI, USA, 21–26 July 2017.
18. Yang, M.; Yu, K.; Zhang, C.; Li, Z.; Yang, K. Denseaspp for semantic segmentation in street scenes. In Proceedings of the IEEE Conference on Computer Vision and Pattern Recognition, Salt Lake City, UT, USA, 18–28 June 2018.
19. Chen, L.C.; Papandreou, G.; Kokkinos, I.; Murphy, K.; Yuille, A.L. Deeplab: Semantic image segmentation with deep convolutional nets, atrous convolution, and fully connected crfs. *IEEE Trans. Pattern Anal. Mach. Intell.* **2017**, *40*, 834–848. [CrossRef] [PubMed]
20. Chen, L.C.; Papandreou, G.; Schroff, F.; Adam, H. Rethinking atrous convolution for semantic image segmentation. *arXiv* **2017**, arXiv:1706.05587. Available online: https://arxiv.org/abs/1706.05587 (accessed on 20 July 2020).
21. Chen, L.C.; Papandreou, G.; Kokkinos, I.; Murphy, K.; Yuille, A.L. Semantic image segmentation with deep convolutional nets and fully connected crfs. *arXiv* **2014**, arXiv:1412.7062. Available online: https://arxiv.org/abs/1412.7062 (accessed on 20 July 2020).
22. Chen, L.C.; Zhu, Y.; Papandreou, G.; Schroff, F.; Adam, H. Encoder-decoder with atrous separable convolution for semantic image segmentation. In Proceedings of the European Conference on Computer Vision (ECCV), Munich, Germany, 8–14 September 2018.
23. Jenkins, M.D.; Carr, T.A.; Iglesias, M.I.; Buggy, T.; Morison, G. A deep convolutional neural network for semantic pixel-wise segmentation of road and pavement surface cracks. In *2018 26th European Signal Processing Conference (EUSIPCO)*; IEEE Institute of Electrical and Electronics Engineers: Piscataway, NJ, USA, 2018.
24. Oliveira, H.; Correia, P.L. Road surface crack detection: Improved segmentation with pixel-based refinement. In *2017 25th European Signal Processing Conference (EUSIPCO)*; IEEE Institute of Electrical and Electronics Engineers: Piscataway, NJ, USA, 2017.
25. Elkerdawy, S.; Zhang, H.; Ray, N. Lightweight monocular depth estimation model by joint end-to-end filter pruning. In *2019 IEEE International Conference on Image Processing (ICIP)*; IEEE Institute of Electrical and Electronics Engineers: Piscataway, NJ, USA, 2019.
26. Perez, L.; Wang, J. The effectiveness of data augmentation in image classification using deep learning. *arXiv* **2017**, arXiv:1712.04621. Available online: https://arxiv.org/abs/1712.04621 (accessed on 20 July 2020).
27. Mikołajczyk, A.; Grochowski, M. *Data Augmentation for Improving Deep Learning in Image Classification Problem*; 2018 International Interdisciplinary Ph.D. Workshop (IIPhDW): Świnouście, Poland, 2018; pp. 117–122.
28. Ning, X.; Gong, K.; Li, W.; Zhang, L. JWSAA: Joint Weak Saliency and Attention Aware for person re-identification. *Neurocomputing* **2020**. [CrossRef]
29. Duta, I.C.; Liu, L.; Zhu, F.; Shao, L. Improved Residual Networks for Image and Video Recognition. *arXiv* **2020**, arXiv:2004.04989. Available online: https://arxiv.org/abs/2004.04989 (accessed on 20 July 2020).
30. Wu, Y.; He, K. Group normalization. In Proceedings of the European Conference on Computer Vision (ECCV), Munich, Germany, 8–14 September 2018.
31. Dauphin, Y.N.; De Vries, H.; Bengio, Y. Equilibrated adaptive learning rates for non-convex optimization. *arXiv* **2015**, arXiv:1502.04390. Available online: https://arxiv.org/abs/1502.04390 (accessed on 10 March 2021).
32. Hou, Q.; Zhang, L.; Cheng, M.M.; Feng, J. Strip Pooling: Rethinking Spatial Pooling for Scene Parsing. In Proceedings of the IEEE/CVF Conference on Computer Vision and Pattern Recognition, virtual conference, Seattle, WA, USA, 14–19 June 2020.
33. Tao, H.; Feng, H.; Xu, L.; Miao, M.; Long, H.; Yue, J.; Li, Z.; Yang, G.; Yang, X.; Fan, L. Estimation of Crop Growth Parameters Using UAV-Based Hyperspectral Remote Sensing Data. *Sensors* **2020**, *20*, 1296. [CrossRef] [PubMed]
34. Ma, Y.; Fang, S.; Peng, Y.; Gong, Y.; Wang, D. Remote estimation of biomass in winter oilseed rape (Brassica napus L.) using canopy hyperspectral data at different growth stages. *Appl. Sci.* **2019**, *9*, 545. [CrossRef]
35. Moeckel, T.; Dayananda, S.; Nidamanuri, R.R.; Nautiyal, S.; Hanumaiah, N.; Buerkert, A.; Wachendorf, M. Estimation of vegetable crop parameter by multi-temporal UAV-borne images. *Remote Sens.* **2018**, *10*, 805. [CrossRef]

entropy

MDPI

Article

Optical Channel Selection Avoiding DIPP in DSB-RFoF Fronthaul Interface

Zbigniew Zakrzewski

Institute of Telecommunications and Computer Science, Bydgoszcz University of Science and Technology, Al. Prof. Sylwestra Kaliskiego 7, 85-796 Bydgoszcz, Poland; zbizak@pbs.edu.pl

Abstract: The paper presents a method of selecting an optical channel for transporting the double-sideband radio-frequency-over-fiber (DSB-RFoF) radio signal over the optical fronthaul path, avoiding the dispersion-induced power penalty (DIPP) phenomenon. The presented method complements the possibilities of a short-range optical network working in the flexible dense wavelength division multiplexing (DWDM) format, where chromatic dispersion compensation is not applied. As part of the study, calculations were made that indicate the limitations of the proposed method and allow for the development of an algorithm for effective optical channel selection in the presence of the DIPP phenomenon experienced in the optical link working in the intensity modulation–direct detection (IM-DD) technique. Calculations were made for three types of single-mode optical fibers and for selected microwave radio carriers that are used in current systems or will be used in next-generation wireless communication systems. In order to verify the calculations and theoretical considerations, a computer simulation was performed for two types of optical fibers and for two selected radio carriers. In the modulated radio signal, the cyclic-prefix orthogonal frequency division multiplexing (CP-OFDM) format and the 5G numerology were used.

Keywords: fronthaul; Xhaul; DSB-RFoF; A-RoF; B5G; 6G; DIPP; optical channel selection

Citation: Zakrzewski, Z. Optical Channel Selection Avoiding DIPP in DSB-RFoF Fronthaul Interface. *Entropy* **2021**, *23*, 1554. https://doi.org/10.3390/e23111554

Academic Editors: Michal Choras, Robert Burduk, Agata Giełczyk, Rafal Kozik and Tomasz Marciniak

Received: 22 September 2021
Accepted: 20 November 2021
Published: 22 November 2021

1. Introduction

The 4G and 5G cellular systems have gained independence in using the domain of packet switching. Bypassing telephone centrals when setting up connections has made it possible to provide broadband services based on wireless access. This kind of approach has resulted in an exponential growth in demand for packet mobile services, which were previously only provided over wired networks. It quickly turned out that the approach to designing mobile systems needed to be changed, as the architecture known from the first generations was no longer effective. As a result of the evolutionary changes, the boundary between the core part of the system and the section with radio resource management and multilevel processing of radio signals from baseband and beyond has been clearly marked.

The original architecture of the network in the radio domain of the distributed radio access network (D-RAN) type, also known as an all-in-one macro base station, provides simplicity, as it consists in constructing base stations as single-device systems for processing and broadcasting the radio signals. The use of such a solution is appropriate when the use of the base station is uniform in time and the peak-load is well below the hardware capabilities of the equipment. As mentioned earlier, the rapidly growing demand for broadband Internet access has changed the paradigm in the approach to designing new wireless communication systems, especially those that can be used to provide more dynamic and sophisticated network services. In order to eliminate the disadvantages of the D-RAN solution, the centralized/cloud radio access network (C-RAN) architecture was introduced. In this solution, the user equipment (UE) is set at the center, which means that the provision of coordinated multi-point (CoMP) services becomes much simpler as spatial access aggregation can take place at the level of one base station. In addition, coverage of the area with a radio signal is much more effective, because the central unit, often located in a computing cloud, has greater coordination possibilities and easier access to the often different needs related to the software signal processing.

The C-RAN concept was used in mobile networks from the second generation (2G), and the architecture related to it was disseminated only at the beginning of the 21st century in order to increase the efficiency of next-generation cellular networks. In all cases, the implementation of this concept required the use of specific media that made it possible to transfer pre-prepared radio signals from a central unit (CU) to a radio unit (RU). With the current advances in transmission technologies, this can be done with a wireless microwave medium or a fiber-optic medium. The microwave radio-link was sufficient for the distribution of signals from 2G and 3G interfaces. The appearance on the global market of 4G systems solely with the domain of packet switching has revolutionized the approach to the methods of modernizing the C-RAN architecture. Optical fiber and networks using its transmission potential appeared in the foreground. In the case of 5G mobile systems, in both the core and radio domains, it is difficult to imagine the functioning of such demanding networks without solutions based on optical fibers. The concept of the Xhaul network [1,2] proposed in the 5GPPP project shows exceptionally well the degree of integration of the fiber-optic and radio domains, which is widely used in 5G solutions [3].

In the paper, will pay particular attention to the use of fiber-optic networks in the radio domain of next generations of mobile systems. Next-generation radio access networks (NG-RANs) are very much based on optical networks, which ensures the construction of efficient and broadband interfaces for backhaul (BH), midhaul (MH) and fronthaul (FH) links. Backhaul networks are a very important broadband link between the core network (5GC) and the NG-RAN domains. They are purely digital networks for the delivery of content for final distribution over the wireless mobile radio links. Midhaul and fronthaul networks are used somewhat differently, as they act as networks distributing digital control signals and radio signals with varying degrees of processing. At this point, the radio-over-fiber (RoF) technique [4] is used, which consists in transmitting radio signals from the baseband (BBoF), intermediate-frequency (IFoF) and radio-frequency (RFoF) over an optical path or optical link, such as an optical fiber [5]. Such a method of increasing the transmission range of a radio signal was first defined in [6], where a unit for introducing a radio signal onto an optical carrier and a unit for inverting the process are defined. Radio signals can be transported over a fiber-optic medium in a digitized version, in the digitized-radio-over-fiber (D-RoF) technique, or in an analog version, in the analog-radio-over-fiber (A-RoF) technique [7].

The D-RoF technique is widely used in radio access systems and in microwave transmission systems. In this respect, the most recognizable is the common public radio interface (CPRI) [8], which has monopolized the market in this field and allows the delivery of digitized radio baseband signals. This interface produces very high bit rates, which is noticeable in the case of antenna installations based on multi-element matrices [9,10]. In order to eliminate the monotype of the D-RoF interface, new solutions were introduced to increase the dynamics of this link. The most recognizable dynamic D-RoF interfaces include eCPRI [11] and the next generation fronthaul interface (NGFI) [12]. An initial outline of these solutions was presented by 3GPP [13] and 5GPPP [3] for applications in the 5G mobile systems. Both solutions were prepared so that radio signal transport services could be implemented in the radio-over-ethernet (RoE) mode. This is a good and optimal complement to the shortcomings of CPRI but requires the use of synchronous packet networks [14–17]. The popularization of the implementation of the NGFI family interfaces in 5G systems initiated the trend of looking for solutions that reduce the costs of transporting the digitized radio signals prepared in the edge cloud to an active antenna unit or remote radio head (AAU/RRH) [18]. This resulted in the emergence of a new concept of an open radio access network (O-RAN) [19]. The main players in this field dealing with standardization include O-RAN Alliance, 3GPP and IEEE. The openness of mobile networks in the radio domain ensures the cooperation of 5G systems with local and regional networks belonging to various local operators but meeting specific requirements at the level of real-time packet transport. Fulfilling these requirements is more difficult the higher the split/option of the supported network interface is [11,12].

The A-RoF technique is still used in the field of research, as it is based on the transport of signals over optical fibers in the original version. Such a solution is particularly useful when there is a need to transmit a highly processed radio signal at the intermediate carrier (IFoF) or radio broadcast carrier (RFoF) level. The technique of analog transport of radio signals can be implemented in the fiber-to-the-antenna (FTTA) architecture or with the use of an all-optical transport network. Proposed solutions enabling the simultaneous operation of D-RoF and A-RoF interfaces are presented in [20,21]. Optical systems using coarse/dense WDM (C/DWDM) techniques with all-optical nodes and elastic optical networks (EONs) provide a very good network structure suitable for transporting A-RoF signals. On this basis, it can be concluded that all-optical networks fit well with the concept of O-RAN because their appearance in the local and regional domains ensures the transport of optical signals modulated in any modern and classic formats, as long as they are compatible with the grid of optical channels of a given link.

The transmission of optical signals in the A-RoF technique is very spectral-effective and allows signals to be built up on the edge of the network or in the edge cloud to the final form (software defined radio (SDR) technique). The mast with antennas operating in AAU mode is, in this case, a simple device for amplifying and radiating the RF signal. This solution also fits very well with digital (DBF), analog (ABF) and hybrid (HBF) beamforming on the wireless side of the network. However, it should be taken into account that the signal coming from the A-RoF interface has the structure of an analog signal and is strongly influenced by phenomena occurring in the optical path. In the work [21], the author proposed the introduction of two new Splits/Options, in relation to the 3GPP model [13] to support transport in the IFoF (Option 9) and RFoF (Option 10) formats (Figure 1). In the future B5G/6G mobile systems, it is planned to use high millimeter frequencies up to 100 GHz. High radio frequency provides mass and broadband radio access, but over a not very large area. Transporting the radio signals in the RFoF format at such a high radio frequency requires special measures, as the optical path is dispersive. We mean sections of links or optical paths, which should not exceed 20 km, but the influence of phenomena, especially related to the relative delay, is noticeable. Massive access and local or regional use of the RFoF interface necessitates the use of simple and the same modulation and demodulation techniques, so the solution based on IM-DD modulation seems to be optimal. During the IM modulation, a signal in DSB format appears. The sidebands carry information, but at the point of the direct detection (DD) receiver, they are subject to interference, significantly lowering the receiver sensitivity. This phenomenon is strictly dependent on the RF radio frequency, the length of the optical fiber path and the chromatic dispersion coefficient occurring in the optical channel.

For the first time, the effect of the dispersion-induced power penalty (DIPP) phenomenon on the reception of an optical signal in the DSB format was presented in [22]. This was confirmed in many subsequent experiments [23–27]. In the meantime, new formats for modulating radio signals have emerged, and the demand for RoF solutions for use in mobile wireless communication systems has increased. Methods and techniques have been presented in many publications to mitigate or compensate for the effect of DIPP on the efficient reception of modulated optical signals carrying microwave radio signals. The publication of [28] proposed a technique to mitigate the influence of DIPP by appropriate biasing of the electrooptical Mach–Zehnder modulator (MZM) either at the maximum or at the minimum transmission bias points of the operation of an external MZM. The authors of [29] proposed the use of an FBG taper, whose task is narrowband compensation of chromatic dispersion. This is a good solution, but in most optical paths, it is implemented in a static manner with dispersion-compensating fiber (DCF). In the case of fiber-optic access networks, chromatic dispersion compensation is usually not used, which requires it to be taken into account when designing the link. An alternative to static compensators can be dynamic tunable dispersion–compensation module (TDCM) compensation [30,31].

Single-sideband (SSB) modulation was proposed in [32,33] as an effective way to avoid the sideband interference effect at the photodetector point. This method, however, increases the complexity of the modulator system and lowers the sensitivity of the direct photodetection. A method of mitigating the influence of the DIPP was also proposed by appropriately selecting the modulation index and introducing phase imbalance in the signal modulated in the dual-sideband optical carrier suppression (DSB-OCS) format [34]. Another way to eliminate the phenomenon of deep sideband interference in the photodetector can be the introduction of interleaving light intensity modulation (IM-DD) with phase modulation (PM-DD) [35,36].

In this paper, we propose a solution based on a deep analysis of the DIPP interference characteristics and the introduction of a tunable optical carrier source, according to the flexible DWDM system grid [37]. Such an approach requires the use of a suitable transceiver insert on the optical transmitter side with the ability to tune the laser. If the signal is introduced into the optical path of the DWDM network, the laser tuning must follow the system's optical channel grid, which fits well with the Xhaul architecture supported by the Open RAN Alliance [38]. At the moment, this solution cannot be classified as cheap, but it should be assumed that the increased demand for the development of laser techniques and the wide implementation of flexible DWDM systems in the Xhaul domain will mean that transceivers/transponders equipped with lasers working in accordance with the DWDM grid will in time cease to be a luxury product. The results of the first studies in this area were presented at the international scientific conference "Optical Fibers and Their Applications", which took place in Poland in 2020 [39].

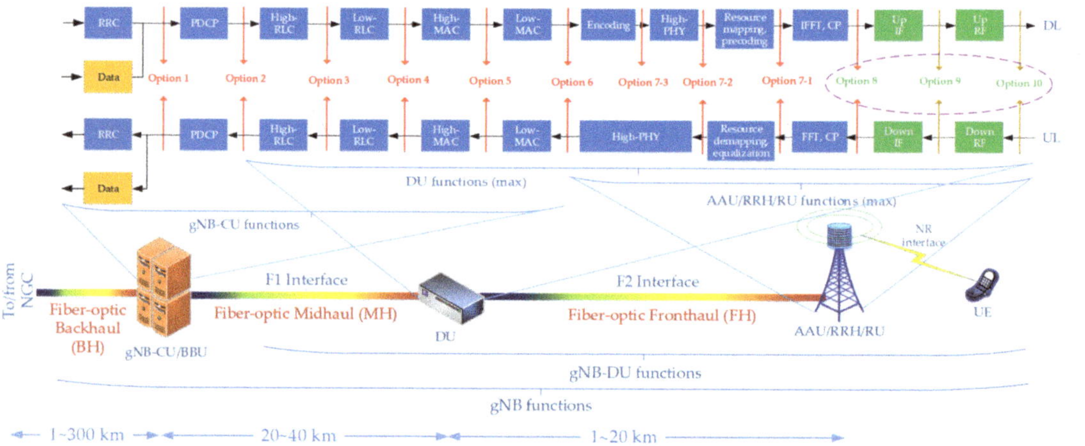

Figure 1. Functional splits proposed by 3GPP [13] in NG-RAN with an example of IF/RF extensions proposed by the author for A-RoF functions introduced into the distributed unit (DU) and the radio unit (RU) (green blocks and options) [21]. Optical BH/MH/FH and their maximal links could be realized in mobile 5G systems [40].

The paper consists of six sections. The first section presents a broad introduction showing the advisability of researching the optical signals modulated in the A-RoF format, with the rapidly developing D-RoF format. Section 2 describes the system, the methods used, and the network components. In this section, Section 2.1 gives a detailed description of the fiber-optic fronthaul path parameters, and Section 2.2 presents the theoretical basis for the DSB–RFoF signal. Section 3 is devoted to the description of the DIPP phenomenon and the analysis of the interference characteristics formed at the photo-receiver side with direct detection. This section is divided into four subsections, where Section 3.1 deals with the calculations related to the DIPP phenomenon occurring in the Option 10 interface, Section 3.2 deals with the calculations related to the selection of the optical channel for the vRF service, Section 3.3 contains the calculations of the relative delay within the CP-OFDM symbol, and Section 3.4 is devoted to the proposed algorithm for selecting an optical channel. Section 4 presents the results of the verification simulations carried out using the VPIphotonics platform, which show the possibilities and limitations of the proposed "Option 10" [21] split working in the DSB–RFoF format. The last two sections present a broad discussion of the obtained results of calculations and simulations, as well as conclusions and predictions for further research.

2. System, Methods, and Network Components

Modern mobile 5G systems are based, in the radio domain, on the C-RAN architecture, strongly promoted by the Open RAN Alliance. It is highly likely that this type of architecture will be the basic determinant of the development of B5G/6G systems and networks. The study below is based on the 3GPP model [13], in which eight basic Splits/Options and three intermediate Splits/Options were defined, which enable the split of functions between network devices due to the processing of information and signals at the level of the NG-RAN domain. The assignment of appropriate functions depends on the current network load and the processing load of individual nodal components. It should be noted here that, in the case of real-time services, e.g., from the ultra-reliable low-latency communication (URLLC) family, it should be ensured that as many stages of information and signal preparation as possible are carried out very close to the place where the radio signal is emitted. On the other hand, the simplicity of the network components located right next to the antenna array provides a flexible approach to network reconfiguration, as most signal processing can be performed in an edge cloud.

Figure 1 shows the functional split of NG-RAN work according to the structure proposed by 3GPP [13]. In order to introduce interfaces from the A-RoF family, two new Splits (Option 9 and Option 10) and the extension of the functions of split number 8 (Option 8) [21] were proposed. When Option 10 is used in the fronthaul link, then the optical signal is in the RFoF format. In Ref. [39], this service is referred to as the virtual radio frequency (vRF), as it provides full virtualization of the site of software building the radio signal to the level of the radio frequency (RF) carrier. Transporting the RF signal over optical fiber is possible only in the fronthaul network link, the length of which cannot exceed 20 km [40]. When we use the Xhaul architecture and the proposed Option 10 split, the RF signal can be transported over various generations of the fiber-optic media. The impact of the accidental transmission parameters of a fiber-optic link, due to the possibility of automatic reconfiguration of the optical path, requires a continuous correction of the use of specific optical resources. In Figure 1, a distributed unit (DU) is responsible for software building a radio signal and inserting it into the optical path, which may be a specialized device or a separate process in the edge computing cloud. On the wireless part, the radio signal is built by the user equipment (UE) as a mobile device. The RU device is responsible for the modulation and demodulation of the optical signal, as well as the amplification of the radio signal and its transmission and reception from the wireless path, by using the antenna array.

2.1. *Fiber-Optic Fronthaul Path*

The fiber-to-the-antenna (FTTA) architecture is used in the basic solutions for connecting the DU and RU. In this case, the optical path is homogeneous, and no active devices are used along the path. In the extended version of the FTTA link, a passive optical network (PON) can be used, which causes a deterioration of the optical power balance due to the presence of high loss optical splitters, but it enables the use of the already existing resources of the optical network. The Xhaul is a universal extension of the optical fiber network with active and reconfigurable devices. The radio signal traveling over the all-optical path will reach optical fiber sections with different propagation properties. At this point, choosing the right optical carrier is crucial.

Three generations of single-mode telecommunications optical fibers are used for the calculations and simulations. The most common single-mode fiber standard is ITU-T

G.652D [41]. It has a wide range of single modality versus wavelength. Its counterpart in the version resistant to macro-bending is the fiber of the ITU-T G.657A standard [42]. This is an SMF fiber that is increasingly replacing the fibers of the G.652D standard. ITU-T G.655 fiber cables are often used in DWDM systems [43]. These fibers have a narrower range of single modality as a function of wavelength, due to adjustments related to DWDM standardization. The standard defined in two tables, i.e., G.655D and G.655E, are used for calculations.

Table 1 shows the standard and maximum attenuation coefficients of single-mode optical fibers, which were used to determine a more precise absorption characteristic as a function of wavelength [41–44] (Figure 2).

Table 1. Empirical and usually measured single-mode optical fiber attenuation coefficients [44].

Fiber	1310 nm	1380 nm	1550 nm	1625 nm
G.652D/G.657A—std	0.33 dB/km	0.31 dB/km	0.20 dB/km	0.25 dB/km
G.652D/G.657A—max [1]	0.40 dB/km	0.4 dB/km	0.3 dB/km	0.4 dB/km
G.655D/E—std	-	-	0.22 dB/km	0.27 dB/km
G.655D/E—max [2]	-	-	0.35 dB/km	0.4 dB/km

[1] According to recommendations [41,42]. [2] According to recommendation [43].

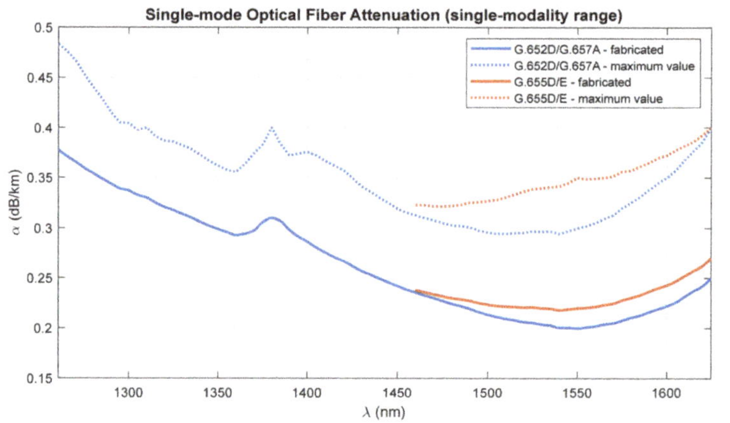

Figure 2. Modeled attenuation characteristics of single-mode optical fibers in range of single-modality limited by the cut-off wavelength [41–43].

Parameters D and S were used to model the delay characteristics of the modulated signals. In order to determine the averaged characteristics of the chromatic dispersion coefficient D of the G.652D and G.657A standards, the Sellmeier polynomials included in the recommendations [42,43] were used:

$$D_{G.652/G.657} = \begin{cases} \frac{\frac{\lambda S_{0max}}{4}\left[1-\left(\frac{\lambda_{0max}}{\lambda}\right)^4\right]+\frac{\lambda S_{0min}}{4}\left[1-\left(\frac{\lambda_{0min}}{\lambda}\right)^4\right]}{2} & \text{for } 1260 \text{ nm } \leq \lambda \leq 1300 \text{ nm} \\ \frac{\frac{\lambda S_{0max}}{4}\left[1-\left(\frac{\lambda_{0max}}{\lambda}\right)^4\right]+\frac{\lambda S_{0max}}{4}\left[1-\left(\frac{\lambda_{0min}}{\lambda}\right)^4\right]}{2} & \text{for } 1300 \text{ nm } < \lambda \leq 1324 \text{ nm} \\ \frac{\frac{\lambda S_{0min}}{4}\left[1-\left(\frac{\lambda_{0max}}{\lambda}\right)^4\right]+\frac{\lambda S_{0max}}{4}\left[1-\left(\frac{\lambda_{0min}}{\lambda}\right)^4\right]}{2} & \text{for } 1324 \text{ nm } < \lambda \leq 1460 \text{ nm} \\ 0.06 \cdot (\lambda - 1460) + 10.5485 & \text{for } 1460 \text{ nm } < \lambda \leq 1625 \text{ nm} \end{cases} \quad (1)$$

where $\lambda_{0min} = 1300$ nm, $\lambda_{0max} = 1324$ nm, $S_{0min} = 0.073$ ps/nm$^2 \cdot$ km, and $S_{0min} = 0.092$ ps/nm$^2 \cdot$ km. The characteristics of the chromatic dispersion coefficient of G.655 fibers were modeled using the polynomials contained in the appropriate tables D and E of the recommendation [43] (Figure 3):

$$D_{G.655D} = \begin{cases} \frac{4.955}{90}(\lambda - 1460) - 0.455 & \text{for } 1460\text{ nm } \leq \lambda \leq 1550\text{ nm} \\ \frac{4.015}{75}(\lambda - 1550) + 4.500 & \text{for } 1550\text{ nm } < \lambda \leq 1625\text{ nm} \end{cases} \quad (2)$$

$$D_{G.655E} = \begin{cases} \frac{5.035}{90}(\lambda - 1460) + 2.650 & \text{for } 1460\text{ nm } \leq \lambda \leq 1550\text{ nm} \\ \frac{3.710}{75}(\lambda - 1550) + 7.685 & \text{for } 1550\text{ nm } < \lambda \leq 1625\text{ nm} \end{cases} \quad (3)$$

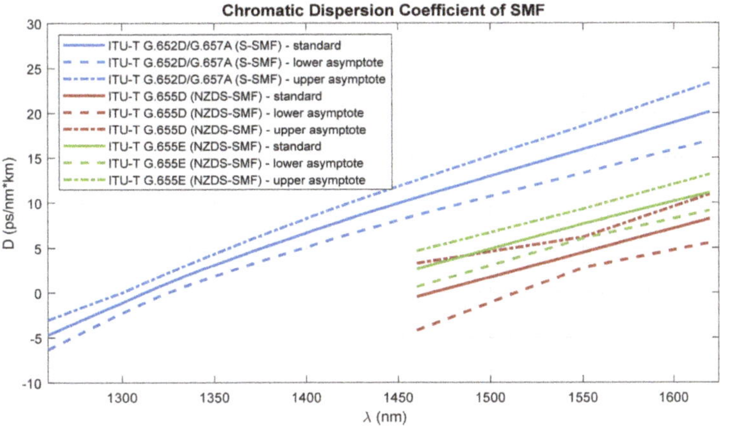

Figure 3. Modeled characteristics of the chromatic dispersion coefficient of the single-mode optical fibers in range of single-modality limited by the cut-off wavelength [41–43].

Due to the specificity of the operation of flexible DWDM systems, as mentioned before, the average characteristics determined in the wavelength range from 1460 nm to 1625 nm are used for further calculations.

Figure 3 shows the characteristics of the chromatic dispersion coefficients (modeled on the basis of (7)–(9)) for three generations of single-mode fibers, which can be found in the optical fiber network working in the Xhaul architecture. The dashed line defines the upper and lower limits of the chromatic dispersion coefficient that can be found in standard SMFs (S-SMFs) and non-zero dispersion-shifted SMFs (NZDS-SMFs).

Optical fibers of the G.652D standard are the basic medium in optical access networks, so it can be assumed that they are present in cities and their surroundings, but also in rural areas. The G.655D/E optical fibers, which are adapted to DWDM links, are often found between cities and between larger regions. We assume that these fibers appear only in the links between optical nodes that support DWDM streams and very often constitute all-optical nodes as the reconfigurable/optical add-drop multiplexers (OADM/ROADM). The fibers of the recently emerged G.657A generation have parameters very similar to those of the G.652D standard but are more resistant to macro-bending events. As a result, fiber-optic cables with G.657A fibers can be installed in locations with more difficult installation conditions. This type of cabling is an excellent solution when connecting AAU/RRH/RU to CU/DU in FTTA architecture.

Figure 4 shows an exemplary slice of the flexible optical Xhaul with fiber access connections. The flexible Xhaul is based on G.655D/E fibers, which guarantees the possibility of providing broadband services with a long range. It should be added that the Xhaul can also provide services in the backhaul domain, which introduces the need to significantly increase the length of optical links (Figure 1). In the access part of the network, on the DU side, standard G.652D fibers are usually used, and on the RRH side, the G.657A standard fibers are preferred. Each of the proposed optical fibers is single mode, so it can carry the traffic of optical signals in accordance with the flexible DWDM channel grid, which is represented in Figure 4 by multi-colored lines connecting network nodes.

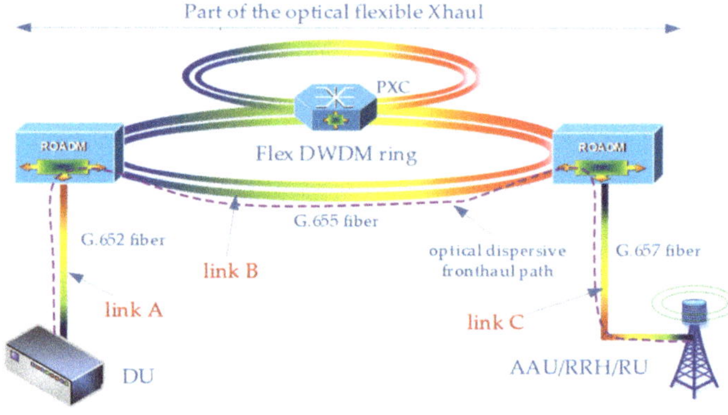

Figure 4. Example of the optical fronthaul path assembled with use of different standards of the single-mode optical fibers.

The optical path, created as a result of connecting several different optical links, can be characterized by a specific attenuation depending on the carrier wavelength (Figure 2), as well as the averaged chromatic dispersion coefficient (Figure 3). The path attenuation depends not only on the length of the optical fiber used, but also on the number of splices or the physical contact (PC) connectors, the number of ROADM/OADM or passive optical splitters (Table 2). In a network using the flexible DWDM system, we assume that optical splitters are not applicable, but in the case of the access part, such optical systems can be used. In this case, PON networks, which currently constitute the basis for the operation of cable networks and fiber-optic Internet access in FTTH format, come to the rescue. In the part of the optical path, on the DWDM link side, an optical amplifier may be used. Such a system is not taken into account during calculations and simulations due to the significant band limitations introduced in the wavelength domain. If there is an amplifier in the optical path, the choice of the optical channel is significantly limited.

Table 2. Example loss/attenuation of passive components in the optical fronthaul path.

Path Component	Fiber (α)	Splice (a_s)	Connector APC/UPC (a_c)	ROADM Add-Drop (a_{ad})	ROADM Passthrough (a_{pt})	Splitter (a_s)
Loss (dB)	-	0.05–0.1	0.1–0.3	1.5–3	1–4	3.5–20
Attenuation (dB/km)	0.2–0.4 [1]	-	-	-	-	-

[1] Changing with wavelength according to Figure 2.

The attenuation of the individual components in the optical path varies with the wavelength of the optical signal carrier. However, the proportions of wavelength-dependent changes are greatest in the case of attenuation of the optical fiber itself. With a short optical path with a length not exceeding 20 km, the number of connections and node devices is not too large. Constant average values provided by the system or device manufacturer may be used in the calculations.

In order to visualize the changes taking place in the averaged chromatic dispersion coefficient, sample calculations were made for five different variants (Table 3) based on the network diagram in Figure 4.

Table 3. Examples of the optical paths with different standards of the single-mode optical fibers used in individual internodal links.

Variant No.	Link A		Link B		Link C	
	Fiber	Length (km)	Fiber	Length (km)	Fiber	Length (km)
1	G.652D	5	G.655D	12	G.652D	3
2	G.652D	3	G.655E	10	G.657A	2
3	G.657A	7	G.655D	8	G.652D	3
4	G.657A	10	G.655E	3	G.657A	7
5	G.652D	1	G.655D	17	G.657A	1

In all variants, it was assumed that link B is equipped with an optical fiber adapted to work in the DWDM environment, while external access cables are based on standard fibers, which have a higher chromatic dispersion coefficient.

The calculation results presented in Figure 5 show that despite the relatively short optical path, the variability of the averaged chromatic dispersion coefficient, with a large variety of the fibers used, may be significant. The nodal devices are not taken into account in the calculations, as they do not introduce dispersive phenomena large enough to be of significant importance in creating the averaged dispersion characteristic.

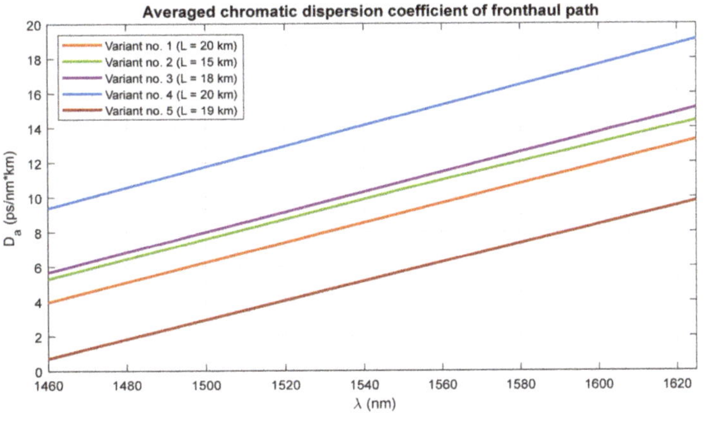

Figure 5. Averaged characteristics of the chromatic dispersion coefficients for the exemplary optical fronthaul paths.

2.2. *Radio-Frequency-over-Fiber in DSB Format*

From the three A-RoF interface formats mentioned in the introduction above, the RFoF format is the most susceptible to the dispersion parameters of a single-mode optical fiber. This type of signal is present at the level of the proposed Option 10 interface (Figure 1). In the case of the classic and at the same time the simplest optical IM-DD link, the RFoF signal contains amplitude modulation (AM) products in the continuous version of the carrier. At the beginning of the analysis, we assume that the radio signal consists of a non-modulated single-tone carrier with a frequency f_{RF} that is a signal modulating an optical carrier with wavelength λ_0. In the case of RF signals, we operate with frequency units and in the case of optical signals, with the wavelength units. At a later stage, we move on to the frequency domain also on the optical side.

The signal modulated in the RFoF format, coming from the optical modulator, can be represented by the formula making the optical carrier signal dependent on the RF signal:

$$E_{IM}(t) = \sqrt{1 + m_{IM}\cos(2\pi f_{RF})}\exp\left(j2\pi\frac{c}{\lambda_0}t\right), \tag{4}$$

where m_{IM} is the intensity modulation index. When m_{IM} is small, the optical field (4) can be written as follows:

$$E_{IM}(t) = (1 + m_{IM}\cos(2\pi f_{RF} t))\exp\left(j2\pi\frac{c}{\lambda_0}t\right). \tag{5}$$

If we transform formula (5) trigonometrically, we obtain a relationship showing the frequency components that create the IM signal:

$$E_{IM}(t) = \exp\left(j2\pi\frac{c}{\lambda_0}t\right) + \frac{m_{IM}}{2}\left\{\exp\left[j2\pi\left(\frac{c}{\lambda_0} + f_{RF}\right)t\right] + \exp\left[j2\pi\left(\frac{c}{\lambda_0} - f_{RF}\right)t\right]\right\}. \tag{6}$$

As we can see in formula (6), the signal takes the form of a DSB, where we can distinguish the optical carrier and two side fringes, which in the case of a modulated RF signal, are sidebands. Now the signal represented by (6) is introduced to the optical path with the dispersive optical fiber, while ignoring the absorption and scattering loss:

$$\begin{aligned} E_{IM}(t,\cdot) = \exp\left(j2\pi\tfrac{c}{\lambda_0}t\right) + \tfrac{m_{IM}}{2}\Big\{\exp\left[j2\pi\left(\tfrac{c}{\lambda_0} + f_{RF}\right)t - \Phi(\cdot)\right] + \\ + \exp\left[j2\pi\left(\tfrac{c}{\lambda_0} - f_{RF}\right)t - \Phi(\cdot)\right]\Big\}. \end{aligned} \tag{7}$$

In the context of the phenomenon taking place in the photodetector working in the direct detection format, it can be assumed that the optical path is a band-pass filter with a flat amplitude response and a linear group delay factor. On this basis, we can assume that the fiber-optic transfer function can be represented by the following [45]:

$$H(f_{RF}, \lambda_0) = \exp(-j\Phi(f_{RF}, \lambda_0)) \tag{8}$$

where

$$\Phi(f_{RF}, \lambda_0) = \pi D(\lambda_0) L \frac{\lambda_0^2}{c} f_{RF}^2 \tag{9}$$

and $D(\lambda_0)$ is the chromatic dispersion coefficient changing with the optical carrier λ_0 (Figure 3). L is the optical path length. After substituting (9) into (7), we obtain the dependence on the electric field strength after the signal passes through the dispersive optical path:

$$\begin{aligned} E_{IM}(t, f_{RF}, \lambda_0) = \exp\left(j2\pi\tfrac{c}{\lambda_0}t\right) + \tfrac{m_{IM}}{2}\Big\{\exp\left[j2\pi\left(\tfrac{c}{\lambda_0} + f_{RF}\right)t - \pi D(\lambda_0) L \tfrac{\lambda_0^2}{c} f_{RF}^2\right] + \\ + \exp\left[j2\pi\left(\tfrac{c}{\lambda_0} - f_{RF}\right)t - \pi D(\lambda_0) L \tfrac{\lambda_0^2}{c} f_{RF}^2\right]\Big\}. \end{aligned} \tag{10}$$

The electrical signal behind the photodetector is proportional to the intensity of the incoming light, so the square law detection applies [26,35,36]:

$$i_{IM}(t) \propto |E_{IM}(t, f_{RF}, \lambda_0)|^2. \tag{11}$$

When we substitute (10) into (11), we obtain a relationship containing a constant component, f_{RF} dependent components and higher-order components. When we assume that the modulation index m_{IM} is small, the higher-order terms become negligibly small. If we also omit the constant component which does not carry information, the signal takes the following approximate form:

$$i_{IM}(t, f_{RF}, \lambda_0) \approx 2m_{IM}\cos\left(\pi D(\lambda_0) L \frac{\lambda_0^2}{c} f_{RF}^2\right)\cos(2\pi f_{RF} t) \tag{12}$$

Consequently, the normalized power response, downstream of the photodetector, on the excitation of the fiber-optic path by the IM signal, is determined by the following relationship [46]:

$$P_{IM}(f_{RF},\lambda_0) \approx \cos^2\left(\pi D(\lambda_0)L\frac{\lambda_0^2}{c}f_{RF}^2\right). \tag{13}$$

Based on formula (13), it can be seen that the power of the received RF radio signal periodically declines to zero. We define this phenomenon as a dispersion induced power penalty (DIPP), and it is the main element of further considerations. After that, we can go to the logarithmic scale to determine the decay ranges of the received RF signal and to combine the DIPP with the absorption loss of the optical path. All components of the optical signal around the λ_0 carrier undergo the same attenuation due to absorption. On this basis, we can determine the degree of fading of the sidebands signal as a result of the DIPP carrier-to-interference ratio (DIPP-CIR) interaction, in relation to the optical carrier as the source of the constant component:

$$\begin{aligned} \mathrm{DIPP}-\mathrm{CIR}_{IM} &= 10\log\tfrac{P_C}{P_{SBs}} = 10\log\cos^2\left(\pi D(\lambda_0)L\tfrac{\lambda_0^2}{c}f_{RF}^2\right) \\ &= 20\log\cos\left(\pi D(\lambda_0)L\tfrac{\lambda_0^2}{c}f_{RF}^2\right). \end{aligned} \tag{14}$$

Given that the optical signal is also subject to absorption loss, the power of the received RF signal downstream of the photodetector is approximately as follows:

$$P_{out}(f_{RF},\lambda_0) \approx P_{in} - \mathrm{DIPP}-\mathrm{CIR}_{IM}(f_{RF},\lambda_0) - A_{pp}(\lambda_0), \tag{15}$$

where $A_{pp}(\lambda_0)$ is the optical path loss.

3. Calculation Results

In the modern 5G-NR radio interface, two frequency ranges are planned to be used, i.e., FR1 [47] to 6 GHz and FR2 [48] to 52.6 GHz. The future B5G/6G wireless systems will also use the higher frequency ranges, i.e., above 60 GHz. For this reason, slightly higher radio frequencies (RFs) are also used in the calculations.

3.1. DIPP-CIR in the Option 10 Fronthaul Path

In fiber-optic communications with optical access, transceivers operating at a wavelength of 1310 nm are very often used. In the case of the Xhaul using optical channels supported by the flexible DWDM system, we move to the S, C and L optical bands. In this case, a transponder should be used in the place of connection to the optical fronthaul path, which will convert the wavelength in accordance with the DWDM grid. In the middle of the C band there is an optimal wavelength of 1550 nm for which preliminary calculations were made to show the influence of the DIPP phenomenon.

Figures 6–8 show that RF dropouts due to direct detection are periodic both as a function of the optical path length and as a function of the radio frequency, which is the modulating signal. This is a well-known phenomenon but has always been considered in the context of a G.652D-based optical path and a wavelength of 1550 nm as the primary optical carrier.

In Figure 6, we can see the significant effect of G.652D fiber dispersion. In this case, decay occurs quickly, and their periodicity is highly densified. The G.655D/E fiber (Figures 7 and 8) has a much smaller impact on the short periodicity of the interference signal fading. Additionally, in the areas of constructive interference, there are quite flat characteristics, which ensure good transmission of broadband radio signals. This issue is discussed later in the paper.

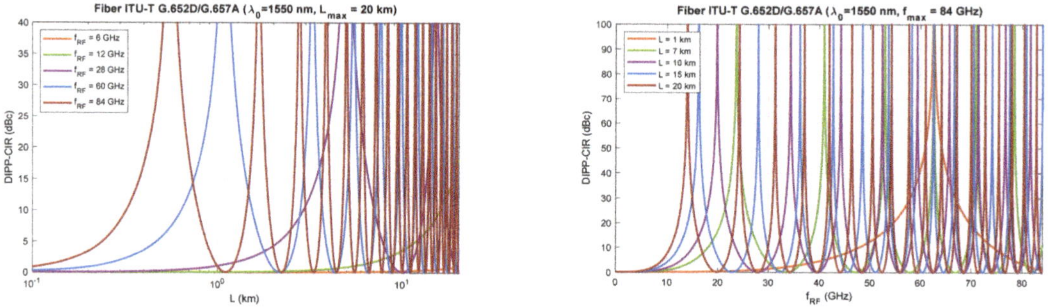

(a) **(b)**

Figure 6. Dispersion induced power penalty—carrier-to-interference ratio obtained after propagation over G.652D or G.657A single-mode fiber and direct detection in photodetector: (**a**) calculated values for the five selected RF carriers as a function of the optical fronthaul path length; (**b**) calculated values for the five selected optical fronthaul path lengths as a function of the RF carrier.

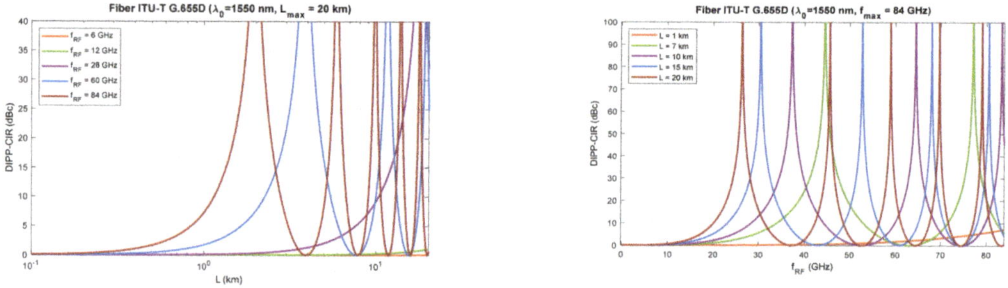

(a) **(b)**

Figure 7. Dispersion induced power penalty—carrier-to-interference ratio obtained after propagation over G.655D single-mode non-zero dispersion-shifted fiber and direct detection in photodetector: (**a**) calculated values for the five selected RF carriers as a function of the optical fronthaul path length; (**b**) calculated values for the five selected optical fronthaul path lengths as a function of the RF carrier.

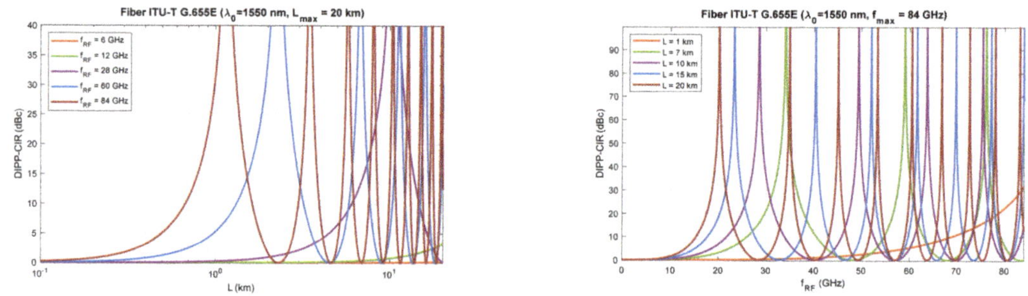

(a) **(b)**

Figure 8. Dispersion-induced power penalty—carrier-to-interference ratio obtained after propagation over G.655E single-mode non-zero dispersion shifted fiber and direct detection in photodetector: (**a**) calculated values for the five selected RF carriers as a function of the optical fronthaul path length; (**b**) calculated values for the five selected optical fronthaul path lengths as a function of the RF carrier.

On the basis of the obtained results, it can also be concluded that the influence of the DSB signal interference phenomenon during direct detection is significant at higher radio frequencies. This mainly applies to the FR2 band [48] intended for use in 5G/B5G systems, which reaches the range of 60 GHz, as well as higher bands, which are adapted in subsequent releases and generations of wireless access systems. The optical path for signal transmission in the RFoF technique should not exceed 20 km (Figure 1), and thus, can be implemented in a passive network, where chromatic dispersion compensation is usually not practiced.

The next stage of calculations is closely related to the dispersion characteristics of single-mode optical fibers of different generations, in relation to the optical channels used in DWDM systems. The single-modality range of G.652D and G.657A fibers is much wider (Figure 3), but due to the limitations of G.655 fibers, the total optical bandwidth (S + C + L) is used for the calculations. This is a practical range of single-mode optical fibers adapted to work with dense wavelength multiplexing systems.

Figures 9–11 show the fixed limits of 3 dB and 10 dB that define the areas in the optical wavelength and path length domains. On this basis, it is possible to determine the relationship that classifies the quality of the optical channel (QoOCh) prepared to carry the DSB signal over the optical path in the RFoF format:

$$\text{QoOCh} = \begin{cases} \text{strong} \;\; for \;\; \text{DIPP} - \text{CIR}_{IM} < 3\text{ dB} \\ \text{weak} \;\; for \;\; 3\text{ dB} < \text{DIPP} - \text{CIR}_{IM} < 10\text{ dB} \\ \text{unusable} \;\; for \;\; \text{DIPP} - \text{CIR}_{IM} > 10\text{ dB} \end{cases} \tag{16}$$

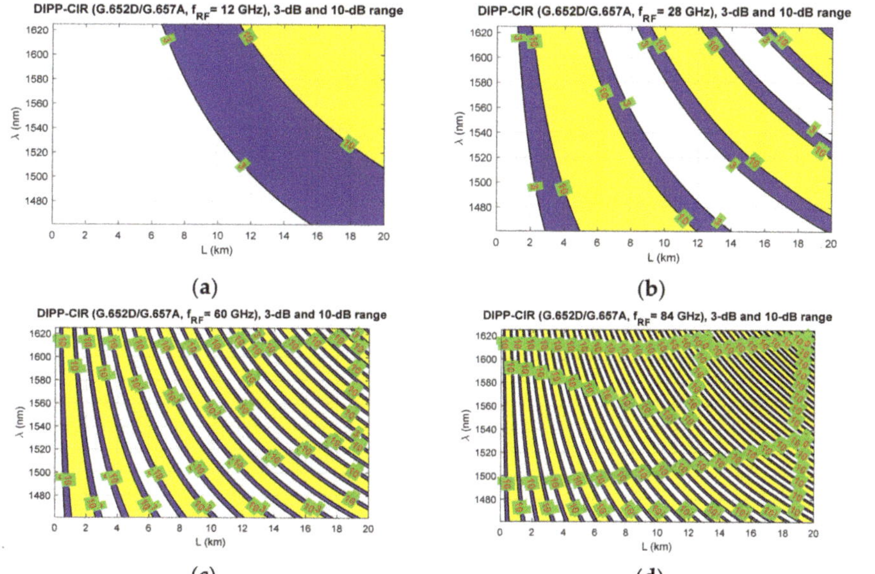

Figure 9. The results of the dispersion-induced power penalty calculations, taking into account the 3 dB and 10 dB thresholds, obtained in the range of the S, C and L optical bands for the fronthaul path based on the G.652D or G.657A optical fiber of variable length: (**a**) $f_{RF} = 12$ GHz; (**b**) $f_{RF} = 28$ GHz; (**c**) $f_{RF} = 60$ GHz; (**d**) $f_{RF} = 84$ GHz.

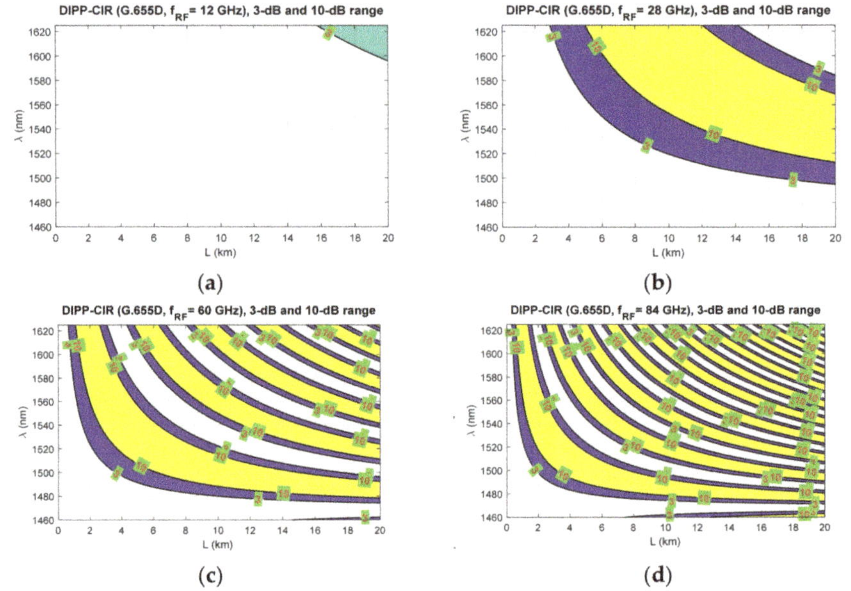

Figure 10. The results of the dispersion-induced power penalty calculations, taking into account the 3 dB and 10 dB thresholds, obtained in the range of the S, C and L optical bands for the fronthaul path based on the G.655D optical fiber of variable length: (**a**) $f_{RF} = 12$ GHz; (**b**) $f_{RF} = 28$ GHz; (**c**) $f_{RF} = 60$ GHz; (**d**) $f_{RF} = 84$ GHz.

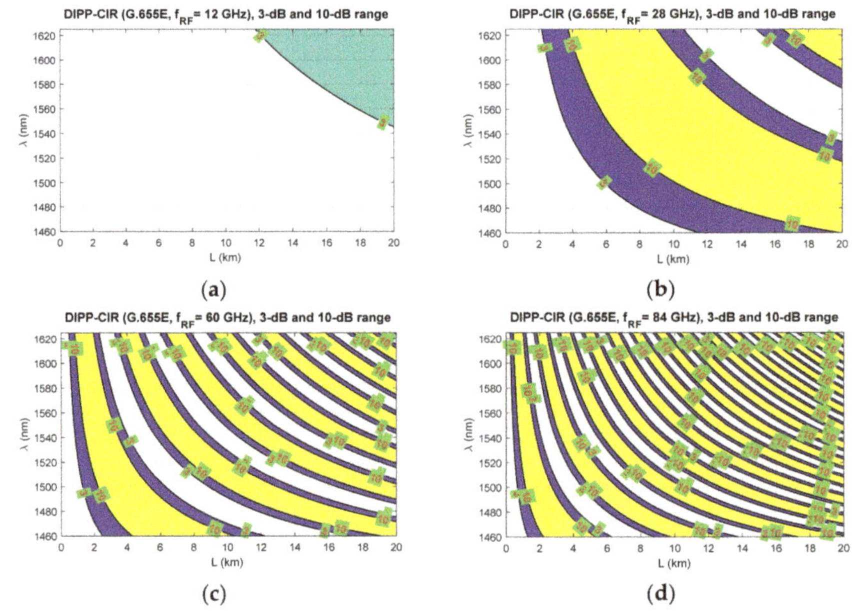

Figure 11. The results of the dispersion-induced power penalty calculations, taking into account the 3 dB and 10 dB thresholds, obtained in the range of the S, C and L optical bands for the fronthaul path based on the G.655E optical fiber of variable length: (**a**) $f_{RF} = 12$ GHz; (**b**) $f_{RF} = 28$ GHz; (**c**) $f_{RF} = 60$ GHz; (**d**) $f_{RF} = 84$ GHz.

The white area visible in Figures 9–11 is the best optical carrier range for a particular optical path length. The blue field is a transition region with a fast quenching of the sidebands in the 3–10 dB range. Yellow areas indicate a DIPP-CIR greater than 10 dB, which, combined with the absorption losses, gives ranges of poor transmission quality. It should be noted, however, that in the case of radio carrier modulation with a low order applied, this range may also be partially used.

Figure 9 shows the results of the calculations concerning the DIPP phenomenon occurring in the photodetector after the modulated optical signal has passed over the G.652D or G.657A optical fiber. In this type of fiber, there are yellow areas that cover the entire range of the S + C + L bands. This mainly concerns short sections of several kilometers for radio frequencies from 28 GHz (Figure 9b–d). In the case of the G.655D fiber, the situation is much better (Figure 10) because we can always find the optical range that lies in the white or blue field. The G.655E standard optical fiber is characterized by an indirect characteristic of the chromatic dispersion coefficient (Figure 2), which allows the lack of availability for most radio carrier frequencies to be avoided. The exception here is the case of frequencies above 80 GHz, where for an optical path with a length of about 1.5 km, there may be problems with selecting the appropriate optical channel.

Figures 12–14 show the results of the DIPP-CIR signals calculated as a function of the optical wavelength. In order to determine the 3 dB and 10 dB optical wavelength range, a cut-off condition was applied. On this basis, we can try to adjust the optical channel for a specific optical path length and for a specific radio carrier frequency. The calculation results presented in Figures 12–14 relate to a path with a length of 20 km. Figure 12 shows the results of the DIPP interaction after passing over the G.652D or G.657A fiber. It is significant that for the case f_{RF} = 12 GHz, no optical band range with a threshold below 3 dB is available (Figure 12b—black color). The only thing left to do here is to select the optical channel in the range between the 3 dB and 10 dB thresholds (Figure 12c). The situation looks much better in the case of the optical path built on the basis of G.655D (Figure 13) and G.655E (Figure 14) fibers. In this case, we can always find optical resources of good quality due to DIPP.

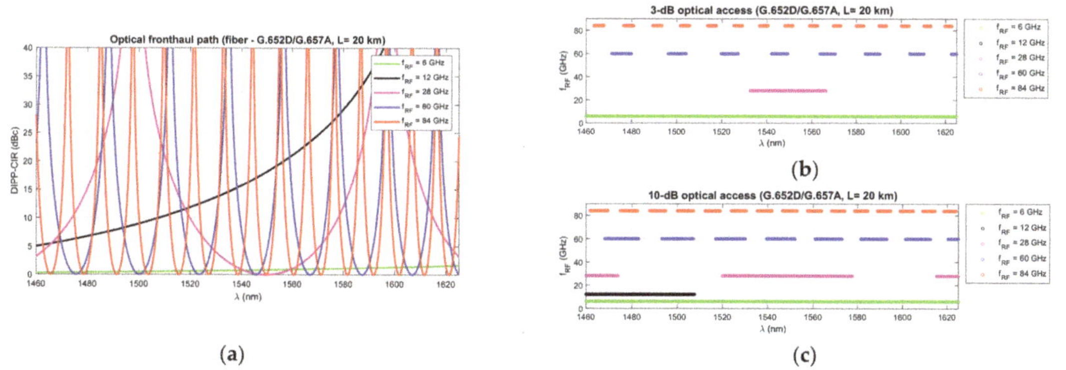

Figure 12. DIPP-CIR as a function of the optical wavelength for selected radio carrier frequencies and the 20 km optical path created on the basis of the G.652D or G.657A optical fiber: (**a**) calculation results without selection; (**b**) 3 dB cut-off optical access ranges; (**c**) 10 dB cut-off optical access ranges.

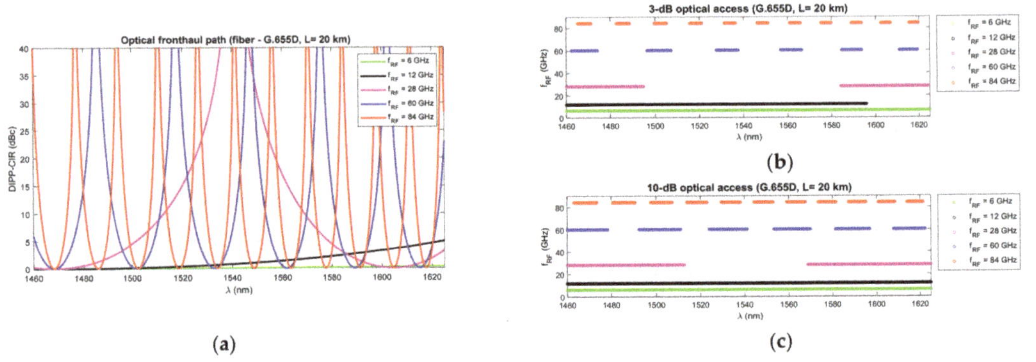

Figure 13. DIPP-CIR as a function of the optical wavelength for selected radio carrier frequencies and the 20 km optical path created on the basis of the G.655D optical fiber: (**a**) calculation results without selection; (**b**) 3 dB cut-off optical access ranges; (**c**) 10 dB cut-off optical access ranges.

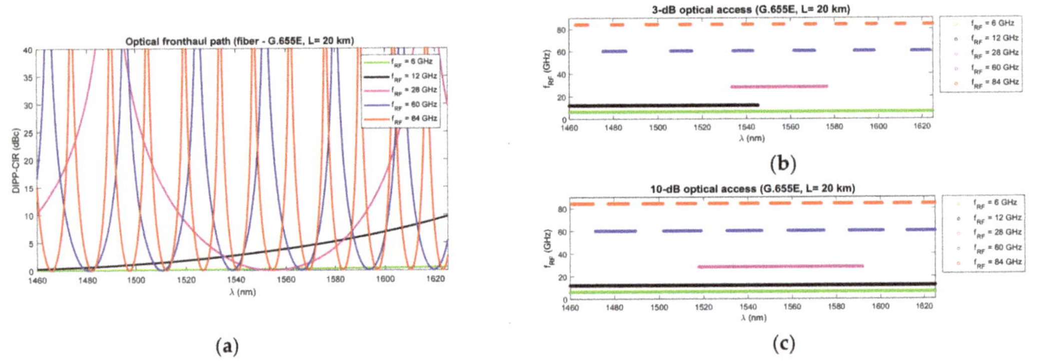

Figure 14. DIPP-CIR as a function of the optical wavelength for selected radio carrier frequencies and the 20 km optical path created on the basis of the G.655E optical fiber: (**a**) calculation results without selection; (**b**) 3 dB cut-off optical access ranges; (**c**) 10 dB cut-off optical access ranges.

3.2. Optical Subband Selection for vRF Service

The bandwidth of the optical channel is strictly dependent on the grid defined in the DWDM system [37]. However, the flexibility of the system allows for uneven reservation of optical resources, which increases the spectral efficiency of the use of the optical path built on the basis of a single-mode optical fiber. The optical carrier can change with a 2 times finer grid (6.25 GHz) in relation to the minimum channel bandwidth, which means that the minimum optical channel bandwidth is 12.5 GHz. In an all-optical fronthaul link, it can be assumed that the nodal devices are insensitive to the modulation format used in the optical channel, but the channel bandwidth and optical carrier must be defined by the optical resource management system (Table 4). ROADM optical access nodes are equipped with filters, which should also be adapted to the supported DWDM grid. Lasers used in optical transmitters are not always characterized by highly stable operation, which causes the phenomenon of optical drift. Hence, the optical modulated signal needs a wider optical channel, both when passing through a filter built into the ROADM node and when transmitting over a DWDM link on the Xhaul side. From here, on the optical side, we also use the unit of frequency, which ensures compliance with the grid given in [37] and makes the specification of the channel width independent of the carrier wavelength.

Table 4. Optical channel bandwidths for selected RF carriers in the vRF service (RFoF interface). $\Delta\lambda_0$ is given for an optical carrier with a wavelength equal to 1550 nm.

f_{RF} (GHz)	6	12	28	60	84
Δf_0 (GHz)	12.5	25	100	200	200
$\Delta\lambda_0$ (nm)	~0.1	~0.2	~0.8	~1.6	~1.6

In the next step, based on relationships (14) and (16), optical subbands (from the S + C + L band range) are determined, in which optical channels can be selected with a step of 6.25 GHz or higher. This depends on the capabilities of the tunable laser, which is built into the optical transmitter, or on the fixed lasers available at the transmitter site. In the case of a low frequency radio carrier where the DIPP interaction is weak, a narrower optical channel should be used (Table 4), and the needs accordingly increase as the frequency of the radio carrier increases. The degree of laser tuning or wavelength selection from the available allows for a much more precise selection of optical resources in the presence of the DIPP. In filling the gaps between the channels, such a dense laser tuning mesh also increases flexibility. Table 5 shows the subbands that provide access to a 20 km fronthaul path based on G.652D or G.657A optical fiber. The 3 dB criterion was not met in the whole wavelength range for the radio signal with f_{RF} = 12 GHz, which resulted in a lack of access to the service for this link. In other cases, optical resources are available. In a situation where a specific optical path does not have optical resources that can be used to transmit a radio signal with a specific radio frequency, then an attempt should be made to look for an alternative path with a different chromatic dispersion factor D. It is advisable that with such a systemic approach, each optical path found should be tested in advance for its accumulated chromatic dispersion. This ensures a quick decision by the algorithm searching for matched optical resources for transporting the RF signal. The tabulated results show that calculations of this kind can be performed for any FH path length and any type of single-mode fiber, also in mixed versions (for example, Figure 5 and Table 3).

Table 5. Calculated availability of optical subbands for two thresholds in a 20-km fronthaul path designed on the basis of G.652D or G.657A single-mode fiber.

f_{RF} (GHz)	**Optical Subband (THz)**	
	3-dB	**10-dB**
6	205.3380–184.4880	205.3380–184.4880
12	-	205.3380–198.8568
28	195.6068–191.4130	205.3380–203.4067; 197.2505–190.0130 [1]; 185.6130–184.4880
60	203.7692–202.5880; 200.3817–199.3067; 197.2817–196.2880; 194.4130–193.4942; 191.7505–190.8942; 189.2692–188.4692; 186.9442–186.1942; 184.7630–184.4880	204.2067–202.1693; 200.7817–198.9193; 197.6442–195.9317; 194.7505–193.1630 [1]; 192.0692–190.5880; 189.5630–188.1817; 187.2255–185.9255; 185.0255–184.4880
84	204.8630–204.2442; 203.0505–202.4567; 201.3130–200.7505; 199.6567–199.1192; 198.0755–197.5567; 196.5505–196.0567; 195.0880–194.6130; 193.6817–193.2255; 192.3255–191.8817; 191.0192–190.5942; 189.7567–189.3442; 188.5380–188.1380; 187.3568–186.9692; 186.2130–185.8380; 185.1068–184.7442	205.0942–204.0192; 203.2630–202.2442; 201.5255–200.5443; 199.8568–198.9255; 198.2630–197.3692; 196.7318–195.8755; 195.2630–194.4380; 193.8505–193.0567; 192.4942–191.7255; 191.1755–190.4380; 189.9130–189.1942; 188.6880–187.9942; 187.5005–186.8317; 186.3505–185.7067; 185.2380–184.6130

[1] selected optical subband for simulation.

Tables 6 and 7 contain the results of searching for the available optical subbands also for two thresholds, but for optical paths built on the basis of optical fibers of the G.655 family. The optical fiber with a non-zero chromatic dispersion-shifted coefficient is characterized by a much lower dispersion coefficient, which makes the available areas of the optical subbands wider. It can be concluded that in the case of reduced dispersion or partially compensated, e.g., in a link belonging to the DWDM transit domain (for example Figure 4, link B), the number of useful optical channels is much greater.

Table 6. Calculated availability of optical subbands for two thresholds in a 20 km fronthaul path designed on the basis of G.655D non-zero dispersion shifted single-mode fiber.

f_{RF} (GHz)	Optical Subband (THz)	
	3-dB	10-dB
6	205.3380–184.4880	205.3380–184.4880
12	205.3380–187.8630	205.3380–184.4880
28	205.3380–200.5755; 189.2442–184.5630	205.3380–198.1880; 191.1068–184.4880 [1]
60	205.0192–203.3630; 200.3505–198.9067; 196.2442–194.9630; 192.5630–191.3817; 189.1817–188.1130; 186.1130–185.1317	205.3380–202.7880; 200.8880–198.3942; 196.7255–194.5130 [1]; 193.0067–190.9630; 189.5817–187.7318; 186.4755–184.7880
84	204.6068–203.7630; 202.1693–201.3880; 199.9005–199.1693; 197.7755–197.0880; 195.7692–195.1192; 193.8817–193.2630; 192.0567–191.4567; 190.3130–189.7505; 188.6567–188.1192; 187.0817–186.5630; 185.5692–185.0817	204.9130–203.4630; 202.4630–201.1068; 200.1693–198.9067; 198.0255–196.8380; 196.0130–194.8880; 194.1068–193.0380; 192.2755–191.2442; 190.5255–189.5443; 188.8568–187.9255; 187.2692–186.3817; 185.7505–184.9005

[1] selected optical subband for simulation.

Table 7. Calculated availability of optical subbands for two thresholds in a 20 km fronthaul path designed on the basis of G.655E non-zero dispersion shifted single-mode fiber.

f_{RF} (GHz)	Optical Subband (THz)	
	3-dB	10-dB
6	205.3380–184.4880	205.3380–184.4880
12	205.3380–194.0192	205.3380–184.4880
28	195.5130–190.1505	197.5317–188.3255
60	203.2130–201.7692; 199.1130–197.8317; 195.4505–194.2880; 191.9880–190.8255; 188.6567–187.5942; 185.6130–184.6380	203.7505–201.2630; 199.5942–197.3755; 195.8755–193.8817; 192.4193–190.4130; 189.0443–187.2192; 185.9692–184.4880
84	204.9317–204.1567; 202.6755–201.9505; 200.5567–199.8755; 198.5630–197.9130; 196.6693–196.0567; 194.8755–194.2943; 193.1442–192.5317; 191.3505–190.7630; 189.6380–189.0817; 188.0067–187.4755; 186.4505–185.9442; 184.9567–184.4880	205.2192–203.8755; 202.9442–201.6880; 200.8130–199.6255; 198.8005–197.6817; 196.8942–195.8380; 195.0942–194.0817; 193.3755–192.3067; 191.5630–190.5505; 189.8442–188.8817; 188.2005–187.2817; 186.6380–185.7567; 185.1380–184.4880

The calculation results presented in the tables and figures above relate to the transmission of unmodulated radio carrier in the links of the optical fiber network, but do not take into account the bandwidth of the radio frequency channel. The modulation of the radio carrier causes the appearance of frequency components that make up a broadband signal that must be within the range of a specific radio channel. The bandwidth of this channel depends on the modulation format, bit rate and the filtering criterion.

3.3. Relative Delay and CP-OFDM Modulation

The CP-OFDM modulation is applied to the 5G-NR downlink, which allows dynamic allocation of radio resources to different user equipment (UE) on the same frequency channel. The smallest unit consists of 12 OFDM subcarriers, and the frequency bandwidth of this resource is dependent on a numerological parameter that defines the frequency spacing between OFDM subcarriers [49,50] (Table 8). As a result of combining all radio resource blocks, a frequency channel is created, which has a width of up to 400 MHz in 5G NR. Of course, the entire channel is not signal covered, as it requires frequency guard intervals [50], which limit the actual frequency bandwidth of the modulated OFDM signal. The maximum channel bandwidths, which are given in Table 8, are taken into account for the calculations. This ensures the consistency of the calculations and better visual analysis resulting from the numerology used in the 5G NR interface.

Table 8. Selected parameters of CP-OFDM modulation used in the 5G New Radio interface [49,50].

5G Numerology (μ)	BB-Bandwidth (Δf_{RF})	Subcarrier Spacing	OFDM Subcarriers	Modulation Order (min–max)	Symbol-Rate	1-Layer Bit Rate (max)	FFT Symbol Duration	CP Duration (T_{CP})	FFT + CP Duration
-	(MHz)	(kHz)	-	-	(Bd)	(Mbps)	(µs)	(µs)	(µs)
0	5	15	300	2–256	14,000	33.60	66.67	4.69	~71.36
0	10	15	624	2–256	14,000	69.89	66.67	4.69	~71.36
0	20	15	1272	2–256	14,000	142.46	66.67	4.69	~71.36
0	50	15	3240	2–256	14,000	362.88	66.67	4.69	~71.36
1	80	30	2604	2–256	28,000	583.30	33.33	2.34	~35.67
1	100	30	3276	2–256	28,000	733.82	33.33	2.34	~35.67
2	200	60	3168	2–256	56,000	1419.26	16.67	1.17	~17.84
3	400	120	3168	2–256	112,000	2838.53	8.33	0.57	~8.90
4	400	240	1536	2–256	224,000	2752.51	4.17	0.29	~4.46

In order to determine the effect of the channel bandwidth on the OFDM subcarrier transfer unevenness in the RFoF link, we introduce a dependence on the degree of variation in the level of the extreme subcarriers belonging to the RF radio channel. For this purpose, we use formula (14), in which the frequency spacing of the extreme OFDM subcarrier from the middle RF carrier is defined as half the width of the frequency channel:

$$\begin{gathered}\Delta \mathrm{DIPP} - \mathrm{CIR}_{(\mathrm{dB})}(f_0, \Delta f_{RF}) = \\ \left| \mathrm{DIPP} - \mathrm{CIR}_{(\mathrm{dBc})}\left(f_0 - \frac{\Delta f_{RF}}{2}\right) - \mathrm{DIPP} - \mathrm{CIR}_{(\mathrm{dBc})}\left(f_0 + \frac{\Delta f_{RF}}{2}\right) \right|.\end{gathered} \tag{17}$$

The value of ΔDIPP-CIR, determined by relationship (17), is related to the first *Th*1 threshold, which determines the allowed DIPP-CIR value for the carrier frequency f_0 of the optical channel. In order to decide whether the optical channel is suitable for carrying a signal in the DSB-RFoF format, a second threshold, *Th*2, is determined that ultimately limits the use of the designated channel by threshold *Th*1. The *Th*1 and *Th*2 thresholds are used in the algorithm for determining the optical channel subband as a fixed value of the impact of the DIPP phenomenon.

Figures 15–17 show the results of the differential DIPP-CIR calculation for the first 3 dB threshold only. The influence of the radio frequency channel bandwidth on the severity of the OFDM subcarrier level differentiation phenomenon is significant. With increasing frequency of the radio carrier, the increase is so large that the selection of the optical channel is possible in very narrow ranges, much narrower than those given in Tables 6–8. It should also be noted that not all channels with 100 MHz bandwidth can be used. In the radio band FR1 [50], it is not possible to create wider channels, due to the high fragmentation and dedicated use of subbands. Wider channels realization is only possible in the higher frequency bands in the FR2 range [50]. Thus, it can be seen that not all the combinations, shown in Figures 15–17, are currently possible and practical to implement. In the case of the G.655D fiber, the results of the calculations show that practically, the full 3 dB range, according to *Th*1, can be accepted for implementation (Figure 16). In the case of the G.655E fiber, the *Th*2 threshold limitation is necessary for 400 MHz bandwidth channels and a radio carrier with a frequency above 60 GHz (Figure 17d). In this case, the differential DIPP-CIR exceeds 3 dB. When analyzing the graph in Figure 15, we can see that significant limitations appear with a 200 MHz channel, obviously with a carrier frequency above 60 GHz. Here, in the optical path, there is a classic G.652D or G.657A access fiber. Thus, a *Th*1 threshold of 10 dB may have practical applications for bands lying in the FR1 range.

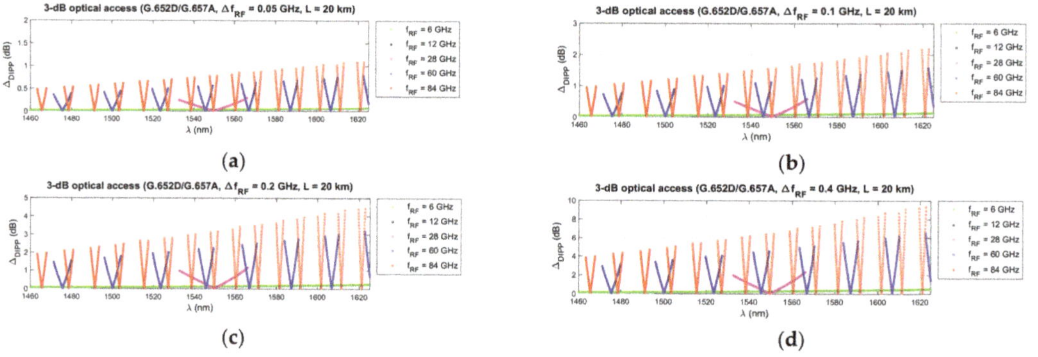

Figure 15. Differential DIPP-CIR as a function of the optical wavelength for selected radio carrier frequencies and the 20 km optical path created on the basis of the G.652D or G.657A optical fiber, and for selected radio channel widths: (**a**) 50 MHz; (**b**) 100 MHz; (**c**) 200 MHz; (**d**) 400 MHz.

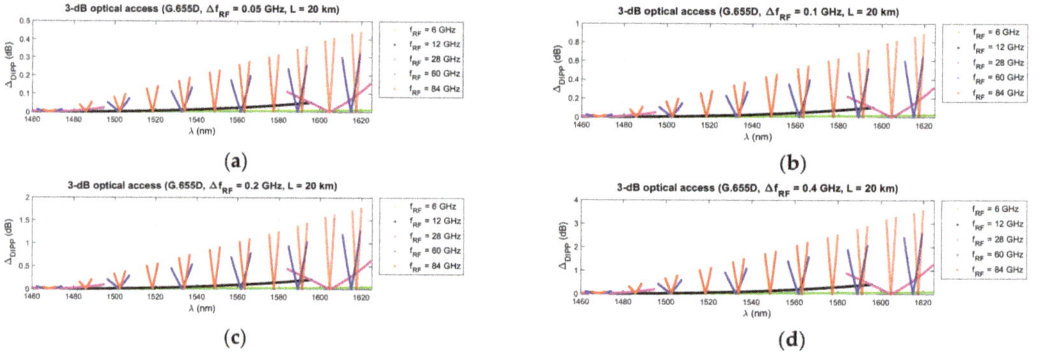

Figure 16. Differential DIPP-CIR as a function of the optical wavelength for selected radio carrier frequencies and the 20 km optical path created on the basis of the G.655D optical fiber, and for selected radio channel widths: (**a**) 50 MHz; (**b**) 100 MHz; (**c**) 200 MHz; (**d**) 400 MHz.

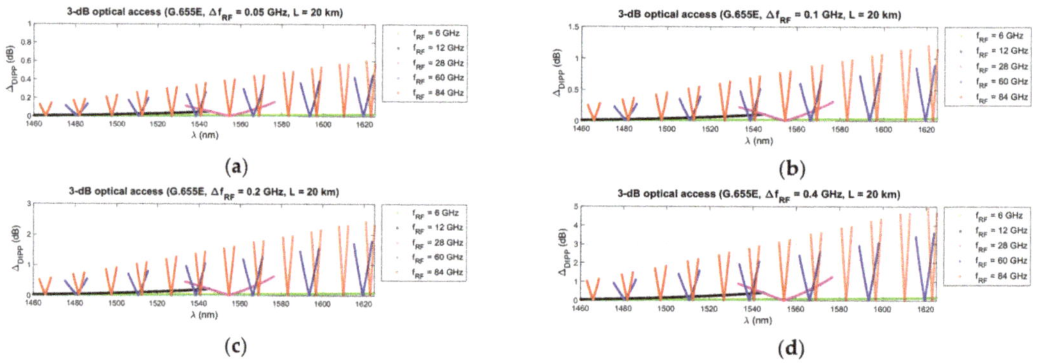

Figure 17. Differential DIPP-CIR as a function of the optical wavelength for selected radio carrier frequencies and the 20 km optical path created on the basis of the G.655E optical fiber, and for selected radio channel widths: (**a**) 50 MHz; (**b**) 100 MHz; (**c**) 200 MHz; (**d**) 400 MHz.

In the next step, we look at the susceptibility of the CP-OFDM symbol to the multipath effect caused by the propagation of two optical sidebands in the DSB format. For this purpose, we determine the dependence on the total width of the optical channel (assuming a very narrow spectral line of the laser in relation to the radio frequency):

$$\Delta f_0 = 2f_{RF} + \Delta f_{RF} \tag{18}$$

where f_{RF} is the central radio frequency of the channel and Δf_{RF} is the radio frequency width of the signal containing OFDM subcarriers and side bands of the outer subcarriers. Then we move on to the wavelength domain, while taking into account the optical carrier wavelength that is used during transmission in a fiber of a certain standard. For this purpose, we use the following approximate relationship:

$$\Delta\lambda_0 = \lambda_2 - \lambda_1 = \frac{c \cdot \Delta f_0}{(f_2 - \Delta f_0) \cdot f_2} = \frac{c \cdot \Delta f_0}{(f_1 + \Delta f_0) \cdot f_1} \cong \frac{c \cdot \Delta f_0}{f_0^2} \cong \frac{\Delta f_0 \cdot \lambda_0}{c} \tag{19}$$

where $f_0 = f_1 + \Delta f_0/2 = f_2 - \Delta f_0/2$, and $f_0 = c/\lambda_0$. After substituting the dependence (18) into (19), we obtain the following:

$$\Delta\lambda_0 = \frac{\lambda_0(2f_{RF} + \Delta f_{RF})}{c} \tag{20}$$

The effective chromatic dispersion coefficient is used to calculate the relative delay as follows:

$$D_{eff}(\lambda_0, \Delta\lambda_0) = D(\lambda_0) + S(\lambda_0) \cdot \left(\frac{\Delta\lambda_0}{2}\right) \tag{21}$$

where S is the chromatic dispersion slope. Now, taking into account (21), the relative delay can be determined using the following equation:

$$\tau_{DSB\max}(\lambda_0) = D_{eff}(\lambda_0)\Delta\lambda_0 L = \left(|D(\lambda_0)| \cdot \Delta\lambda_0 + \frac{1}{2}S \cdot \Delta\lambda_0^2\right) \cdot L. \tag{22}$$

After substituting (20) into (22) we obtain the following:

$$\tau_{DSB\max}(\lambda_0) = \left(|D(\lambda_0)|(2f_{RF} + \Delta f_{RF}) + \frac{S\lambda_0(2f_{RF} + \Delta f_{RF})^2}{2c}\right)\frac{L\lambda_0}{c}. \tag{23}$$

Formula (23) can be used to calculate, with high accuracy, the relative delay in the optical channel, especially when the optical chromatic dispersion coefficient of the path is characterized by a small value below 1 ps/nm*km.

The results of the relative delay of the sidebands are shown in Figure 18. The calculations were made for the fronthaul optical path with the maximum allowable length of 20 km. The optimal channel bandwidth was taken into account, i.e., 100 MHz, which is available in all bands from the FR1 and FR2 ranges, intended to be used in the 5G NR interface. The calculation results show that the relative delay is several orders less than the allowable relative multi-path delay over the radio link. The allowed relative delay value in the CP-OFDM modulated signal is governed by the duration of the cyclic prefix. In this case $T_{CP_{\min}} \gg \tau_{\max}$, so it can be concluded that the OFDM symbol used in the RFoF link is safe (Table 8). Figure 18 shows the calculation results for the optical carrier wavelength of 1550 nm and the radio carrier frequency of 84 GHz (this frequency is currently outside the range of FR2 but will likely be used in B5G/6G solutions). However, by testing the link parameters in the radio domain and by extended setting of the channel state information (CSI), this quantity should be taken into account.

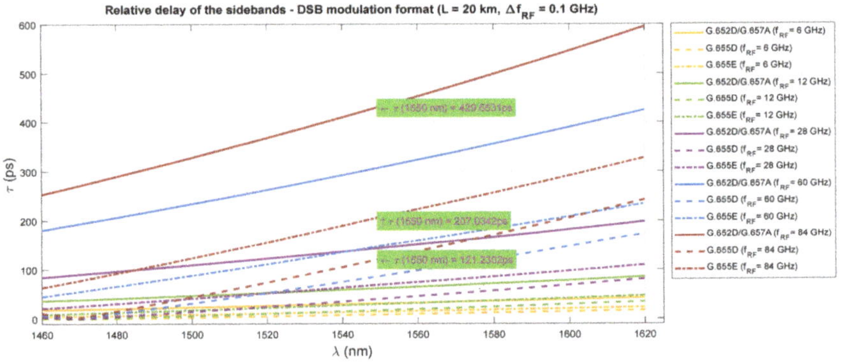

Figure 18. Calculation results of relative sideband delay in the optical channel (length of optical path/link is equal to 20 km and the radio channel frequency bandwidth for CP-OFDM modulation format is equal to 100 MHz).

3.4. Optical Channel Selection Algorithm

The proposed algorithm (Figure 19) assumes that the routing protocol working in the Xhaul will find available optical paths connecting the DU and RU. Thus, a set of possible connections on the optical layer should be available. It should be assumed that these paths are all-optical and transparent to signals modulated in the RFoF format. The first criterion excluding the path is its length L, which must be smaller than $Lmax$, because the permissible delay in the fronthaul link is of key importance. Further criteria are based on scanning free optical resources for the $Dmax$, $Th1$ and $Th2$ thresholds.

A first decision threshold, $Th1$, relates to the determination of an optical subband range that can be used for carrying the signal in the RFoF format. Two values were adopted here, i.e., 3 dB and 10 dB. In extreme cases, especially in the presence of a high radio frequency and a wideband frequency channel, an even stricter 1 dB criterion can be adopted. The second threshold, $Th2$, relates to the slope of the DIPP-CIR characteristic, which indicates how much they differ in the blanking level of the two outside subcarriers contained in the OFDM radio signal or physical resource block (PRB). The second threshold is especially important in cases where a wide frequency channel is used on a high carrier frequency with a high coefficient of chromatic dispersion present in the optical path.

The analysis of the phenomena occurring in the optical fronthaul path allowed for the construction of an algorithm (Figure 19) for the effective determination of the optimal optical channel for signal transmission in the DSB-RFoF format. Before the application created on the basis of the algorithm is launched, a lot of initial data must appear, which indicate not only the parameters of the radio signal, but also the properties of the optical path selected for transmission. The properties of the optical path include the chromatic dispersion coefficient D (or cumulative chromatic dispersion), path length L and its attenuation A_{pp}. Due to the DIPP phenomenon occurring in a receiver operating in the DD mode, the most important parameter is the chromatic dispersion coefficient. This measurement can be performed with the use of the built-in OTDR system with the CD measurement function or with the use of an algorithm using optical AM modulation [51].

The introduction of the option to select an optical channel that meets the condition of minimum dispersion $D_{\min}$ greatly reduces the occurrence of cross-phase modulation (XPM) and four-wave mixing (FWM) phenomena. The value of the minimum chromatic dispersion coefficient is variable, as it depends on the type of optical signals multiplexed in the DWDM Xhaul link. In the case of a constant wavelength spacing between the channels in which radio signals with the same RF radio frequency are carried, the requirement for $D_{\min}$ is much higher. The paths/links of the DWDM Xhaul can transport the streams from the core network, backhaul, midhaul and fronthaul. Only on the fronthaul links can the RFoF signal be used. In other cases, it is a digital transmission in the OOK or coherent format. Interleaving in the wavelength domain of RFoF and OOK signals, with a flexible DWDM grid, further reduces the mutual influence of signals transported in adjacent optical channels. Please note that in the case of OOK modulation, the signal is concentrated close to the optical carrier, while in the RFoF signal, the sidebands containing the information are located at the periphery of the channel.

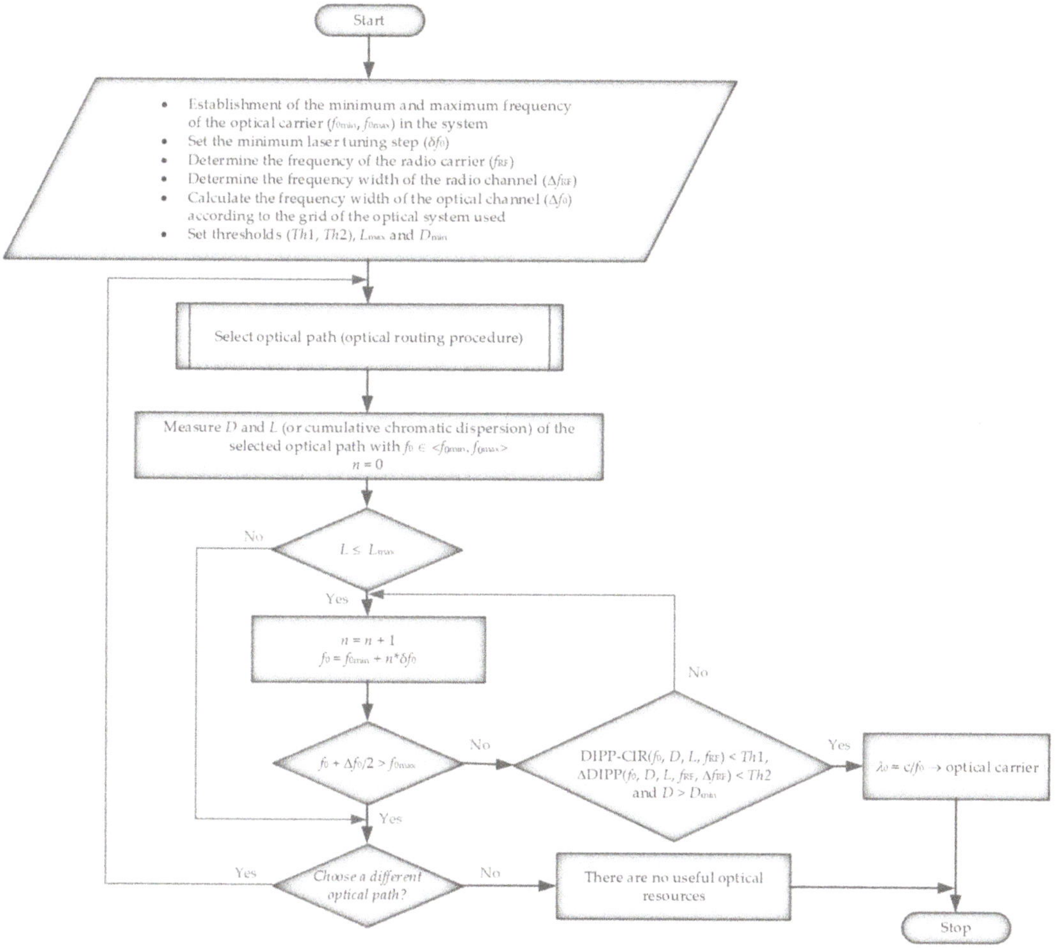

Figure 19. Optical channel selection algorithm based on DIPP-CIR calculations at two decision thresholds.

The capabilities of the transmitting laser indicate a jump in the tuning of the optical carrier and the maximum range of wavelength change. The main limitation here is the operating range of the optical system, where $f_0 \in \langle f_{0min}, f_{0max} \rangle$, including the ROADM nodal devices and optical fibers belonging to the selected optical path. During the calculations, it was assumed that in the DWDM system we do not exceed the narrowest single-modality range of optical fibers with non-zero chromatic dispersion shifted. The speed of tuning or switching the laser is decisive in terms of the final speed of selecting and setting up an optical channel. If no suitable optical channel is detected (Figure 12, Table 5), the process of selecting a new optical path must take place. This is related to the need to run a routing procedure, which was also included in the algorithm (Figure 19).

An EON or flexible DWDM network, e.g., an algorithm including routing, modulation level, and spectrum allocation (RMLSA) techniques [52–54], can be equipped with an additional computational-decision-making component based on the algorithm shown in Figure 19. This operation not only ensures precise selection of the optical path parameters in the case of a signal in the RFoF format, but can significantly improve the classical transmission in the on–off keying (OOK) format because fast streams also experience the DIPP phenomenon at the location of the photodetector, which lowers the sensitivity of the direct receiver.

4. Simulation Results

The above calculations can be confirmed by hardware experiments or computer simulation. In practice, the quality of the CP-OFDM radio signal transmitted over the optical path depends on many other factors that were not taken into account in the theoretical calculations. These include, among others, laser, modulator and photodetector parameters. Additionally, there is phase noise resulting from the influence of chromatic dispersion on the current phase state of individual OFDM subcarriers.

In order to confirm the correctness of the theoretical selection of optical subbands suitable for carrying the RFoF signal in the DSB format, the VPIphotonics Design Suite 11.1 simulation platform was used. This allowed for point-modeling of the quality of the CP-OFDM signal, carried by a specific radio and optical carrier, then demodulated in the direct detection format on the optical side, and in the coherent format on the radio side. The simulation scheme is shown in Figure 20.

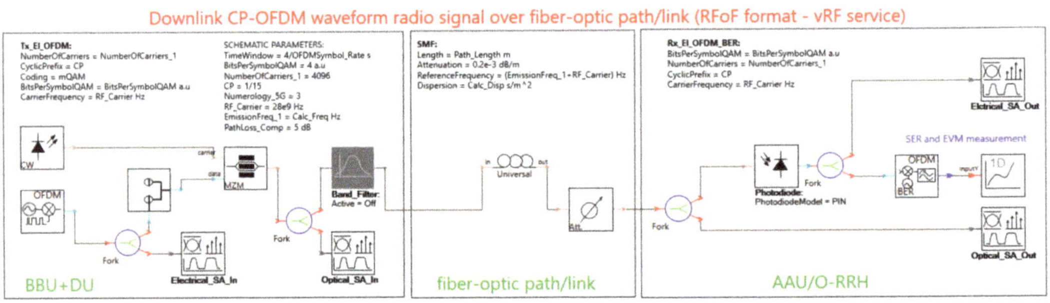

Figure 20. Simulation diagram prepared on the VPIphotonics Design Suite 11.1 platform, presenting three fronthaul parts, configured for downlink transmission: BBU + DU cloud side, fiber-optic path/link and AAU/O-RRH as an antenna side.

The CP-OFDM signal was generated with the OFDM transmitter module. The signal structure was adopted according to the 3GPP-5G parameters given in Table 8, except for the number of subcarriers. A maximum number of 4096 ("NumberOfCarriers" parameter in Figure 20) was adopted to simplify the configuration options and to standardize the signal format for all radio frequency ranges. With this assumption, more difficult detection conditions arose due to the broadband signals. In practice, the simulator allows for switching off the subcarriers, but the variability of the planned guard intervals for individual radio bands would violate the constancy of the reference parameters.

The simulator performed calculations in the time of four CP-OFDM symbols ("TimeWindow" parameter—Figure 20). The value of this parameter was established by a compromise between the accuracy of the SER obtained and the time of the calculation. It should be noted that a fairly low level of the total loss of the 20 km path (0.2 dB/km) was established on the level of 9 dB with the attenuation components ("PathLoss_Comp" = 5 dB), which resulted in minimizing the impact of photodetector noise on the symbol condition. Thanks to this assumption, it was possible to demonstrate the characteristics of the SER dependent mainly on the DIPP phenomenon. It can be concluded here that any appearance of an additional attenuation component, e.g., a ROADM node in the passive optical path, will worsen these conditions. Additionally, it should be noted that the OFDM symbol is very energetic because its duration is due to the very low symbol rate (Table 8) occurring in the interfaces of cellular systems based on the OFDM modulation technique. The filling of the constellation is possible due to the large number of OFDM subcarriers, each of which is modulated separately, using modulation from the n-QAM family.

The CP-OFDM signal is upconverted to the RF carrier. Two carrier frequencies were selected for the simulation, i.e., 28 GHz and 60 GHz. To generate the optical carrier, a DFB laser emitting light with a power of 10 dBm and a spectral line of 10 kHz was used. During the simulation, the laser used was tuned in 100 GHz increments. The optical zero reference channel is located in the middle of the C band, and its central frequency is equal to 193.1 THz. The optical modulation process is carried out with the use of an external MZI-based modulator, the operating parameters of which were selected so that the RF radio signal is introduced onto the optical carrier in a linear manner. Light intensity modulation in MZI is performed correctly when the system operates in quadrature bias point (QP) configuration. This means that the optical MZM operates at a fixed operating point in the middle of the transmission characteristic. The modulated signal goes to a Gaussian bandpass filter, which is only turned on when the signal needs to be created in SSB format, for comparison purposes. The signal in the DSB format goes to the SMF, the length of which was set as the maximum, i.e., 20 km. In the fronthaul network, the optical path also cannot exceed this value, as reserved in the 3GPP recommendation.

Two types of optical fibers were selected for the simulation, i.e., G.652D/G.657A and G.655D. The calculation script developed on the basis of (1) and (2) determines the chromatic dispersion coefficient according to the current optical channel. The attenuation coefficient of the optical fiber is 0.2 dB/km. Additionally, an optical attenuator is used, which is a substitute for the optical loss components of the optical path according to Table 2. The default value of the attenuator is set to 5 dB. The optical signal is received by a PIN-type photo-receiver operating in the direct detection mode. This is where optical demodulation takes place. The RF radio signal goes to a coherent receiver, where it is demodulated to the baseband. This process takes place in an OFDM demodulator circuit, which is equipped with a software DSP. The options of automatic synchronization with the radio carrier and automatic normalization of the n-QAM constellation were enabled in the DSP. In practice, a separate synchronization line is needed, which provides an unmodulated radio carrier. The DSP is equipped with components enabling the determination of the SER and EVM. The SER reading, estimated by the Monte Carlo (MC) method, was used to evaluate the quality of the transmitted CP-OFDM signal.

In order to be able to view the spectral characteristics, four spectrum analyzers were connected: two for monitoring the input and output optical signals and two for monitoring the RF electrical signals. Exemplary spectrograms of the optical side are shown in Figure 21. The spectrograms indicate the DSB format of the modulated optical signal, where the center frequency of the radio channel is 60 GHz. The optical carrier is 195.7 THz; it is the center of the optical channel number 26, which is located in the center of the optical subband to be simulated.

Figure 22 shows the spectrograms of radio signals that take place around a radio carrier of 60 GHz. The signal in this case was formed according to a numerological parameter of 2. Using 4096 subcarriers, a signal with a frequency width of approximately 247.7 MHz was created. This signal transports a 1.835 Gbps stream. Figure 22a shows the spectrum of the signal input to the optical modulator, and Figure 22b,c shows the spectra of the signal output from the photo-detector. Figure 22b shows the signal transported on the optical channel located in the center of the determined 10 dB subband according to $Th1$, and Figure 22c shows the spectrum of the signal transported on the optical channel located on the border of the determined 10 dB subband.

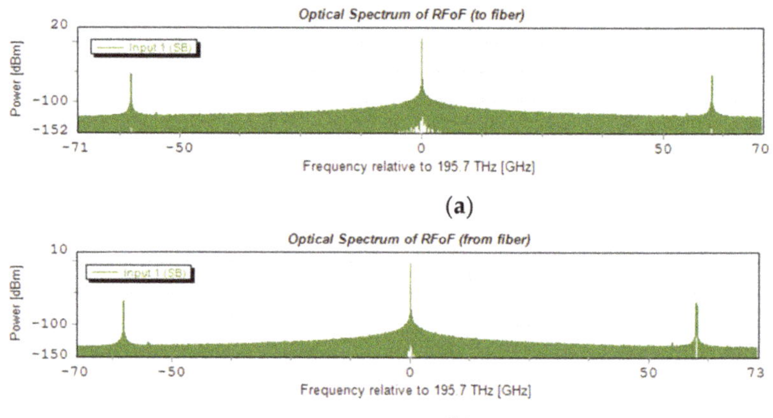

Figure 21. Example optical power spectrum of RFoF signal (f_{RF} = 60 GHz, G.655D fiber, optical channel no. 26 (195.7 THz), 4096 subcarriers, μ = 2): (**a**) inserted into the optical single-mode fiber/path, (**b**) at the photodetector input.

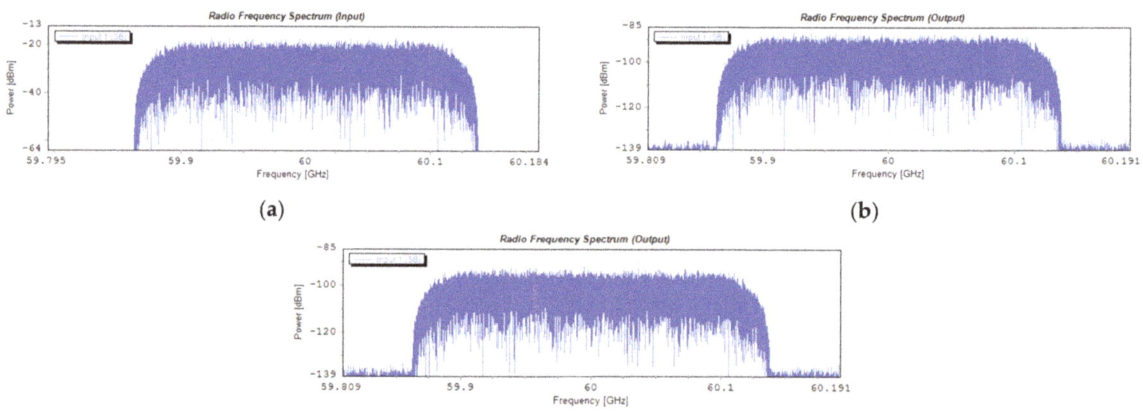

Figure 22. Example radio power spectrum of RF signal (f_{RF} = 60 GHz, 4096 subcarriers, μ = 2, 1.835 Gbps, Δf_{RF} = 247.7 MHz): (**a**) input signal; (**b**) output signal transported over G.655D fiber in optical channel no. 26 (195.7 THz); (**c**) output signal transported over G.655D fiber in optical channel no. 14 (194.5 THz).

Figure 23 shows exemplary OFDM subcarrier constellations modulated in 256-QAM format. As in the case of Figure 22, Figure 23a shows the constellation of the radio signal carried on the optical channel no. 26, and Figure 23b, the constellation of the signal carried on the 10 dB outermost channel no. 14.

Figure 24 shows the spectrograms of the radio signals introduced into the G.655D fiber and led out from the fiber. In this case, the radio signal was created on the basis of a numerological parameter with a value of 3, which resulted in the channel width increasing twice as compared to the signal presented in Figure 22.

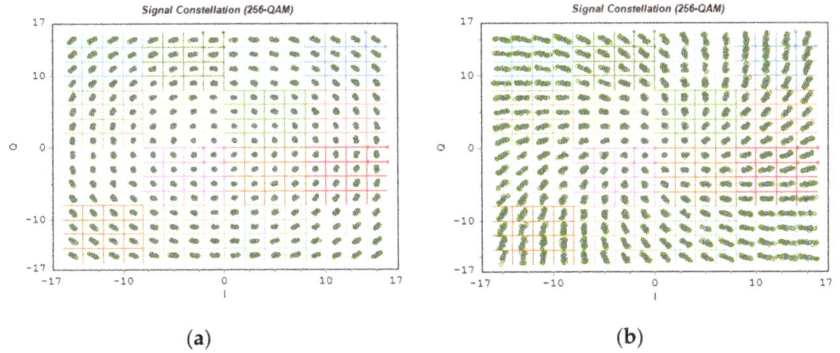

Figure 23. Example constellations of 256-QAM signal (f_{RF} = 60 GHz, G.655D fiber, μ = 2): (**a**) demodulated output signal transported over G.655D fiber in optical channel no. 26 (195.7 THz); (**b**) demodulated output signal transported over G.655D fiber in optical channel no. 14 (194.5 THz).

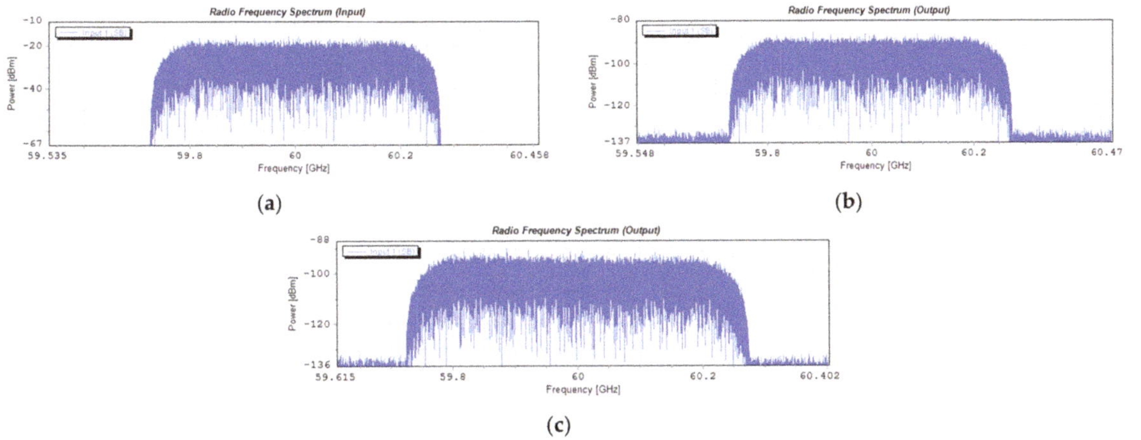

Figure 24. Example radio power spectrum of RF signal (f_{RF} = 60 GHz, 4096 subcarriers, μ = 3, R_b = 2.752 Gbps, Δf_{RF} = 491.5 MHz): (**a**) input signal; (**b**) output signal transported over G.655D fiber in optical channel no. 26 (195.7 THz); (**c**) output signal transported over G.655D fiber in optical channel no. 14 (194.5 THz).

Figure 25 shows the constellations of a 64-QAM modulated signal on each OFDM subcarrier in the baseband. In this case, the signal comes from an OFDM demodulator to which 4096 subcarriers were provided on a frequency channel with a width of 491.5 MHz. Such a channel width indicates that a numerological parameter with a value of 3 was used. The propagation conditions in a single-mode fiber, in this case, are much more difficult. This manifests itself in the rapidly changing position of the constellation points in the interval, in the case of the optical channel located on the edge of the 10 dB optical band (Figure 25b).

In order to fully verify the correctness of the selection in the computational part of 10 dB optical subbands, suitable for transmitting radio signals in the RFoF format in the DSB version, simulations were carried out with the selected, previously mentioned, optical fibers and system parameters. The simulation results are presented in Figures 26 and 27.

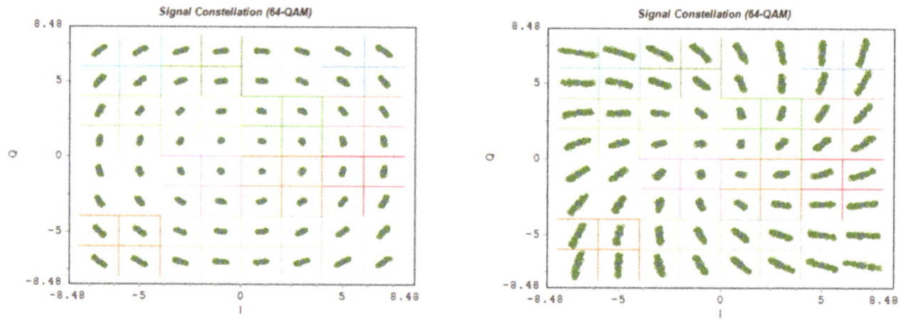

(a) (b)

Figure 25. Example constellations of 64-QAM signal (f_{RF} = 60 GHz, G.655D fiber, μ = 3): (**a**) demodulated output signal transported over G.655D fiber in optical channel no. 26 (195.7 THz); (**b**) demodulated output signal transported over G.655D fiber in optical channel no. 14 (194.5 THz).

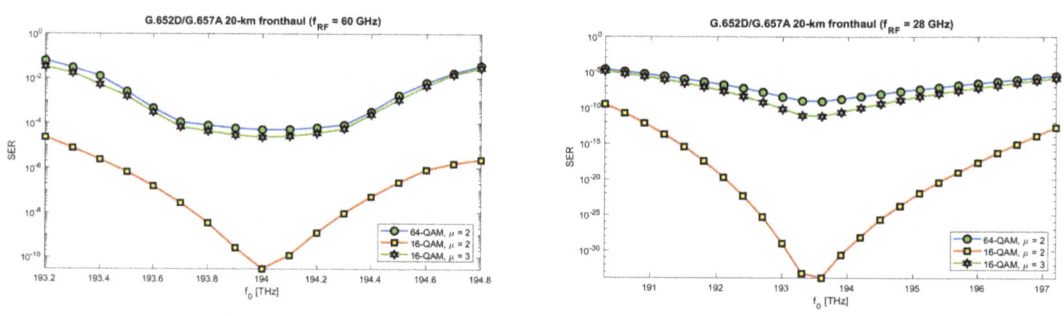

(a) (b)

Figure 26. Simulation results presenting SER as a quality parameter of the selected CP-OFDM signals transported over the G.652D/G.657A single-mode fiber fronthaul (L = 20 km): (**a**) for radio carrier f_{RF} = 60 GHz and the calculated optical 10 dB subband: 193.1630–194.7505 THz; (**b**) for radio carrier f_{RF} = 28 GHz and the calculated optical 10 dB subband: 190.0130–197.2505 THz.

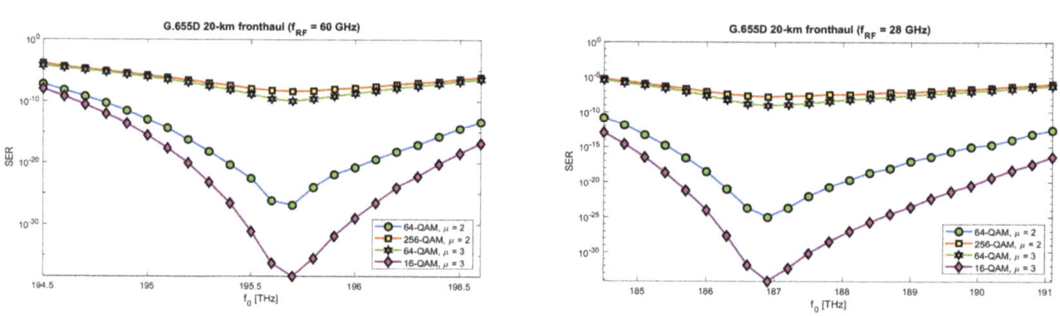

(a) (b)

Figure 27. Simulation results presenting SER as a quality parameter of the selected CP-OFDM signals transported over the G.655D non-zero dispersion-shifted single-mode fiber fronthaul (L = 20 km): (**a**) for radio carrier f_{RF} = 60 GHz and the calculated optical 10 dB subband: 194.5130–196.7255 THz; (**b**) for radio carrier f_{RF} = 28 GHz and the calculated optical 10 dB subband: 184.4880–191.1068 THz.

In the case of the optical path based on the G.652D/G.657A fiber, the radio signal was used in three modulation configurations with 16-QAM and 64-QAM modulation orders and a numerological parameter with a value of 2 and 3. Figure 27 shows the results of the simulation of radio signal transmission over an optical path based on the G.655D standard fiber. In this case, the tests were performed for 16-QAM, 64-QAM and 256-QAM modulation orders, with numerological parameters of 2 and 3. Not all combinations were tested, but the results obtained are sufficient to confirm the correct selection of optical subbands during the theoretical calculations.

5. Discussion

The Analog Option 10 interface, proposed to be introduced in the mobile fronthaul, working in the RFoF format (Figure 1), is a big challenge, but it can allow for a significant increase in the efficiency of using the network's optical resources. This issue was widely analyzed and discussed in [21]. The idea of selecting an appropriate optical channel in order to transport the signal in analog form comes from the fact that the optical transmitter and receiver usually work in a DSB format. This means the work simplicity of both the modulator and the optical demodulator. The only component increasing the cost of fabrication of the transceiver is the option to tune the laser to the carrier wavelength selected using the proposed algorithm (Figure 19). To the best of the author's knowledge, this type of approach in relation to the RFoF solutions has not been presented in the literature on the subject.

An external modulator is used in the simulation; however, in the case of a radio carrier with a lower frequency, the direct modulation technique can be used. This choice depends entirely on the laser frequency response. This results in an additional reduction in the costs of fabrication of the optical transmitter device. It is known that in the current 5G and future B5G/6G radio systems, MIMO techniques are and will be used, which means there is a need to provide more streams to the RRH antenna mast in order to properly radiate the beam (ABF and HBF techniques) and to use the spatial multiplexing technique. If we assume that the RF signal intended for a single antenna module is built in the cloud, then sending more such signals over the network will require a greater broadband. Fiber-optic networks make this possible because they use the xWDM techniques more and more often. The aforementioned optical Xhaul is also based on this.

The results of the calculations performed, presented in Section 3, show that this method may not be useful in all cases. Therefore, it is important to have a larger number of optical paths managed by network routing protocols (Figure 19). The optical sidebands interference phenomenon at the location of the photodetector is very strongly dependent not only on the chromatic dispersion of the optical path, but also on its length and the radio carrier frequency. Due to such a complex relationship, there are more dimensions that can be used to accurately select the optical resources. In Figures 6–8, we can see how the fading effect of the sidebands signal increases with respect to the optical carrier as the carrier radio frequency increases. The periodicity of the DIPP interaction is also dependent on the optical fiber type used in the optical path. It should be assumed that the path may consist of various optical fibers (Table 3, Figure 5). Therefore, it is important to perform a test for its length and its averaged chromatic dispersion coefficient in accordance with the algorithm. Exemplary calculations were made for three types of optical fibers with the assumption that the fronthaul path reaches the maximum allowable length of 20 km and is made of one type of optical fiber. The results of the calculations shown in Figures 9–11 are much more representative and illustrative. In addition to the three types of optical fibers, four radio carrier frequencies were used for the calculations. Not all the selected radio carrier frequencies are in the range FR1 or FR2. It is expected that the ranges indicated by 3GPP will be extended in the future. It is all about adding bands that are currently free from concession. In Figures 6a, 7a and 8a, we can see that the impact of DIPP on the 6 GHz frequency, after passing over the 20 km optical path, is negligibly small. On this basis, it was decided that the FR1 range would not be taken into account during further calculations, as very good conditions for receiving the DSB signal exist here. Carrier frequencies of 12, 28, 60 and 84 GHz were used for further calculations. Figures 9a, 10a and 11a show that only in the case of a long optical path based on the G.652D/G657A fiber may the transmission quality be lowered. Here, the choice of a longer wavelength of the optical carrier allows us to go in the range between 3 dB and 10 dB. For 28 GHz and 60 GHz radio frequencies, there is moderate decay periodicity, allowing us to find the right resources. However, we can find cases where we find ourselves on the border of the 10 dB range. Such a situation can, for example, be observed in the case of the G.652D path with a length of 5 km and the radio frequency of 28 GHz. In this case, all the optical carriers lying in the system band are in the range below the 10 dB threshold (Figure 9b). Such cases should be caught by an algorithm that requests an optical path change. Another problem is the frequent periodicity of fading. This is especially characteristic of the 84 GHz frequency that is transmitted over the G.652D/G.657A fiber. The range of white clearances is small, indicating a low number of optical channels that ensures high quality DSB signal transmission.

In the next stage of calculations, the process of determining the optical subbands for selected radio frequency carriers and the fronthaul path with a maximum length of 20 km was performed (Figures 12–14). These types of charts allow us to see when there are no matching optical resources. Such a situation occurred, for example, in the case of the G.652D/G657A fiber and the 12 GHz frequency in the 3 dB access range (Figure 12b). On the basis of the performed calculations, the ranges of subbands for individual types of optical fibers and selected radio frequencies were generated. The results are listed in Tables 5–7. The values were recalculated and given in frequency to comply with the provisions used in the recommendations for the optical channel grid of the flexible DWDM system [37]. The analysis of the results presented in Tables 5–7 shows that in the case of optical fibers with a lower chromatic dispersion coefficient, the ranges are wider, which allows for the selection of an optical channel with a flat transmission characteristic of broadband signals. This is confirmed by the calculations, the results of which are presented in Figures 15–17. Earlier calculations were performed for an unmodulated carrier. In the next stage of calculations, it was assumed that the radio signal is composed of a larger number of subcarriers, which is characteristic of the CP-OFDM format. The dispersiveness of the fiber causes different subcarriers to experience different relative delays. In the case of a large number of OFDM subcarriers in the radio signal, the channel bandwidth may reach 400 MHz. This value can be quickly achieved at higher values of the numerological parameter (Table 8). As can be expected, the greatest differential delay occurs in the case of optical fibers with a high chromatic dispersion coefficient, i.e., G.652D/G.657A (Figure 15). Here, we can observe ΔDIPP of a few dB. It is an important parameter, as it influences the correct functioning of the radio receiver in the scope of normalization of the constellation of the modulated signal in the n-QAM format in the area of the radio resource unit. A high value of this parameter results in the exclusion of a specific modulation order in the case of the need to transmit a radio signal in a specific optical channel based on a dispersive optical fiber.

In addition, calculations for the so-called multipath were made. The phenomenon of differential propagation of the optical sidebands of the DSB signal causes two replicas of the OFDM symbol to reach the receiver. The calculations are only a formality because the relative delay is several orders lower than the duration of the guard interval (Table 8). The results of the calculations are shown in Figure 18. However, when determining the parameters of the radio link between the DU and UE, this value should be taken into account as an addition to the channel state information (CSI) in a radio link.

In order to verify the possibility of transmitting the modulated RF signal over the optical path and the correct selection of optical subbands, a simulation was performed. Two types of single-mode optical fibers (G.652D/G.657A and G.655D) were used in the simulation and two radio frequency carriers were selected, i.e., 28 and 60 GHz. The simulation results showed that the optical subbands were calculated correctly (Figures 26 and 27). Optical transmission of the DSB signal in the RFoF technique is burdened with many distortions, which was partially shown in the computational part of this study. The first limitation is the significant loss in the places where the signal is converted from electrical-to-optical (E/O) and optical-to-electrical (O/E). The results of the passive path loss can be seen in the pictured sample spectrograms (Figures 22 and 24). It is a key component determining the overall level of the signal reaching the radio demodulator. The noise floor of the radio receiver and the photo-detector constitute the basic signal-to-noise ratio, which determines the quality of the received useful signal. It should be noted that there is no optical amplifier on the optical side, which is typical for short-range links. It should also be taken into account that, at the location of the RF signal receiver, before the signal is delivered to the antenna module, a radio amplifier should be used. However, the signal-to-noise ratio is crucial here. The simulation did not use linear and redundant information coding. The Monte Carlo (MC) method was used to assess the SER parameter. However, the results obtained are relative, as the SER value is strongly dependent on the parameters of optoelectronic systems and the cumulative loss of optical components in the optical path. Such components may include, for example, OADMs/ROADMs and splitters (Table 2). The simulation assumed that the total additional attenuation of optical components is 5 dB. An increase in this value proportionally increases the SER value over the entire range, as it lowers the signal-to-noise ratio, which is not due to the DIPP effect. This assumption is due to the fact that the classic optical components used in the optical path are not dispersive, with the exception of dedicated chromatic dispersion compensators.

The constellations of the signals modulated in the 256-QAM and 64-QAM formats are shown in Figures 23 and 25, respectively. Figures 23a and 25a show constellations of the signal transported over the optical fiber in the most optimal optical channel in the designated optical subband. These constellations show the presence of phase noise that results from phase shifts due to optical sideband interference. This phenomenon intensifies with an increase in the chromatic dispersion coefficient and an increase in the radio carrier frequency. In this case, the increase in the radio frequency causes the frequency sidebands to recede, which in turn causes rapid changes in the resultant phase. The depicted constellations also show that, with low loss of the optical path (together with the optical components), the effect of the thermal noise of the photodetector is minimal. Figures 23b and 25b show the constellations of the modulated signal carried in the optical channel at the edge of the 10 dB subband. In this case, the differences in the levels of adjacent subcarriers start to differ substantially. We observe this in the form of increasing amplitude noise in the presence of constant phase noise. External symbols (of greater amplitude) are most exposed to the increasing qualitative SER parameter, which is easy to observe. Figures 26 and 27 show the simulation results as a compendium. In the case of the G.652D/G.657A fiber, the simulation of the modulated signal in the 256-QAM format was abandoned, as phase noise causes too much reduction in the transmission quality. It is possible but with a significant reduction in the radio carrier or the bandwidth of the frequency channel. The simulations were performed for frequencies within the FR2 range and above this range, where it was not planned to use a numerological parameter with a value below 2. The simulation results also show that the transmission quality is significantly influenced by the bandwidth of the frequency channel, and more specifically, the numerological parameter of the CP-OFDM signal. For $\mu = 3$, the subcarrier spacing increases to 120 kHz, which greatly increases the distance differences between the levels of the same numbered interfering subcarriers in the optical sidebands. Additionally, with the increasing value of the parameter μ, the length of the OFDM symbol is shortened, which reduces its energy. The total consequences can be seen in Figure 26, where we compare the SER value for 16-QAM modulation, with the μ parameter being 2 and 3. Increasing the μ parameter by one degree results in a SER similar to 64-QAM for $\mu = 2$. This applies to both the 60 GHz and 28 GHz radio carriers.

The same phenomenon is observed in the case of the G.655D fiber, although in this case, the results are much better, due to the lower chromatic dispersion coefficient. Here, 256-QAM modulation for $\mu = 2$ was introduced, due to its much better dispersion parameters. Confusingly, Figure 27a,b shows swapped values. In the case of the 28 GHz radio carrier, the SER values are slightly higher than in the case of the 60 GHz carrier. However, one should look at the subband in which the simulations were performed. The optical subband for 60 GHz is characterized by higher frequencies, i.e., shorter wavelengths, so here, there is a lower chromatic dispersion coefficient. In the case of 28 GHz, the optical frequencies are much lower, so the optical wave is longer at a higher value of the chromatic dispersion coefficient.

It is undisputed that the introduction of redundant coding will significantly increase the quality of the transmitted digital information. However, the strict quality characteristics presented in Figures 26 and 27 clearly show which parameters of the radio signal must be taken into account when we are selecting an optical channel. For lower radio frequencies, the available optical subbands are much wider. This results in a lot of freedom in the selection of optical channels, and in such a channel, the amplitude characteristics are much flatter.

The discussion presented above concerned the research performed on the basis of the well-known standards of single-mode optical fibers, which are designed and manufactured with the use of silica. These types of optical fibers have become very entrenched in telecommunications fiber-optic networks, so breaking this barrier will certainly not be easy. A very significant disadvantage of standard single-mode optical fibers is the small diameter of the core, in which optical signals appearing in the adjacent optical channels very quickly increase the power density. This fact causes the appearance of the above-mentioned non-linear effects, which can significantly deteriorate the transmission. The chromatic dispersion occurring in, and in the vicinity of, the core, which is the essential limiter of DSB-RFoF signal transport, is important from the point of view of limiting the Kerr effect, which results in XPM and FWM distortions. In conclusion, these two phenomena should be balanced so that communication is effective. If we want to stick to this technology, we need to popularize multi-core fibers much faster [55]. Fibers based on a greater number of cores are more difficult to couple with the transmitter and receiver systems, but they allow for a significant spatial increase in cable capacity. In the B5G/6G wireless access systems, the massive-MIMO technique will be the norm. Connecting a large number of antenna modules, located on the RU/RRH mast, to the BBU, also via the DU, requires the use of a large number of optical fibers heavily loaded with high optical power density. This is the specificity of the C-RAN architecture described in the introduction. Therefore, it is indisputable that the popularization of multi-core optical fibers in distribution and access networks will significantly simplify the management of often terabit fronthaul traffic.

6. Conclusions and Future Work

The proposed new analog interface, with the acronym Option 10, could apply to next-generation wireless systems. The condition for introducing this solution is the development of optical Xhaul networks working in the flexible grid DWDM format. Optical nodal devices must be transparent due to the possibility of carrying RFoF signals. The A-RoF technique is more spectrally effective than D-RoF solutions and allows the use of simple AAU/RRH devices. However, the transmission of analog signals over optical fibers causes many problems. The attenuation of silica fibers is very low, but they have dispersion that must be taken into account. The paper presents a method of using the optical path to transport the DSB-RFoF radio signal in which no chromatic dispersion compensation is used. However, it is indisputable that static or dynamic compensation of chromatic dispersion is an unrivaled but costly solution. Chromatic dispersion pre-compensation can improve the total dispersion balance, which will allow the transport of RFoF signals with a higher radio carrier frequency, also exceeding 84 GHz. This may be particularly important when the fiber-optic access path lengths are long in relation to the links between the Xhaul nodes. The access fibers are usually of the G.652D or G.657A standard, which in the DWDM system operating band, introduces a high chromatic dispersion coefficient (examples of variants are given in Table 3 and Figure 5). SSB transmission can also be used, which, however, is associated with the need to use tunable filters or dedicated single-sideband modulation systems. The proposed method of selection of an optical channel based on the flexible DWDM grid for transporting RF radio signals between the DU and RU/RRH, over a dispersive optical fiber, may be a good complement to the existing techniques. This can be especially useful when we need to mass-produce transceivers for optical fronthaul communications.

The calculations and simulations were carried out on the assumption that only one modulated RF signal is transported in the selected optical channel. In practice, in an optical Xhaul link working in the flexible DWDM format, there may be more optical channels located close to each other. There may be a high power density in the core of the optical fiber. In such a case, one should take into account the possibility of the above-mentioned non-linear distortions causing the interaction of signals located in adjacent optical channels. This is one of the issues that will be discussed in further research. In the case of multi-channel, also we need to pay attention to the guard spacing between channels. This will entail the need to disable OFDM subcarriers as required by the system requirements of the 3GPP recommendation. In this paper, this element was omitted in order to simplify the structure of the CP-OFDM signal.

Funding: This research received no external funding.

Institutional Review Board Statement: Not applicable.

Informed Consent Statement: Not applicable.

Data Availability Statement: All data are included within manuscript.

Conflicts of Interest: The author declares no conflict of interest.

References

1. Camps-Mur, D.; Gutierrez, J.; Grass, E.; Tzanakaki, A.; Flegkas, P.; Choumas, K.; Giatsios, D.; Beldachi, A.F.; Diallo, T.; Zou, J. 5G-XHaul: A novel wireless-optical SDN transport network to support joint 5G backhaul and fronthaul services. *IEEE Commun. Mag.* **2019**, *57*, 99–105. [CrossRef]
2. 5G-XHaul, D2.3. Architecture of Optical/Wireless Backhaul and Fronthaul and Evaluation. 2017. Available online: https://www.5g-xhaul-project.eu/download/ (accessed on 4 June 2021).
3. 5GPPP, Architecture Working Group. View on 5G Architecture. 2020. Available online: https://5g-ppp.eu/wp-content/uploads/2020/02// (accessed on 4 June 2021).
4. Cooper, A.J. Fiber/radio for the provision of cordless/mobile telephony services in the access network. *Electron. Lett.* **1990**, *26*, 2054–2056. [CrossRef]
5. Lee, C.H. *Microwave Photonics*, 2nd ed.; CRC Press, Taylor & Francis Group: Boca Raton, FL, USA, 2013.
6. ITU-R, F.1332-l. Radio-Frequency Signal Transport through Optical Fibres. 1999. Available online: https://www.itu.int/rec/R-REC-F.1332/ (accessed on 4 June 2021).
7. ITU-T, Series G, Supplement 55. Study Group 15. Radio-over-Fibre (RoF) Technologies and Their Applications. 2015. Available online: https://www.itu.int/rec/T-REC-G.Sup55/ (accessed on 5 June 2021).
8. CPRI Industry Forum (Ericsson, Huawei, NEC, and Nokia). CPRI Specification 7.0. 2015. Available online: www.cpri.info (accessed on 5 June 2021).
9. Pfeiffer, T. Next generation mobile fronthaul and midhaul architectures. *J. Opt. Commun. Netw.* **2015**, *7*, B38–B45. [CrossRef]
10. Zakrzewski, Z. Fronthaul optical networks working with use of the hybrid analog and digital radio-over-fiber techniques. In Proceedings of the International Society for Optical Engineering, 17th Conference on Optical Fibers and Their Applications, Supraśl, Poland, 10 February 2017; Volume 10325. [CrossRef]
11. CPRI Industry Forum (Ericsson, Huawei, NEC, and Nokia). eCPRI Specification 2.0. 2019. Available online: www.cpri.info (accessed on 5 June 2021).
12. IEEE 1914.1. Standard for Packet-Based Fronthaul Transport Networks. 2019. Available online: https://standards.ieee.org/standard/1914_1-2019.html (accessed on 5 June 2021).
13. 3GPP TR 38.801 v14.0.0. 03.2017. Study on New Radio Access Technology: Radio Access Architecture and Interfaces. 2017. Available online: https://www.3gpp.org/ftp//Specs/archive/38_series/38.801/38801-e00.zip (accessed on 5 June 2021).
14. IEEE, 1588-2019. IEEE Standard for a Precision Clock Synchronization Protocol for Networked Measurement and Control Systems. PNCS—Precise Networked Clock Synchronization Working Group. 2019. Available online: https://standards.ieee.org/standard/1588-2019.html (accessed on 5 June 2021).

15. ITU-T, G.8261. Timing and Synchronization Aspects in Packet Networks. 2019. Available online: https://www.itu.int/rec/T-REC-G.8261 (accessed on 5 June 2021).
16. ITU-T, G.8262. Timing Characteristics of Synchronous Ethernet Equipment Slave Clock. 2018. Available online: https://www.itu.int/rec/T-REC-G.8262 (accessed on 5 June 2021).
17. IEEE, P802.1CM. IEEE Standard for Local and Metropolitan Area Networks. *Time-Sensitive Networking for Fronthaul.* 2018. Available online: https://standards.ieee.org/standard/802_1CM-2018.html (accessed on 5 June 2021).
18. Ng, B.L. Fulfilling the promise of massive MIMO with 2D active antenna array. In Proceedings of the IEEE Globecom Workshops, Anaheim, CA, USA, 3–7 December 2012.
19. O-RAN Alliance. Use Cases and Overall Architecture Workgroup. O-RAN Architecture Description 3.0. Technical Specification ORAN-WG1. 2020. Available online: https://www.o-ran.org/specification-access (accessed on 8 June 2021).
20. Zakrzewski, Z. Optical RRH working in an all-optical fronthaul network. In Proceedings of the 8th International Conference on Photonics, Devices and Systems, SPIE—The International Society for Optical Engineering, Prague, Czech Republic, 1 December 2017; Volume 10603, pp. 1–12. [CrossRef]
21. Zakrzewski, Z. D-RoF and A-RoF interfaces in an all-optical fronthaul of 5G mobile systems. *Appl. Sci.* **2020**, *10*, 1212. [CrossRef]
22. Meslener, G.J. Chromatic dispersion induced distortion of modulated monochromatic light employing direct detection. *IEEE J. Quantum Electron.* **1984**, *20*, 1208–1216. [CrossRef]
23. Ih, C.S.; Gu, W. Fiber induced distortions in a subcarrier multiplexed lightwave system. *IEEE J. Select. Areas Commun.* **1990**, *8*, 1296–1303.
24. Schmuck, H. Comparison of optical millimeter-wave system concepts with regard to chromatic dispersion. *Electron. Lett.* **1995**, *31*, 1848–1849. [CrossRef]
25. Elrefaie, A.F.; Lin, C. Clipping Distortion and Chromatic Dispersion Limitations for 1550 nm Video Trunking Systems. In Proceedings of the IEEE Symp. Computers and Communications, Alexandria, Egypt, 27–29 June 1995; pp. 328–337. Available online: https://www.computer.org/csdl/proceedings-article/iscc/1995/70750328/ (accessed on 4 July 2021).
26. Gliese, U.; Nielsen, S.N.; Nielsen, T.N. Limitations in distance and frequency due to chromatic dispersion in fiber-optic microwave and millimeter-wave links. In Proceedings of the IEEE MTT-S Int. Microwave Symp. Dig., San Francisco, CA, USA, 17–21 June 1996; pp. 1547–1550. Available online: https://backend.orbit.dtu.dk/ws/portalfiles/portal/4537651/ (accessed on 4 July 2021).
27. Park, J.; Elrefaie, A.F.; Lau, K.Y. Fiber chromatic dispersion effects on multichannel digital millimeter-wave transmission. *IEEE Photon. Technol. Lett.* **1996**, *8*, 1716–1718. [CrossRef]
28. Fuster, J.M.; Marti, J.; Corral, J.L.; Polo, V.; Ramos, F. Generalized study of dispersion-induced power penalty mitigation techniques in millimeter-wave fiber-optic links. *J. Lightwave Technol.* **2000**, *18*, 933–940. [CrossRef]
29. Marti, J.; Fuster, J.M.; Laming, R.I. Experimental reduction of chromatic dispersion effects in lightwave microwave/millimetre-wave transmissions using tapered linearly chirped fibre gratings. *Electron. Lett.* **1997**, *33*, 1170–1171. [CrossRef]
30. Ilgaz, M.A.; Batagelj, B. Using tunable dispersion-compensated modules to overcome the power penalty of a millimeter-wave opto-electronic oscillator signal that is distributed via a passive optical network for 5G networks. In Proceedings of the 11th International Symposium on Communication Systems, Networks & Digital Signal Processing (CSNDSP), Budapest, Hungary, 18–20 July 2018. [CrossRef]
31. Ilgaz, M.A.; Baliž, K.V.; Batagelj, B. A flexible approach to combating chromatic dispersion in a centralized 5G network. *Opto-Electron. Rev.* **2020**, *28*, 35–42. [CrossRef]
32. Smith, G.H.; Novak, D.; Ahmed, Z. Technique for optical SSB generation to overcome fiber dispersion penalties in fiber-radio systems. *Electron. Lett.* **1997**, *33*, 74–75. [CrossRef]
33. Won, P.; Zhang, W.; Williams, J. Self-phase modulation dependent dispersion mitigation in high power SSB and DSB + dispersion compensated modulated radio-over-fiber links. In Proceedings of the MTT-S International Microwave Symposium Digest, San Francisco, CA, USA, 11–16 June 2006. [CrossRef]
34. Yaakob, S.; Mahmood, R.M.; Zan, Z.; Rashidi, C.B.M.; Mahmud, A.; Anas, S.B.A. Modulation index and phase imbalance of dual-sideband optical carrier suppression (DSB-OCS) in optical millimeter-wave system. *Photonics* **2021**, *8*, 153. [CrossRef]
35. Ishimura, S.; Kim, B.G.; Tanaka, K.; Nishimura, K.; Kim, H.; Chung, Y.C.; Suzuki, M. Broadband IF-over-fiber transmission with parallel IM/PM transmitter overcoming dispersion-induced RF power fading for high-capacity mobile fronthaul links. *IEEE Photonics J.* **2018**, *10*, 1–9. [CrossRef]
36. Ishimura, S.; Bekkali, A.; Tanaka, K.; Nishimura, K.; Suzuki, M. 1.032-Tb/s CPRI-equivalent rate IF-over-fiber transmission using a parallel IM/PM transmitter for high-capacity mobile fronthaul links. *J. Lightwave Technol.* **2018**, *36*, 1478–1484. [CrossRef]
37. ITU-T, G.694.1. Spectral Grids for WDM Applications: DWDM Frequency Grid. 2020. Available online: https://www.itu.int/rec/T-REC-G.694.1/ (accessed on 8 July 2021).
38. O-RAN Alliance. O-RAN Open X-haul Transport WG9. WDM-based Fronthaul Transport 1.0. Technical Specification ORAN-WG9. 2021. Available online: https://www.o-ran.org/specification-access (accessed on 8 July 2021).
39. Zakrzewski, Z. Transport of Rel-15/16 waveform radio signals over optical 5G fronthaul path. In Proceedings of the International Society for Optical Engineering, 19th Conference on Optical Fibers and Their Applications, Białowieża, Poland, 27–31 January 2020; Volume 11456. [CrossRef]
40. ITU-T, GSTR-TN5G. Transport Network Support for IMT2020/5G. 2018. Available online: https://www.itu.int/dms_pub/itu-t/opb/tut/ (accessed on 9 July 2021).

41. ITU-T, G.652. Characteristics of a Single-Mode Optical Fibre and Cable. 2016. Available online: https://www.itu.int/rec/T-REC-G.652/ (accessed on 16 July 2021).
42. ITU-T, G.657. Characteristics of a Bending-Loss Insensitive Single-Mode Optical Fibre and Cable. 2016. Available online: https://www.itu.int/rec/T-REC-G.657/ (accessed on 16 July 2021).
43. ITU-T, G.655. Characteristics of a Non-Zero Dispersion-Shifted Single-Mode Optical Fibre and Cable. 2009. Available online: https://www.itu.int/rec/T-REC-G.655/ (accessed on 16 July 2021).
44. ITU-T, G.650.1. Definitions and Test Methods for Linear, Deterministic Attributes of Single-Mode Fibre and Cable. 2020. Available online: https://www.itu.int/rec/T-REC-G.650.1/ (accessed on 14 July 2021).
45. Elrefaie, A.F.; Wagner, R.E.; Atlas, D.A.; Daut, D.G. Chromatic dispersion limitations in coherent lightwave transmission systems. *J. Lightwave Technol.* **1988**, *6*, 704–709. [CrossRef]
46. Lim, C.; Nirmalathas, A.; Bakaul, M.; Lee, K.L.; Novak, D.; Waterhouse, R. Mitigation strategy for transmission impairments in millimeter-wave radio-over-fiber networks. *J. Opt. Netw.* **2009**, *8*, 201–213. [CrossRef]
47. 3GPP TR 38.101-1 v17.2.0. 06.2021. User Equipment (UE) Radio Transmission and Reception. Part 1: Range 1 Standalone (Release 17). 2021. Available online: https://www.3gpp.org/ftp/Specs/archive/38_series/38.101-1/38101-1-h20.zip (accessed on 23 July 2021).
48. 3GPP TR 38.101-2 v17.2.0. 06.2021. User Equipment (UE) Radio Transmission and Reception. Part 2: Range 2 Standalone (Release 17). 2021. Available online: https://www.3gpp.org/ftp/Specs/archive/38_series/38.101-2/38101-2-h20.zip (accessed on 23 July 2021).
49. 3GPP TR 38.211 v16.6.0. 06.2021. NR. Physical Channels and Modulation (Release 16). 2021. Available online: https://www.3gpp.org/ftp/Specs/archive/38_series/38.211/38211-g60.zip (accessed on 30 July 2021).
50. 3GPP TR 38.104 v17.2.0. 07.2021. NR. Base Station (BS) Radio Transmission and reception (Release 17). 2021. Available online: https://www.3gpp.org/ftp/Specs/archive/38_series/38.104/38104-h20.zip (accessed on 30 July 2021).
51. Devaux, F.; Sorel, Y.; Kerdiles, J.F. Simple measurement of fiber dispersion and of chirp parameter of intensity modulated light emitter. *J. Lightwave Technol.* **1993**, *11*, 1937–1940. [CrossRef]
52. Christodoulopoulos, K.; Tomkos, I.; Varvarigos, E. Elastic bandwidth allocation in flexible OFDM-based optical networks. *J. Lightwave Technol.* **2011**, *29*, 1354–1366. [CrossRef]
53. Christodoulopoulos, K.; Tomkos, I.; Varvarigos, E. Spectrally/bitrate flexible optical network planning. In Proceedings of the 36th Eur. Conf. Exhibit. Opt. Commun., Turin, Italy, 19–23 September 2010. [CrossRef]
54. Tomkos, I.; Azodolmolky, S.; Solé-Pareta, J.; Careglio, D.; Palkopoulou, E. A tutorial on the flexible optical networking paradigm: State of the art, trends, and research challenges. *Proc. IEEE* **2014**, *102*, 1317–1337. [CrossRef]
55. Zakrzewski, Z. Microwave-photonic networks based on single-mode multi-core optical fibers. *Photonics Lett. Pol.* **2013**, *5*, 161–163. [CrossRef]

entropy

Article

Towards an Efficient and Exact Algorithm for Dynamic Dedicated Path Protection

Ireneusz Szcześniak [1,*], Ireneusz Olszewski [2] and Bożena Woźna-Szcześniak [3]

[1] Department of Computer Science, Częstochowa University of Technology, 42-200 Częstochowa, Poland
[2] Institute of Telecommunications, UTP University of Sciences and Technology, 85-796 Bydgoszcz, Poland; irek@utp.edu.pl
[3] Department of Mathematics and Computer Science, Jan Długosz University, 42-200 Częstochowa, Poland; b.wozna@ujd.edu.pl
* Correspondence: ireneusz.szczesniak@pcz.pl

Abstract: We present a novel algorithm for dynamic routing with dedicated path protection which, as the presented simulation results suggest, can be efficient and exact. We present the algorithm in the setting of optical networks, but it should be applicable to other networks, where services have to be protected, and where the network resources are finite and discrete, e.g., wireless radio or networks capable of advance resource reservation. To the best of our knowledge, we are the first to propose an algorithm for this long-standing fundamental problem, which can be efficient and exact, as suggested by simulation results. The algorithm can be efficient because it can solve large problems, and it can be exact because its results are optimal, as demonstrated and corroborated by simulations. We offer a worst-case analysis to argue that the search space is polynomially upper bounded. Network operations, management, and control require efficient and exact algorithms, especially now, when greater emphasis is placed on network performance, reliability, softwarization, agility, and return on investment. The proposed algorithm uses our generic Dijkstra algorithm on a search graph generated "on-the-fly" based on the input graph. We corroborated the optimality of the results of the proposed algorithm with brute-force enumeration for networks up to 15 nodes large. We present the extensive simulation results of dedicated-path protection with signal modulation constraints for elastic optical networks of 25, 50, and 100 nodes, and with 160, 320, and 640 spectrum units. We also compare the bandwidth blocking probability with the commonly-used edge-exclusion algorithm. We had 48,600 simulation runs with about 41 million searches.

Keywords: dynamic dedicated path protection; generic Dijkstra algorithm; elastic optical network; modulation constraints

Citation: Szcześniak, I.; Olszewski, I.; Woźna-Szcześniak, B. Towards an Efficient and Exact Algorithm for Dynamic Dedicated Path Protection. *Entropy* **2021**, *23*, 1116. https://doi.org/10.3390/e23091116

Academic Editor: Ernestina Menasalvas

Received: 1 August 2021
Accepted: 24 August 2021
Published: 27 August 2021

1. Introduction

Optical networks, which are the backbone of communication networks, need to provide protection for the carried traffic to prevent large-scale disruptions due to fiber cuts, human errors, hardware failures, power outages, natural disasters or attacks [1–3]. From among the various ways of protecting traffic in optical networks, dedicated path protection (DPP) is the simplest, and the most effective, albeit the most expensive. In DPP, there are two paths established for a single demand: the working one, and the protecting one. When the working path fails, the protecting path delivers the traffic. DPP has been commonly used and studied for decades.

In a wavelength-division multiplexed (WDM) network, if a client signal does not fully utilize the fixed spectrum of the assigned wavelength, the spectrum of the precious erbium window is wasted, a problem addressed by elastic optical networks (EONs) which divide the spectrum into fine *frequency slot units* (of, e.g., 12.5 GHz width), or just *units*, and then allocating *contiguous units* to form a *slot* tailored to a specific demand [4].

Routing in WDM networks with the spectrum continuity constraint is called routing and wavelength assignment (RWA). Routing in EONs with the spectrum contiguity constraint added is called routing and spectrum assignment (RSA), and with the signal modulation constraint added is called routing, modulation, and spectrum assignment (RMSA). These routing problems can be dynamic or static. In dynamic (aka online) routing, a single demand is routed in a loaded network, as opposed to *static* (aka offline) routing, where many demands are routed in an unloaded network.

When finding an exact solution for a dynamic routing problem in optical networks, some path cost is minimized, and the spectrum and modulation constraints are met. The path cost can be defined in various ways, e.g., the path length, the number of edges, some signal quality measure, monetary cost, or a measure related to availability. The path cost can take into account the cost of traversing not only an edge, but a vertex, too. In routing with DPP, the cost of a path pair is minimized, and the spectrum and modulation constraints must be met for both paths.

Whether routing along paths of lowest cost leads to optimal network performance (as measured, for instance, with the bandwidth blocking probability over a series of established and terminated connections) is, to the best of our knowledge, an open research problem, which we do not address in this work. We research the problem of an exact algorithm, one which finds an optimal (i.e., of lowest cost) solution, and not of an optimal algorithm. An algorithm optimality could imply optimal network performance, or optimal computational complexity, and we address neither of these.

Our novel contribution is an algorithm which solves the dynamic routing problems *with DPP* for WDM networks and EONs without signal regeneration and spectrum conversion. The algorithm can take into account various spectrum allocation policies. With extensive simulations, we demonstrate the computational performance of the proposed algorithm, which can be polynomial, not exponential. We corroborate the optimality of the results found for networks up to 15 nodes. Finally, we provide under a liberal license our free and open-source implementation of the proposed algorithm [5].

The article is organized as follows. In Section 2, we review related works, in Section 3, we state the research problem, in Section 4, we describe the algorithm, and, in Section 5, we report on the simulation results. Finally, Section 6 concludes the article.

2. Related Works

The proposed algorithm is based on the *generic* Dijkstra algorithm recently proposed [6]. The generic Dijkstra algorithm is a generalization of the Dijkstra algorithm, which takes into account the spectrum continuity and contiguity constraints by introducing the *incomparability relation* between solutions. Specifically, we modify the generic Dijkstra algorithm to work on a search graph, which is built using the input graph, and represents the possible ways of finding path pairs. Furthermore, we introduce the incomparability relation between pairs of paths.

To the best of our knowledge, no efficient and exact algorithm (at least demonstrated by simulations) for solving the dynamic routing problem with DPP in optical networks has been published. In [7], the authors offered a proof that the dynamic RWA with DPP is nondeterministic polynomial time complete (NP-complete). In Reference [8], the authors offered a proof that the dynamic RMSA with DPP is NP-complete. In contrast, we propose an algorithm with the efficiency and exactness demonstrated by simulations, which suggests the problem may be tractable.

Dynamic routing without DPP is simpler, but its status seemed unclear. In References [9,10], the problem was solved with exponential worst-case time and memory complexities. However, the dynamic routing problems in EONs can be solved exactly in polynomial time with the spectrum scan method [11], introduced in Reference [12]. The exact-routing concept of the spectrum scan method was introduced earlier for WDM networks in Reference [13] but was called a *heuristic* greedy algorithm. That concept was used under the name of the spectrum window planes [14], and the filtered-graphs algorithm [6].

In Reference [15], the authors solved efficiently (in polynomial time) and exactly the dynamic routing problem in WDM networks with their interconnected-layered-graph algorithm. That *exact* algorithm was later improved and applied to EONs in Reference [16] but was called *heuristic*.

In a very broad sense, dynamic routing problems are multicriteria shortest path problems, which, in turn, are multiobjective combinatorial optimization problems with a set of constraints given to define the combinatorial structure of the problem [17]. Whether a specific routing problem is tractable or not depends on the number and type of criteria (or objective functions) and constraints. Routing problems are defined for discrete or continuous criteria: discrete for, e.g., optical networks [9] and networks capable of advance resource reservation [18], continuous for, e.g., networks with quality-of-service requirements [19,20]. In Reference [21], ten bicriteria shortest path problems were studied, some of them were proven NP-complete, others were solved in polynomial time with a novel multilabeling algorithm. That bicriteria multilabeling algorithm was generalized to any number and type of criteria in Reference [22], which is now called the *Martins algorithm*.

The Martins algorithm is the basic algorithm for exactly solving any multicriteria shortest path problem, but with exponential worst-case memory and time complexities [23]. To use the Martins algorithm for dynamic routing in optical networks, we could consider available spectrum units as discrete criteria, but that would lead to exponential worst-case time and memory complexities. The generic Dijkstra is similar to the Martins algorithm in that it is also a multilabeling algorithm, but the generic Dijkstra algorithm is a single criterion shortest path algorithm, where the ordering between solutions (labels) is partial.

The efficient and exact algorithms for finding a shortest pair of edge-disjoint paths in a graph are: the Suurballe algorithm [24], the Bhandari algorithm [25], and any minimum-cost, maximum-flow algorithm (e.g., the successive shortest path algorithm) with edge capacities set to one [26], all of which use the path augmentation technique. These algorithms cannot be used for solving the stated problem because they do not consider the spectrum continuity and contiguity constraints.

In our simulations, we also used two well-known algorithms for solving the problem: the heuristic *edge-exclusion* algorithm, and the exact *brute-force* algorithm.

The edge-exclusion algorithm is a simple and commonly-used algorithm for finding a pair of edge-disjoint paths: find a shortest path, then remove from the graph the edges found, and then find a shortest path again. This heuristic performs quite well, but it often finds suboptimal solutions, and it can fail even when a solution exists (e.g., for the so-called trap topology).

The edge-exclusion algorithm usually employs the limited (i.e., with a limited K, e.g., $K = 10$) K-shortest path (KSP) algorithm to find a shortest path, which is a heuristic algorithm whose blocking probability depends on the value of K. However, the edge-exclusion algorithm can perform better if an algorithm of lower blocking probabilities is used. The blocking probabilities of the generic Dijkstra algorithm can be even twice as low as the blocking probabilities of the limited KSP algorithm [27]. Therefore, in the edge-exclusion algorithm, we used the generic Dijkstra algorithm.

The brute-force algorithm enumerates the path pairs using a priority queue that sorts the pairs in increasing-cost order. After we retrieve a pair from the queue, we produce new path pairs by extending one of the paths in the pair with an available edge that was not used before because the two paths should be edge-disjoint and without loops. We put a new path pair into the queue, if its paths meet the spectrum continuity and contiguity constraints. We keep looking for path pairs until we find one whose paths end at the destination node, provided we have enough time and memory. We successfully used the brute-force algorithm only for very small networks (15 nodes), since this algorithm is very inefficient.

3. Problem Statement

Given:

- directed multigraph $G = (V, E)$, where $V = \{v_i\}$ is a set of vertexes, and $E = \{e_i\}$ is a set of edges,
- available units function $\mathrm{AU}(e_i)$, which gives the set of available units of edge e_i, which do not have to be contiguous,
- s and t are the source and target vertexes of the demand,
- a cost function $\mathrm{cost}(p)$, which returns the cost of path p,
- a monotonically nondecreasing cost function $\mathrm{COST}(l)$, which returns the (real or integer) cost of path pair l,
- a decision function $\mathrm{decide}(p)$ of monotonically increasing requirements, which returns true if path p can support the demand, otherwise false,
- the set of all units Ω on every edge.

Find:

- a cheapest (i.e., of the lowest cost) pair of edge-disjoint paths (a path is a sequence of edges), the cheaper being the working path, and the more expensive the protecting path,
- continuous and contiguous units for each of the two paths separately: the working path and the protecting path (i.e., each path can have different spectrum).

We denote a set of contiguous units (CU) which start at index a and end at index b inclusive as $[a \,..\, b]$. For instance, $[0 \,..\, 2]$ denotes units 0, 1, and 2. We can treat a set of units as a set of CUs. For instance, $\{0, 1, 3, 4, 5\}$ and $\{[0 \,..\, 1], [3 \,..\, 5]\}$ are the same. Two CUs are *incomparable*, when one is not included in the other. For instance, $[0 \,..\, 2]$ and $[2 \,..\, 3]$ are incomparable, which we denote with the $\|$ relation, e.g., $[0 \,..\, 2] \parallel [2 \,..\, 3]$.

To state the problem generically, we intentionally introduced the cost, COST, and decide functions to consider the RWA, RSA, and RMSA problems with DPP at once. For example, for RWA, the $\mathrm{cost}(p)$ function for path p could return the length of the path, for RSA, the product of the path length and the number of units requested by the demand, and, for RMSA, the product of the path length and the number of units required by the demand for the given path.

We require that the $\mathrm{COST}(l)$ function for a path pair l be monotonically nondecreasing, i.e., for any path pair l' derived from l by appending an edge to one of the paths, $\mathrm{COST}(l) \leq \mathrm{COST}(l')$. This requirement implies the proposed algorithm cannot be used for networks with regeneration, when the path cost is defined as the product of the path length and the required number of units. Regeneration would reduce the number of required units, and the cost of the path pair would be reduced, thus violating this requirement.

We also assume that an optimal path pair has the optimal substructure, i.e., it is built of optimal path pairs, which is required by the dynamic programming principle the proposed algorithm relies on. In our simulations, the defined problem meets this assumption: the path cost is the product of the path length and the number of units required, while the cost of a path pair is the sum of the costs of the two paths.

The decide function accepts or rejects a candidate path, and lets the user define an acceptable path. We require the function to have monotonically increasing requirements, i.e., if the function rejects path p, then any path derived from p by appending an edge should also be rejected. For RWA, the function should make sure that the CU has at least one unit (wavelength), for RSA, that the CU has at least the number of units requested by the demand, and, for RMSA, that a CU has at least the number of units required for the demand for the given path length.

The bitrate of a demand is not a given of the stated problem and, if needed, should be relegated to the decision function as an implementation detail. Likewise, the cost, COST, and decide functions remain undefined in the problem statement. In Section 5, to solve the RMSA problem with DPP, we define the cost, COST, and decide functions in

Section 5.1.3. The decision function defined there by (7) checks for the required number of units, which depends on the path length.

4. Proposed Algorithm

We run the generic Dijkstra algorithm on a *search graph*. Searching for a cheapest solution in the search graph corresponds to searching for a pair of paths of lowest cost in the input graph. The algorithm grows *the search tree* for the search graph.

4.1. Preliminaries

Below, we describe the search graph, the search tree, the priority queue, and the related concepts of the solution, the path trait, and the solution label.

4.1.1. Search Graph

The search graph has a set of vertexes $X = \{x = (v_{x,1}, v_{x,2})\}$, where vertex indexes $v_{x,1}$, and $v_{x,2}$ of the input graph satisfy $v_{x,1} \leq v_{x,2}$. For vertex x, we find a set of solutions, where the *solution* is a pair of paths: one path leads to vertex $v_{x,1}$, and the other to vertex $v_{x,2}$. Which of the paths could eventually (when vertex t is reached by both paths) become working or protecting is unknown and unimportant at this stage.

An edge in the search graph from vertex x to vertex x' represents finding a solution for vertex x' based on a solution for vertex x by taking edge e' in the input graph from either vertex $v_{x,1}$ or $v_{x,2}$. Therefore, the edge in the search graph connects vertex x to some other vertex x' which has one of the vertex indexes $v_{x,1}$ or $v_{x,2}$ taken from x. The other vertex index of x' is the index of the target vertex of edge e'. Vertex x' becomes either $(v_{x',1}, v_{x,2})$, or $(v_{x,1}, v_{x',2})$, with its vertex indexes swapped if necessary, because we require the first one be smaller than or equal to the second one.

Taking a single edge in the input graph is the simplest, and the only one needed, way of producing a new solution in the search graph. Taking at once two edges of the input graph, one edge for each of the two paths, should also work, but this would lead to a more complicated and less efficient algorithm. This is more complicated because we cannot always take two edges, and less efficient because, by taking two edges, we can reach a suboptimal solution, which we would avoid if we took one of those edges first.

4.1.2. Path Trait

A *path trait* p is a pair of a cost and a CU, which describes a path in the input graph. For a path trait p, function $\mathrm{cost}(p)$ gives the path cost, and function $\mathrm{CU}(p)$ the path CU. For example, assuming the cost is the path length, a path trait $p = (500\ \mathrm{km}, [0\,..\,10])$ says the path is $\mathrm{cost}(p) = 500$ km long and has the CU of $\mathrm{CU}(p) = [0\,..\,10]$.

Path trait p_i is better than or equal to path trait p_j, denoted by $p_i \leq p_j$, when the cost of p_i is smaller than or equal to the cost of p_j, and the CU of p_i includes the CU of p_j, i.e., $(\mathrm{cost}(p_i) \leq \mathrm{cost}(p_j)) \wedge (\mathrm{CU}(p_i) \supseteq \mathrm{CU}(p_j))$. If $p_i \leq p_j$, then we drop p_j, since it offers no better path in comparison with p_i, and so we perform the search more efficiently.

This definition of the path trait comparison allows for *incomparability* of path traits, which is needed when searching for paths with the spectrum continuity and contiguity constraints. For instance, path trait $p_1 = (1, [0\,..\,2])$ is incomparable with $p_2 = (2, [0\,..\,3])$ because neither $p_1 \leq p_2$ nor $p_2 \leq p_1$ is true. We are interested in path trait p_2, even though its cost is higher than the cost of p_1, because p_2 has a CU that is incomparable with the CU of p_1.

4.1.3. Solution Label

Solution label $l_x = (p_{x,1}, p_{x,2})$ for vertex x is a pair of path traits $p_{x,1}$, $p_{x,2}$, where the first path which ends at $v_{x,1}$ has trait $p_{x,1}$, and the other path which ends at $v_{x,2}$ has trait $p_{x,2}$.

We compare solution labels to drop those solutions which offer nothing better than we already have, thus limiting the search space, and performing the search more efficiently.

Label l_i is better than or equal to label l_j, denoted by $l_i \leq l_j$, when both path traits of l_i are better than or equal to both path traits of l_j, i.e., $(p_{i,1} \leq p_{j,1}) \wedge (p_{i,2} \leq p_{j,2})$. If $l_i \leq l_j$, then we should not be interested in l_j because it offers no better solution in comparison with l_i.

A solution label for vertex x with the same vertexes in the input graph, i.e., $v_{x,1} == v_{x,2}$, should have its path traits ordered with the $\leq$ relation, i.e., $p_{x,1} \leq p_{x,2}$, so that, when we compare labels of two solutions for vertex x, we compare the working-path traits first, and the equal to or worse protecting-path traits next.

This definition of the label comparison allows for *incomparability* of labels, which is needed when searching for a pair of paths, when these paths can have incomparable traits. For instance, label l_1 of path traits $p_{1,1} = (1, [0..2])$, and $p_{1,2} = (2, [10..12])$ is incomparable with label l_2 of path traits $p_{2,1} = (2, [0..3])$, and $p_{2,2} = (10, [10..12])$ because neither $l_1 \leq l_2$ nor $l_2 \leq l_1$ is true.

4.1.4. Search Tree

The result of the search is the *search tree*, which is organized according to the dynamic-programming principle of reusing data from previous computation. Search-tree node $n_{x'} = (x', l_{x'}, e', n_x)$ represents a solution found for the search-graph vertex $x' = (v_{x',1}, v_{x',2})$ based on the solution found for node x. The solution is described by label $l_{x'} = (p_{x',1}, p_{x',2})$: the first path of the solution which ends at $v_{x',1}$ has trait $p_{x',1}$, and the other path which ends at $v_{x',2}$ has trait $p_{x',2}$. We get the solution from the previous search-tree node n_x for vertex x by taking edge e' in the input graph. For a search-tree node n, function label(n) returns its label, and function vertex(n) returns its search-graph vertex.

A tree node represents a solution which is either *permanent* or *tentative*. A permanent-solution node stays in the tree for good, while a tentative-solution node can be discarded. A tentative-solution node is always a leaf. A tentative solution wants to become permanent, but instead it can be discarded or never processed.

To ensure that a solution is edge-disjoint, we do not add to the search tree a solution node if its edge was already used by its ancestor in the search tree.

4.1.5. Priority Queue

The optimality of the solutions found is achieved with the priority queue, which provides the cheapest solutions. The priority queue stores pairs, where a pair has a cost and a reference to a search-tree node n_x of a tentative solution. The cost in the pair is the cost of the solution, i.e., COST(label(n_x)). The queue sorts the solutions in the increasing-cost order, with the cheapest solution at the top.

A tentative solution is waiting in the queue to be processed, but it also can be either discarded, if we find a better solution, or never processed, if the search finishes sooner. A tentative solution becomes permanent when it is retrieved from the queue.

4.2. Algorithm

The proposed algorithm has the main loop listed in Algorithm 1, and the relax procedure listed in Algorithm 2. The main loop iterates over the permanent solutions retrieved from the priority queue, while the relax procedure pushes tentative solutions to the priority queue.

The solutions for vertex x are maintained in the set P_x of permanent solutions with incomparable labels, and the set T_x of tentative solutions with incomparable labels. The set of all permanent solutions is P, and the set of all tentative solutions is T. Permanent solutions are optimal.

We start the search at vertex $x_s = (s, s)$. We create the tentative solution n_{x_s} (the root of the search tree) of two empty paths starting at vertex s with 0 costs and the CUs of Ω. We insert n_{x_s} into the set of tentative labels for vertex x, and push the pair of $(0, n_{x_s})$ to the priority queue Q.

To cover the maximal part of the search space, we look for the paths with the maximal CU, which satisfy the requirements of the decision function decide used by the relax procedure. For this reason, we start the search with the CU of Ω.

We stop searching when the priority queue is empty, or when we find a permanent solution for vertex $x_t = (t, t)$. If we need a complete (i.e., for all vertexes x) search tree, we should let the algorithm run until the priority queue is empty.

In each iteration of the main loop, we process the cheapest of all tentative solutions, and make it permanent. When we retrieve a pair from the queue, we have to ensure the tentative solution was not discarded by the relax procedure, i.e., that the reference to n_x is not null.

Next, we relax the out edges of vertex x in the search graph. An edge in the search graph represents taking an edge in the input graph from either vertex $v_{x,1}$ or $v_{x,2}$, and so we iterate over the edges leaving vertex $v_{x,1}$ first, and over the edges leaving vertex $v_{x,2}$ next.

Algorithm 1 Dedicated Path Protection Algorithm
In: graph G, source vertex s, target vertex t
Out: a cheapest pair of paths, and their CUs
Here, we concentrate on permanent solutions n_x.

```
x_s = (s, s)
x_t = (t, t)
l_{x_s} = ((0, Ω), (0, Ω))
n_{x_s} = (x_s, l_{x_s}, e_∅, null)
T_{x_s} = {n_{x_s}}
push(Q, (0, n_{x_s}))
while Q is not empty do
  n_x = pop(Q)
  if n_x == null then
    continue the main loop
  x = (v_{x,1}, v_{x,2}) = vertex(n_x)
  // Remove n_x from the set of tentative solutions for x.
  T_x = T_x − {n_x}
  // Add n_x to the set of permanent solutions for x.
  P_x = P_x ∪ {n_x}
  if x == x_t then
    break the main loop
  l_x = (p_{x,1}, p_{x,2}) = label(n_x)
  for each out edge e' of vertex v_{x,1} in G do
    relax(e', v_{x,2}, p_{x,2}, p_{x,1}, n_x)
  for each out edge e' of vertex v_{x,2} in G do
    relax(e', v_{x,1}, p_{x,1}, p_{x,2}, n_x)
return trace(P, x_t, x_s)
```

The relax procedure relaxes a single edge in the search graph, which is described in the procedure parameters: the taken edge e' in the input graph, vertex v_1 and the corresponding path trait p_1 which both do not change, the other trait p_2 of the path to which we try to append edge e', and the previous search-tree node n_x.

The relaxation can find a number of tentative solutions, which would differ only by the CU of C', because there may be a number of spectrum fragments available AU(e') on edge e' which we can use for a tentative solution.

We build (if necessary, we swap the elements of pairs x' and $l_{x'}$ with the swap function) and add a tentative solution $n_{x'}$ to $T_{x'}$ and Q, only when there is no solution with a better or equal label already found. Adding $n_{x'}$ can make some tentative solutions invalid (since $n_{x'}$ is better), so we discard them.

Algorithm 2 relax
In: edge e', const vertex v_1, const trait p_1, other trait p_2,
previous search-tree node n_x
Here, we concentrate on tentative solutions $n_{x'}$.

```
v' = target(e')
c' = cost(path of p2 with e' appended)
for each CU C' in CU(p2) ∩ AU(e') do
  x' = (v1, v')
  p' = (c', C')
  if decide(p') then
    l_x' = (p1, p')
    if v' < v1 then
      swap(x')
      swap(l_x')
    else if v1 == v' and not p1 ≤ p' then
      swap(l_x')
    // Make sure we should be interested in l_x'.
    if ∄n ∈ P_x' : label(n) ≤ l_x' then
      if ∄n ∈ T_x' : label(n) ≤ l_x' then
        // Make sure we are not reusing e'.
        if e' not used by ancestors then
          // Discard worse tentative solutions.
          T_x' = T_x' − {n ∈ T_x' : l_x' ≤ label(n)}
          n_x' = (x', l_x', e', n_x)
          // Add n_x' to the tentative solutions for x'.
          T_x' = T_x' ∪ {n_x'}
          push(Q, (COST(l_x'), n_x'))
```

Spectrum allocation policy should be taken into account in two places. First, the priority queue should choose the solution of the preferred spectrum allocation policy from among the tentative solutions of the same cost. Second, once the solution for the destination vertex is found, the preferred CUs should be allocated from the CUs found.

The trace function, using P, traces back the tree nodes from the node for vertex x_t to the node for vertex x_s. For each tree node there is an edge, which the function appends to one of the two paths. When appending an edge to a path, we have to ensure that not only the cost matches, but the CU matches, too. We have to consult the other path trait of the tree node, to ensure that we are not appending the edge to the wrong path, which coincidentally meets the conditions. The function returns the less expensive path as the working path, and the more expensive path as the protecting. For each of the paths, the function allocates the minimal CU, with the required number of units, from the maximal CU found for the permanent solution for vertex x_t.

4.3. Example

We demonstrate how the algorithm works by finding a solution from the source vertex s to the destination vertex t for a single unit in the trap topology shown in Figure 1, where an edge label gives not only the name of the edge, but also its cost and available units. Not to complicate the example further, the signal modulation constraints are not considered, especially since they are not crucial to the algorithm as they only discard paths.

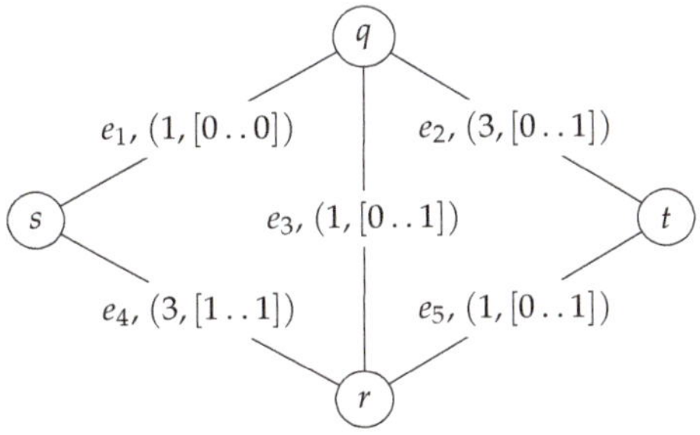

Figure 1. A sample input graph: the trap topology.

The trap topology is well-known since the edge-exclusion algorithm fails for it: when the edges of the shortest path from s to t through vertexes q and r are removed, a second path does not exist. However, the *optimal solution* does exist: one path goes through vertex q with the CU of $[0..0]$, and the other through vertex r with $[1..1]$.

Figure 2 shows the search graph generated, where the edge label gives the name of an input-graph edge to take to make a transition between the vertexes in the search graph. For example, the transition from vertex (s, s) to vertex (q, s) requires edge e_1. Most edges are undirected, since the algorithm examines transitions in both directions. However, there are some directed edges (e.g., from (q, s) to (s, t)), since their reverse transitions (e.g., from (s, t) to (q, s)) are not examined. Because paths in the input graph have to be edge-disjoint, some paths in the search graph are disallowed, e.g., $(s, s) - (q, s) - (q, q)$.

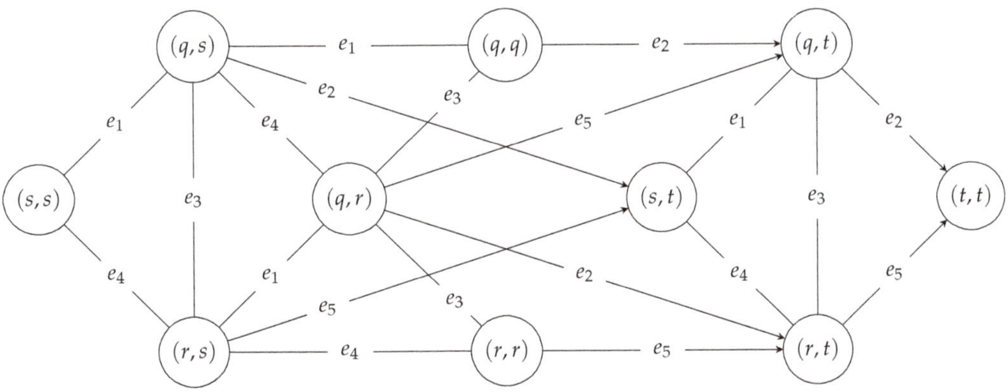

Figure 2. The search graph.

Figure 3 shows the search tree generated, where only the permanent (and not tentative) solutions are represented. The tree is rooted at node n_1 for vertex (s, s). The solution found is represented by node n_{14} for vertex (t, t). A search-graph vertex can have a set of permanent incomparable solutions, which are represented by a set of search-tree nodes, e.g., for vertex (q, s) there are two search-tree nodes: n_2 and n_8.

The algorithm processes solution labels by taking actions on them, as reported in Table 1. A label can be produced and pushed into the queue, as in, e.g., action #0. A row represents an action on a label which was produced for the given search-graph vertex by making a transition with the given edge. For instance, action #2 reports a label that is pushed into the queue, and which was produced for the search-graph vertex (q, s) by making a transition with edge e_1 (from vertex (s, s)).

A label can be retrieved from the queue, and made permanent, as in, e.g., action #1. A search-tree node for a permanent label has a name reported, e.g., n_1 as in action #1. A row for a label made permanent is colored gray to mark the beginning of a sequence of rows that report the actions of the relaxation based on the label from that gray row. For instance, the row for action #1 is gray, and the subsequent rows for actions from #2 to #5 report the actions of the relaxation that ensued.

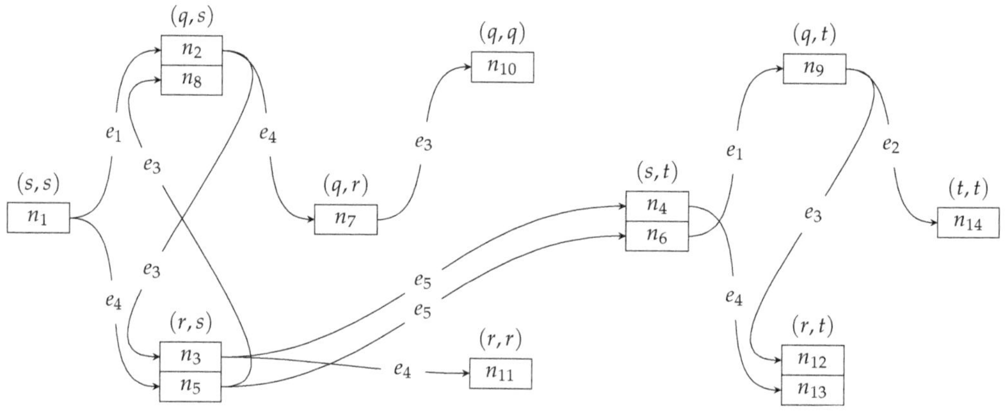

Figure 3. The search tree.

A label can be pushed into the queue, or it can be dropped for two reasons: it is worse than or equal to an existing label (e.g., in action #4, a label is dropped because an equal label of action #2 exists), or it uses an edge twice (e.g., in action #7, a label is dropped because edge e_1 is used twice). A label can also be discarded from the queue if a better label is found (e.g., in action #17, a label is discarded because a better label was found in action #16).

As reported in Table 1, the search is booted with action #0. There are 14 permanent labels found for 10 vertexes of the search graph. The algorithm terminates, when the destination node (t, t) is reached with the search-tree node n_{14}. We can trace back from n_{14} to n_1 to get the aforementioned optimal solution.

Table 1. Solution labels processed.

Action Number	Solution Cost	Search-Tree Node Name	Search-Graph Vertex	Solution Label	Edge	Action
0	0		(s, s)	$((0, [0..1]), (0, [0..1]))$	$e_\emptyset$	push into queue
1	0	n_1	(s, s)	$((0, [0..1]), (0, [0..1]))$	$e_\emptyset$	make permanent
2	1		(q, s)	$((1, [0..0]), (0, [0..1]))$	e_1	push into queue
3	3		(r, s)	$((3, [1..1]), (0, [0..1]))$	e_4	push into queue
4	1		(q, s)	$((1, [0..0]), (0, [0..1]))$	e_1	drop (worse or equal)
5	3		(r, s)	$((3, [1..1]), (0, [0..1]))$	e_4	drop (worse or equal)
6	1	n_2	(q, s)	$((1, [0..0]), (0, [0..1]))$	e_1	make permanent
7	2		(q, q)	$((1, [0..0]), (1, [0..0]))$	e_1	drop (edge reuse)
8	4		(q, r)	$((1, [0..0]), (3, [1..1]))$	e_4	push into queue
9	2		(s, s)	$((0, [0..1]), (2, [0..0]))$	e_1	drop (worse or equal)
10	4		(s, t)	$((0, [0..1]), (4, [0..0]))$	e_2	push into queue
11	2		(r, s)	$((2, [0..0]), (0, [0..1]))$	e_3	push into queue
12	2	n_3	(r, s)	$((2, [0..0]), (0, [0..1]))$	e_3	make permanent
13	3		(q, r)	$((1, [0..0]), (2, [0..0]))$	e_1	drop (edge reuse)
14	5		(r, r)	$((2, [0..0]), (3, [1..1]))$	e_4	push into queue
15	3		(q, s)	$((3, [0..0]), (0, [0..1]))$	e_3	drop (worse or equal)
16	3		(s, t)	$((0, [0..1]), (3, [0..0]))$	e_5	push into queue
17	4		(s, t)	$((0, [0..1]), (4, [0..0]))$	e_2	discard from queue

Table 1. *Cont.*

Action Number	Solution Cost	Search-Tree Node Name	Search-Graph Vertex	Solution Label	Edge	Action
18	3	n_4	(s,t)	$((0,[0..1]),(3,[0..0]))$	e_5	make permanent
19	4		(q,t)	$((1,[0..0]),(3,[0..0]))$	e_1	drop (edge reuse)
20	6		(r,t)	$((3,[1..1]),(3,[0..0]))$	e_4	push into queue
21	3	n_5	(r,s)	$((3,[1..1]),(0,[0..1]))$	e_4	make permanent
22	4		(q,r)	$((1,[0..0]),(3,[1..1]))$	e_1	drop (worse or equal)
23	6		(r,r)	$((3,[1..1]),(3,[1..1]))$	e_4	drop (edge reuse)
24	4		(q,s)	$((4,[1..1]),(0,[0..1]))$	e_3	push into queue
25	6		(s,s)	$((0,[0..1]),(6,[1..1]))$	e_4	drop (worse or equal)
26	4		(s,t)	$((0,[0..1]),(4,[1..1]))$	e_5	push into queue
27	4	n_6	(s,t)	$((0,[0..1]),(4,[1..1]))$	e_5	make permanent
28	5		(q,t)	$((1,[0..0]),(4,[1..1]))$	e_1	push into queue
29	7		(r,t)	$((3,[1..1]),(4,[1..1]))$	e_4	drop (edge reuse)
30	4	n_7	(q,r)	$((1,[0..0]),(3,[1..1]))$	e_4	make permanent
31	5		(r,s)	$((3,[1..1]),(2,[0..0]))$	e_1	drop (worse or equal)
32	7		(r,t)	$((3,[1..1]),(4,[0..0]))$	e_2	drop (worse or equal)
33	5		(r,r)	$((2,[0..0]),(3,[1..1]))$	e_3	drop (worse or equal)
34	5		(q,q)	$((1,[0..0]),(4,[1..1]))$	e_3	push into queue
35	7		(q,s)	$((1,[0..0]),(6,[1..1]))$	e_4	drop (worse or equal)
36	5		(q,t)	$((1,[0..0]),(4,[1..1]))$	e_5	drop (worse or equal)
37	4	n_8	(q,s)	$((4,[1..1]),(0,[0..1]))$	e_3	make permanent
38	5		(q,q)	$((1,[0..0]),(4,[1..1]))$	e_1	drop (worse or equal)
39	7		(q,r)	$((4,[1..1]),(3,[1..1]))$	e_4	drop (edge reuse)
40	7		(s,t)	$((0,[0..1]),(7,[1..1]))$	e_2	drop (worse or equal)
41	5		(r,s)	$((5,[1..1]),(0,[0..1]))$	e_3	drop (worse or equal)
42	5	n_9	(q,t)	$((1,[0..0]),(4,[1..1]))$	e_1	make permanent
43	6		(s,t)	$((2,[0..0]),(4,[1..1]))$	e_1	drop (worse or equal)
44	8		(t,t)	$((4,[0..0]),(4,[1..1]))$	e_2	push into queue
45	6		(r,t)	$((2,[0..0]),(4,[1..1]))$	e_3	push into queue
46	5	n_{10}	(q,q)	$((1,[0..0]),(4,[1..1]))$	e_3	make permanent
47	6		(q,s)	$((4,[1..1]),(2,[0..0]))$	e_1	drop (worse or equal)
48	8		(q,t)	$((4,[1..1]),(4,[0..0]))$	e_2	push into queue
49	6		(q,r)	$((4,[1..1]),(2,[0..0]))$	e_3	drop (edge reuse)
50	8		(q,t)	$((1,[0..0]),(7,[1..1]))$	e_2	drop (worse or equal)
51	6		(q,r)	$((1,[0..0]),(5,[1..1]))$	e_3	drop (worse or equal)
52	5	n_{11}	(r,r)	$((2,[0..0]),(3,[1..1]))$	e_4	make permanent
53	6		(q,r)	$((3,[0..0]),(3,[1..1]))$	e_3	drop (worse or equal)
54	6		(r,t)	$((3,[1..1]),(3,[0..0]))$	e_5	drop (worse or equal)
55	6		(q,r)	$((4,[1..1]),(2,[0..0]))$	e_3	drop (edge reuse)
56	8		(r,s)	$((2,[0..0]),(6,[1..1]))$	e_4	drop (worse or equal)
57	6		(r,t)	$((2,[0..0]),(4,[1..1]))$	e_5	drop (worse or equal)
58	6	n_{12}	(r,t)	$((2,[0..0]),(4,[1..1]))$	e_3	make permanent
59	7		(q,t)	$((3,[0..0]),(4,[1..1]))$	e_3	drop (worse or equal)
60	7		(t,t)	$((3,[0..0]),(4,[1..1]))$	e_5	drop (edge reuse)
61	6	n_{13}	(r,t)	$((3,[1..1]),(3,[0..0]))$	e_4	make permanent
62	7		(q,t)	$((4,[1..1]),(3,[0..0]))$	e_3	drop (edge reuse)
63	9		(s,t)	$((6,[1..1]),(3,[0..0]))$	e_4	drop (worse or equal)
64	7		(t,t)	$((3,[0..0]),(4,[1..1]))$	e_5	drop (edge reuse)
65	8	n_{14}	(t,t)	$((4,[0..0]),(4,[1..1]))$	e_2	make permanent

4.4. Worst-Case Analysis

We argue the size of the search space is polynomially upper bounded. We derive the upper bound L of the number of incomparable labels (i.e., the size of the search space) by considering the worst case where every vertex of all $|X|$ search graph vertexes has the maximum number S of incomparable labels. Therefore, $L = |X|S$. The problem is to derive $|X|$ and S.

The number of vertexes in the search graph is given by (1), since the input graph has $|V|$ vertexes, and since vertexes $x = (v_{x,1}, v_{x,2})$ of the search graph satisfy $v_{x,1} \leq v_{x,2}$. The number of vertexes with $v_{x,1} = v_{x,2}$ is $|V|$, and the number of vertexes with $v_{x,1} < v_{x,2}$ is the number of combinations of two elements from the set of $|V|$ elements.

$$|X| = |V| + \binom{|V|}{2} = \frac{|V|(|V|+1)}{2}. \quad (1)$$

The maximum number S of incomparable labels a vertex can have is given by (2). Since a label describes a solution made up of two independent paths, S is the maximum number of incomparable path traits squared.

The maximum number of incomparable path traits depends only on Ω. We get the largest set of incomparable path traits when the cost of path traits increases as the size of their CUs increases. The largest set has $|\Omega|$ subsets: the first subset has $|\Omega|$ traits with CUs of a single unit and the lowest cost, the second subset has $|\Omega| - 1$ traits with CUs of two units and a higher cost, ..., and the last subset has a single trait with the CU of $|\Omega|$ units and the highest cost. The largest set has $1 + 2 + \ldots + |\Omega| = (|\Omega| + 1)|\Omega|/2$ incomparable path traits.

$$S = \left(\frac{(|\Omega| + 1)|\Omega|}{2}\right)^2. \quad (2)$$

Therefore, the size of the search space is polynomially upper bounded, since $O(L) = O(|V|^2|\Omega|^4)$.

5. Simulations

The simulations had two goals: the optimality corroboration, and the performance evaluation. We had 48,600 simulation runs: 32,400 corroborative runs, and 16,200 performance evaluation runs.

We corroborated the optimality of the results of our algorithm by comparing them with the results of the brute-force enumeration algorithm: for a single search either both algorithms returned results of the same cost, or both algorithms returned no results. Since there are billions of feasible solutions even in small networks, and since the brute-force algorithm enumerates them all, we were able to corroborate the results only for small networks.

We ran the corroborative simulations for networks of 10, 11, 12, 13, 14, and 15 nodes; for 160, 320, and 640 units; for offered loads ranging from light to heavy; and for demands requesting on average from 10 to 64 units. In total, we had 32,400 simulation runs, out of which 183 runs were killed, because they requested more than 120 GB of operating memory, which we did not have. In total, we carried out 17,590,624 searches, all successfully corroborated.

The remainder of this section is about the performance evaluation.

5.1. Simulation Setting

Below, we describe how we model the network, the traffic, and the signal modulation.

5.1.1. Network Model

A network model has an undirected graph, and $|\Omega|$. We randomly generated three groups of network graphs with 25, 50, and 100 vertexes, where each group had one hundred graphs. We generated Gabriel graphs because they have been shown to model the properties of the transport networks very well [28]. The vertexes were uniformly distributed over a square area with the density of one vertex per 10 thousand square km.

We used three spacings of 25 GHz, 12.5 GHz, and 6.25 GHz for the erbium band, which translated to three values for $|\Omega|$: 160, 320, and 640 units.

We present the results only for the first-fit spectrum allocation policy. The first-fit policy allocates units in the first CU that can support the demand, i.e., the CU with the units of the lowest indexes. We also considered the best-fit and random-fit policies. The best-fit policy performed comparably to the first-fit policy, and the random-fit policy performed markedly worse than the first-fit policy. We do not present the results for these alternative policies because they add little to the main results.

5.1.2. Traffic Model

We evaluate the algorithm performance as a function of the *network utilization*, which we define as the ratio of the number of units in use to the total number of units on all edges. We measure the network utilization in response to offered load a, which expresses the desired network utilization.

Demands arrive according to the exponential distribution with rate λ per day. The end nodes of a demand are different and chosen at random. The number of units a demand requests is described by distribution (Poisson$(\gamma - 1) + 1$) with the mean of γ, i.e., a shifted Poisson distribution, so that we do not get a zero. Parameter γ_p expresses the mean number of units that demands request relative to the number $|\Omega|$ of all units on every edge, and so $\gamma = \gamma_p|\Omega|$. We model the connection holding time with the exponential distribution with the mean of τ days. A connection is bidirectional: the same CU is allocated in both directions for a path.

We express λ as a function of a. The offered load is the ratio of the number of demanded units to the total number of units on all edges. For traffic intensity $\lambda\tau$, the number of units demanded is $2\lambda\tau\gamma\alpha$, since a demand requests two paths, and we estimate they require γ units, and α edges each, where α is the average number of edges of a shortest path between the end nodes of the demand in the network being simulated. Therefore, $a = 2\lambda\tau\gamma\alpha/|E||\Omega|$, from which (3) follows.

$$\lambda(a) = \frac{a|E||\Omega|}{2\tau\gamma\alpha}. \tag{3}$$

Equation (3) underestimates the value of $\lambda(a)$ because we assume that every demand has a connection established. For this reason, $a = 1$ does not yield a full network utilization.

5.1.3. Signal Modulation Model

We use the signal modulation model from Reference [6], with M modulations available. For a demand requesting g units for the most spectrally-efficient modulation, the number of units needed to establish a connection of length d is given by (4), where r_1 is the reach of the least spectrally-efficient modulation, and r_M is the reach of the most spectrally-efficient modulation.

$$u(g,d) = \begin{cases} g & \text{if } d \leq r_M \\ \infty & \text{if } r_1 < d \\ \lceil g \cdot log_2(2d/r_M) \rceil & \text{otherwise} \end{cases}. \tag{4}$$

We describe a demand with the number of units g, instead of bitrate b, because the algorithm works with units, not bitrates. If the bitrate is given, we can calculate the number of units using (5), where R is a technology-dependent bitrate (e.g., 2.5 Gb/s), and G is the number of guard-band units [29].

$$g(b) = \lceil b/(R \cdot M) \rceil + G. \tag{5}$$

In the simulations, we assumed $M = 4$, and the reach of the least-spectrally efficient modulation r_1 equals to one and a half lengths of the longest path from among all the shortest paths (i.e., for every source-destination combination) in the network being simulated, which allows us to consider paths much longer than an average shortest path. Following Reference [6], we calculated $r_M = r_1/2^{M-1}$.

5.1.4. The Cost and Decision Functions

The cost and decision functions for path p are given by (6) and (7), where function length returns the length of path p as the sum of positive lengths of the edges used. The cost function for path pair l is given by (8).

$$\text{cost}(p) = \text{length}(p) \cdot u(g, \text{length}(p)), \tag{6}$$

$$\text{decide}(p) = u(g, \text{length}(p)) \leq |\text{CU}(p)|, \tag{7}$$

$$\text{COST}(l = (p_1, p_2)) = \text{cost}(p_1) + \text{cost}(p_2). \tag{8}$$

5.2. Runs and Populations

A simulation run simulated 150 days of a network in operation, with the results from the first 50 days discarded. The parameters of a simulation run were: the network size, $|\Omega|$, γ, a, and τ. A simulation run reported the mean network utilization, the mean and maximum times taken, and the mean and maximum number of 64-bit memory words used by a search for a single demand.

We averaged the mean simulation results to calculate the *sample* mean results, which estimate the *population* mean results, and the average algorithm performance. We took the maximum of the maximum simulation results to get the *sample* maximum results, which estimate the population maximum results, and the worst-case algorithm performance.

In a given population, there were 100 simulation runs whose parameters differed only with the network model. We had 162 populations because we varied 3 network sizes (25, 50, 100 nodes), 3 values of $|\Omega|$ (160, 320, 640 units), 9 values of a (0.05, 0.1, 0.15, 0.2, 0.45, 0.65, 1, 1.5, 2), and two runs for $\gamma = 10$ units, and $\gamma_p = 10\%$ of units available (i.e., 16 units for the case with 160 units, 32 for 320, and 64 for 640). For all populations, the mean connection holding time $\tau = 10$ days was constant. In total, we carried out 16200 simulation runs (162 populations $\times$ 100 samples) with 24,043,157 searches. The sample means credibly estimate the population means, since their relative standard error was below 5%.

5.3. Simulation Results

Figures 4 and 5 show the sample means and the sample maxima of the time taken and memory used by a search, regardless of whether the search was successful or not. The results are shown on a logarithmic scale as a function of network utilization. The curves are plotted dotted for 160 units, dashed for 320 units, and solid for 640 units. The sample means are plotted thin, and the sample maxima thick. Each curve is drawn using 9 data points for different values of a. For the means, we do not report the error bars representing the standard error, since they were too small to plot.

Figures 4 and 5 have three rows and two columns of subfigures. The first row shows the results for the networks with 25 nodes, the second for 50 nodes, and the third for 100 nodes. The first column shows the results for $\gamma = 10$, and the second column the results for $\gamma_p = 10\%$.

The mean times range from 10^{-3} s (for 25 nodes, and 160 units) to 10^2 s (100 nodes, 640 units). While the difference in scale is 10^5, we also note that the problem size increased 16 times. The mean time increases about ten times as we increase the network size by a factor of two. For $\gamma = 10$, the mean time increases about five times as the number of units increases twice (from 160 to 320, and from 320 to 640 units). Interestingly, the time for $\gamma_p = 10\%$ is roughly the same for 160, 320, and 640 units, which suggests the time complexity depends on the number of units requested relative to the number of available units, and indirectly on the spectrum fragmentation. The mean time decreases as the network utilization increases, since the search space gets smaller. As for the sample maximum results, they were usually a hundred times larger than the mean results.

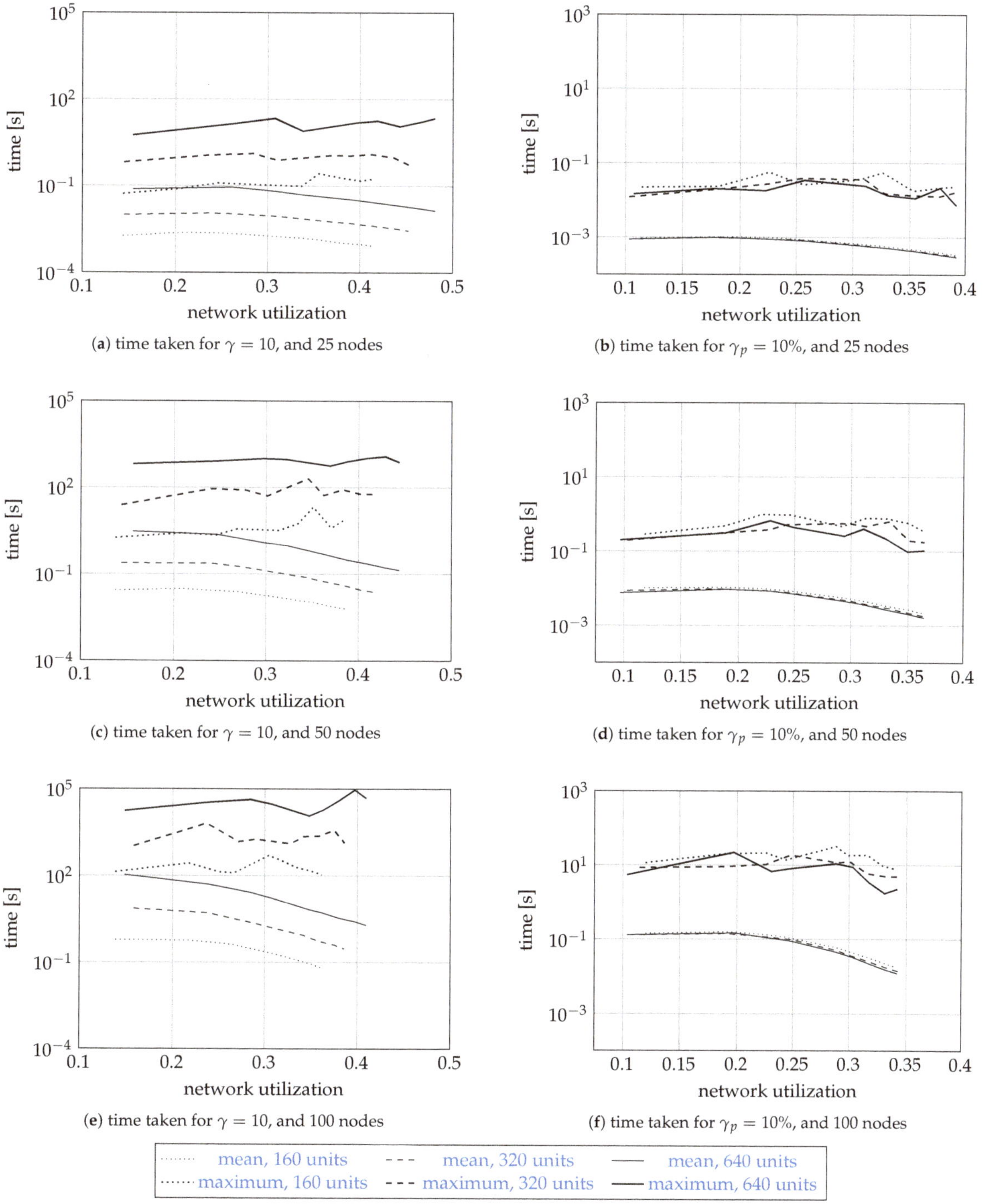

(a) time taken for $\gamma = 10$, and 25 nodes

(b) time taken for $\gamma_p = 10\%$, and 25 nodes

(c) time taken for $\gamma = 10$, and 50 nodes

(d) time taken for $\gamma_p = 10\%$, and 50 nodes

(e) time taken for $\gamma = 10$, and 100 nodes

(f) time taken for $\gamma_p = 10\%$, and 100 nodes

Figure 4. The sample means and maxima of the time taken by the proposed algorithm.

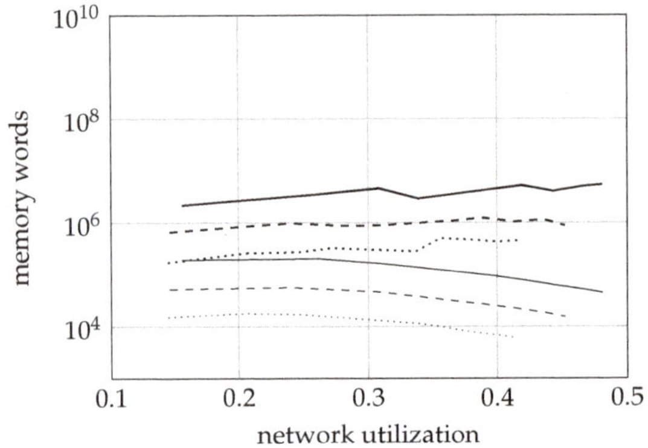

(a) memory used for $\gamma = 10$, and 25 nodes

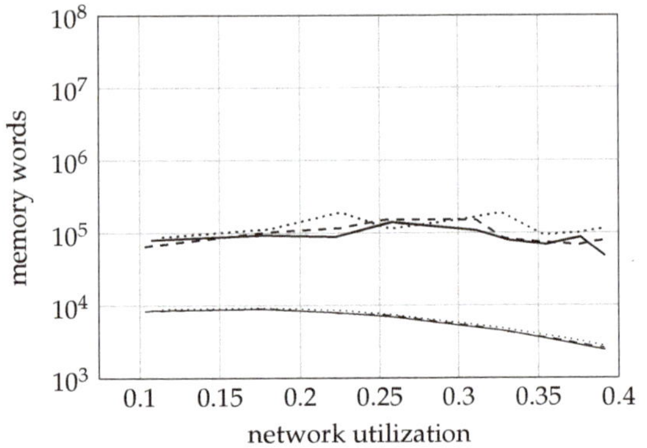

(b) memory used for $\gamma_p = 10\%$, and 25 nodes

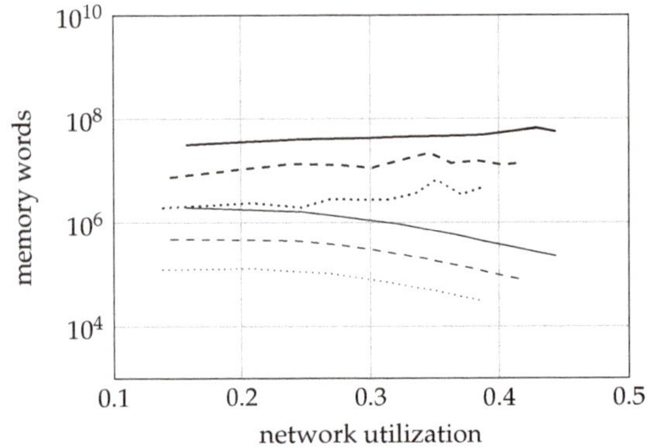

(c) memory used for $\gamma = 10$, and 50 nodes

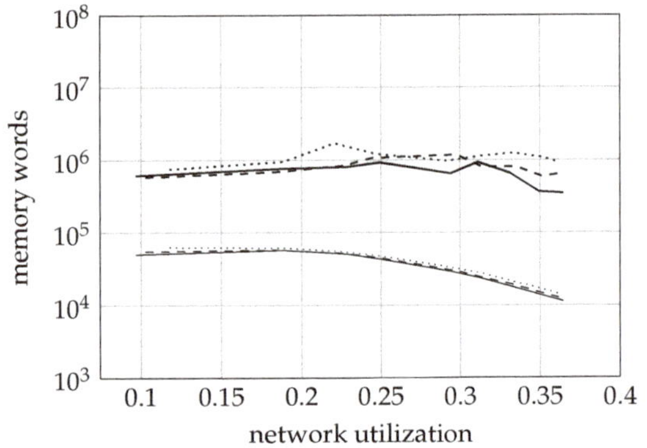

(d) memory used for $\gamma_p = 10\%$, and 50 nodes

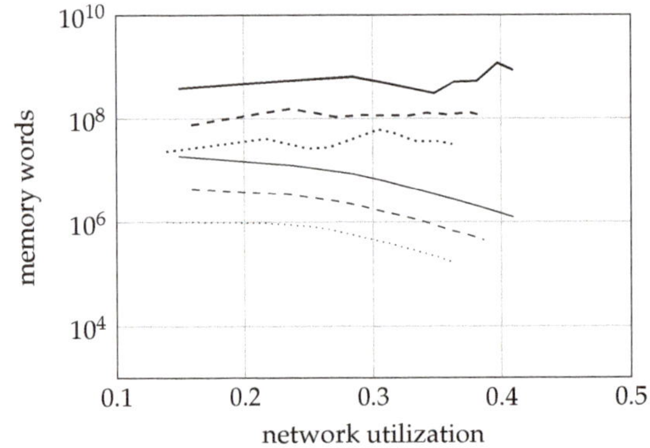

(e) memory used for $\gamma = 10$, and 100 nodes

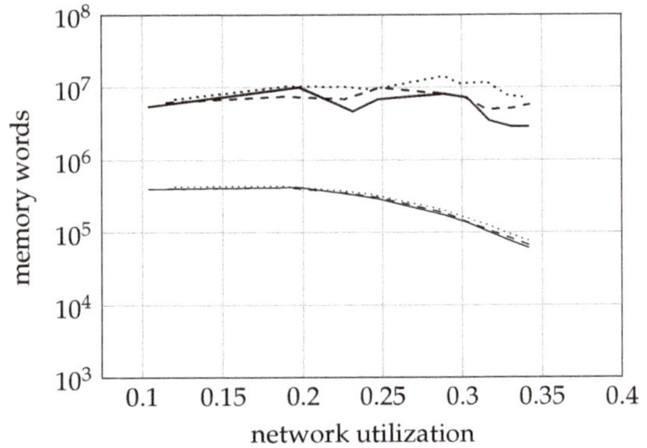

(f) memory used for $\gamma_p = 10\%$, and 100 nodes

mean, 160 units	mean, 320 units	mean, 640 units
maximum, 160 units	maximum, 320 units	maximum, 640 units

Figure 5. The sample means and maxima of the memory used by the proposed algorithm.

The memory results report the number of 64-bit memory words used by the permanent solutions, the tentative solutions, and the priority queue. The network size, $|\Omega|$, and γ affected the memory results similar to how they affected the time results. For the networks with 25 nodes, the mean number of words was about 10^5, while, for the networks with 100 nodes, about 10^9.

The memory used for $\gamma = 10$ is far larger than for $\gamma_p = 10\%$ because the spectrum (the available units) is more fragmented (since it is allocated in smaller fragments), and the algorithm finds more solutions as the search space is larger. Finding more solutions requires more time: simulations for $\gamma = 10$ take more time that the simulations for $\gamma_p = 10\%$.

To examine how the incomparable permanent and tentative labels, and the elements of the priority queue contribute to the memory usage, Figure 6 shows as stack plots the maximal memory used by the proposed algorithm for the networks of 25, 50, and 100 nodes with 320 units, and $\gamma = 10$.

The permanent labels take about 80% of memory, the tentative labels about 20%, and the elements of the priority queue take only a small fraction. We assumed that a label takes 15 64-bit words (implementation details: 4 words for a shared pointer, 1 word for a vertex pair, 4 words for a path trait, 2 words for an edge, 4 words for a shared pointer to the parent node in the search tree). An element of a priority queue is two words long (1 word for cost, 1 word for a weak pointer to its tentative label).

Most of the memory required by the algorithm is consumed by the permanent labels because of the large search space. A permanent label stores an optimal solution, and the results suggest that there are many of them for large networks. Furthermore, that large number of permanent labels helps to keep the number of tentative labels relatively much smaller through the edge relaxation.

To further validate the proposed algorithm, Figure 7 shows, for all populations of interest, the mean bandwidth blocking probabilities of the proposed algorithm as thin curves and of the edge-exclusion algorithm as thick curves. The figure has two rows and three columns of subfigures. The first row shows the results for $\gamma = 10$, and the second for $\gamma_p = 10\%$. The first column shows the results for the networks with 25 nodes, the second for 50 nodes, and the third for 100 nodes. We do not plot the error bars representing the standard error, since they were too small to plot.

Since the proposed algorithm can be exact, and the edge-exclusion algorithm is heuristic, the proposed algorithm should perform better, and indeed this is so. Interestingly, the edge-exclusion algorithm (which uses the generic Dijkstra algorithm) performs very well, at only about 5% worse.

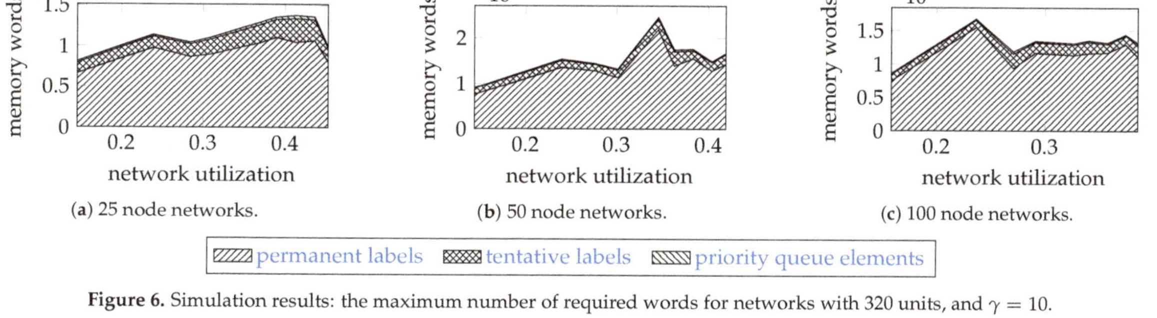

Figure 6. Simulation results: the maximum number of required words for networks with 320 units, and $\gamma = 10$.

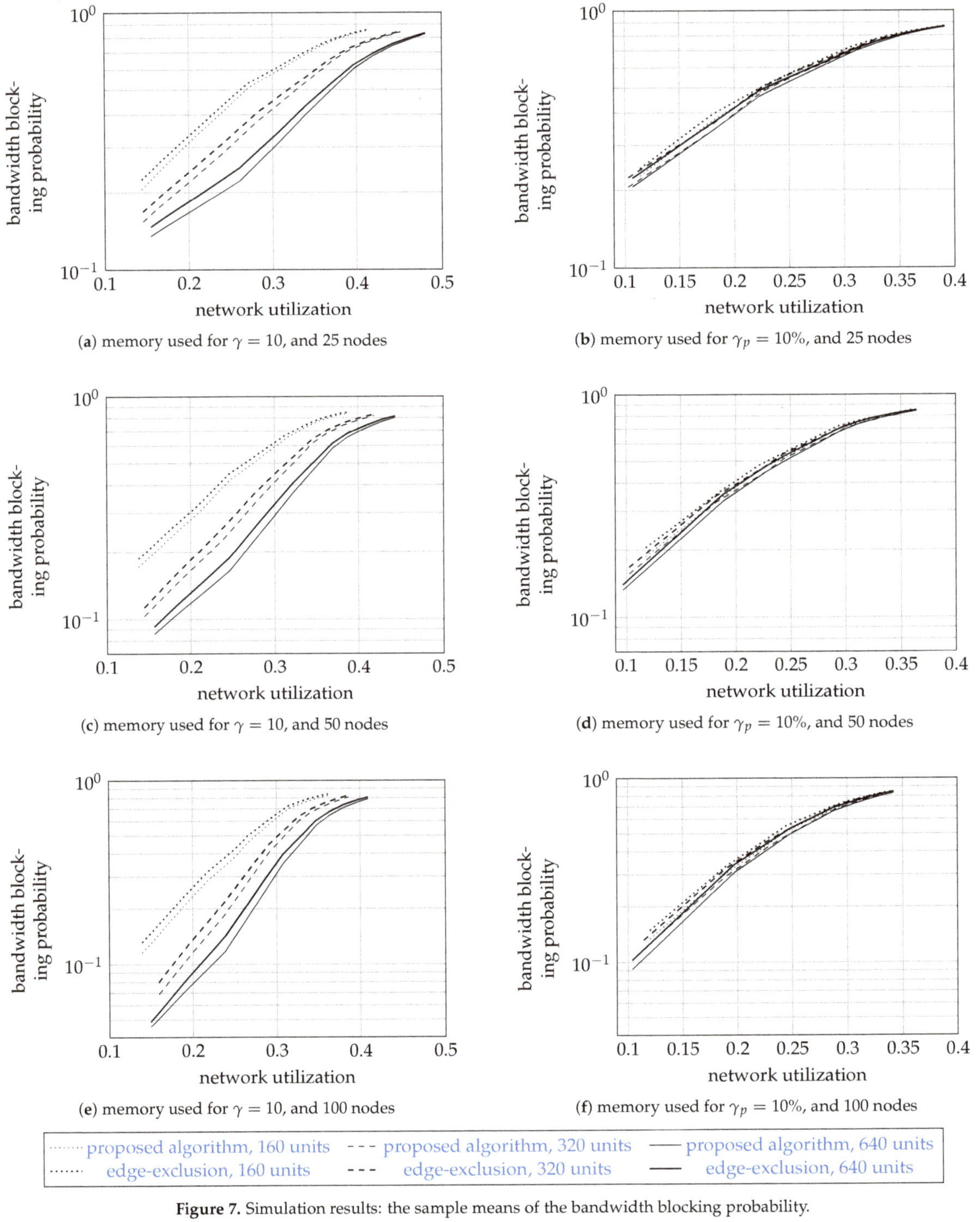

(**a**) memory used for $\gamma = 10$, and 25 nodes

(**b**) memory used for $\gamma_p = 10\%$, and 25 nodes

(**c**) memory used for $\gamma = 10$, and 50 nodes

(**d**) memory used for $\gamma_p = 10\%$, and 50 nodes

(**e**) memory used for $\gamma = 10$, and 100 nodes

(**f**) memory used for $\gamma_p = 10\%$, and 100 nodes

Figure 7. Simulation results: the sample means of the bandwidth blocking probability.

We did not add the edge-exclusion algorithm to the time and memory performance comparison, since it is a heuristic algorithm with the worst-case computational complexity of the dynamic routing problem without DPP, i.e., $O(|\Omega|^2|V|log|V|)$ [6]. In addition, we were unable to add the brute-force algorithm to the comparison because of its exponential complexity.

Admittedly, the reported time and memory consumption of the proposed algorithm seems large: for a network of 100 nodes and 640 units, the algorithm can run even for thousands of seconds, and use even 10 GBs of operating memory. However, to put these results in perspective, we report that, for a *far smaller* network of 15 nodes and 640 units, the brute-force algorithm ran for thousands of seconds, and it requested more than 120 GBs of operating memory in the corroborative runs.

6. Conclusions

The proposed algorithm is capable of solving various dynamic routing problems with dedicated path protection in optical networks, but not all of them, e.g., the algorithm cannot be applied when signal regeneration or spectrum conversion are used. However, the proposed algorithm can solve those routing problems that meet the minimal requirements of the stated research problem. If not, then perhaps the proposed algorithm and its novel principles could be used as a basis for devising more capable algorithms.

The proposed algorithm can also be used to find a pair of paths to different (primary, and secondary) data centers. The algorithm could even be useful in routing with inverse multiplexing, and multipath routing.

We are unable to compare the performance results of the proposed algorithm to some efficient and exact algorithm because, to the best of our knowledge, *no such competing algorithm exists*. For that large problem size, we could not have used the existing exact methods (e.g., the brute-force algorithm, integer linear programming) since they are inefficient, nor could we have used the existing efficient methods (e.g., the edge-exclusion algorithm, the tabu search), since they are suboptimal.

Dedicated path protection can be implemented at the multiplex section (fiber protection), the optical layer (optical signal protection), or the digital layer (the digital signal protection). We presented our algorithm in the setting of the optical signal protection, where we take into account the spectrum continuity and contiguity constraints, but the same principles could be used for other layers (e.g., the Internet protocol layer with multiprotocol label switching) and networks, too.

The algorithm can be adapted for further constraints, e.g., node-disjoint paths or the same spectrum fragments on both paths. Node-disjoint paths can be found if we do not relax the edges of the search graph that leave vertex $(v_{x,1}, v_{x,2})$ when $v_{x,1} == v_{x,2}$. The same spectrum fragment on both paths can be enforced by making sure during edge relaxation that the intersection of two fragments meet the requirements of a demand.

Future work could concentrate on applying the algorithm to related problems (e.g., establishing protected content-oriented connections to data centers), and further improving its performance with parallel computing.

Furthermore, perhaps the principle of the incomparability of solutions could be applied to the path augmentation technique, thus making, e.g., the Suurballe algorithm, even faster than the proposed algorithm.

The provided implementation does not require proprietary software, is implemented in modern C++ using the Boost Graph Library, and with modern functionality, such as concepts, smart pointers, in-place object creation, and move semantics. The implementation can be used to replicate the presented results, as well as stress-test the proposed algorithm.

Author Contributions: Conceptualization, I.S. and I.O.; methodology, I.S., I.O. and B.W.-S.; software, I.S.; validation, I.S., I.O. and B.W.-S., writing—original draft preparation, I.S.; writing—review and editing, I.S., I.O. and B.W.-S. All authors have read and agreed to the published version of the manuscript.

Funding: This research was funded by the Polish Ministry of Science and Higher Education grant number 020/RID/2018/19.

Institutional Review Board Statement: Not applicable.

Informed Consent Statement: Not applicable.

Acknowledgments: We ran the simulations using PL-Grid, the Polish supercomputing infrastructure.

Conflicts of Interest: The authors declare no conflict of interest.

References

1. Shen, G.; Guo, H.; Bose, S.K. Survivable elastic optical networks: Survey and perspective. *Photonic Netw. Commun.* **2016**, *31*, 71–87. [CrossRef]
2. Goścień, R.; Walkowiak, K.; Klinkowski, M.; Rak, J. Protection in elastic optical networks. *IEEE Netw.* **2015**, *29*, 88–96. [CrossRef]
3. Simmons, J.M. *Optical Network Design and Planning*; Optical Networks; Springer: New York, NY, USA, 2014.
4. Gerstel, O.; Jinno, M.; Lord, A.; Yoo, S. Elastic optical networking: A new dawn for the optical layer? *IEEE Commun. Mag.* **2012**, *50*, s12–s20. [CrossRef]
5. Szcześniak, I. The Implementation of the Efficient and Optimal Algorithm for the Dynamic Dedicated Path Protection. 2019. Available online: http://www.irkos.org/ddpp (accessed on 26 August 2021).
6. Szcześniak, I.; Jajszczyk, A.; Woźna-Szcześniak, B. Generic Dijkstra for optical networks. *IEEE/OSA J. Opt. Commun. Netw.* **2019**, *11*, 568–577. [CrossRef]
7. Andersen, R.; Chung, F.; Sen, A.; Xue, G. On disjoint path pairs with wavelength continuity constraint in WDM networks. In Proceedings of the IEEE INFOCOM 2004, Hong Kong, China, 7–11 March 2004; pp. 524–535.
8. Kishi, Y.; Kitsuwan, N.; Ito, H.; Chatterjee, B.C.; Oki, E. Modulation-Adaptive Link-Disjoint Path Selection Model for 1 + 1 Protected Elastic Optical Networks. *IEEE Access* **2019**, *7*, 25422–25437. [CrossRef]
9. Christodoulopoulos, K.; Kokkinos, P.; Varvarigos, E.M. Indirect and direct multicost algorithms for online impairment-aware RWA. *Trans. Netw.* **2011**, *19*, 1759–1772. [CrossRef]
10. Wang, X.; Kuang, K.; Wang, S.; Xu, S.; Liu, H.; Liu, G.N. Dynamic routing and spectrum allocation in elastic optical networks with mixed line rates. *J. Opt. Commun. Netw.* **2014**, *6*, 1115–1127. [CrossRef]
11. Yang, L.; Nan, H.; Xiaoping, Z.; Hanyi, Z.; Bingkun, Z. Polynomial-time adaptive routing algorithm based on spectrum scan in dynamic flexible optical networks. *China Commun.* **2013**, *10*, 49–58. [CrossRef]
12. Liu, Y.; Hua, N.; Wan, X.; Zheng, X.; Liu, Z. A spectrum-scan routing scheme in flexible optical networks. In Proceedings of the 2011 Asia Communications and Photonics Conference and Exhibition, Shanghai, China, 13–16 November 2011; pp. 1–6.
13. Shen, G.; Bose, S.; Cheng, T.; Lu, C.; Chai, T. Efficient heuristic algorithms for light-path routing and wavelength assignment in WDM networks under dynamically varying loads. *Comput. Commun.* **2001**, *24*, 364–373. [CrossRef]
14. Wang, C.; Shen, G.; Bose, S.K. Distance adaptive dynamic routing and spectrum allocation in elastic optical networks with shared backup path protection. *J. Light. Technol.* **2015**, *33*, 2955–2964. [CrossRef]
15. Chen, C.; Banerjee, S. A new model for optimal routing in all-optical networks with scalable number of wavelength converters. In Proceedings of the GLOBECOM '95, Singapore, 14–16 November 1995; Volume 2, pp. 993–997.
16. Hsu, C.F.; Chang, Y.C.; Sie, S.C. Graph-model-based dynamic routing and spectrum assignment in elastic optical networks. *J. Opt. Commun. Netw.* **2016**, *8*, 507–520. [CrossRef]
17. Ehrgott, M.; Gandibleux, X. A survey and annotated bibliography of multiobjective combinatorial optimization. *OR Spektrum* **2000**, *22*, 425–460. [CrossRef]
18. Varvarigos, E.M.; Sourlas, V.; Christodoulopoulos, K. Routing and scheduling connections in networks that support advance reservations. *Comput. Netw.* **2008**, *52*, 2988–3006. [CrossRef]
19. Mieghem, P.V.; Kuipers, F.A. Concepts of exact QoS routing algorithms. *IEEE/ACM Trans. Netw.* **2004**, *12*, 851–864.
20. Wang, Z.; Crowcroft, J. Quality-of-service routing for supporting multimedia applications. *IEEE J. Sel. Areas Commun.* **1996**, *14*, 1228–1234. [CrossRef]
21. Hansen, P. Bicriterion path problems. In *Multiple Criteria Decision Making Theory and Application*; Lecture Notes in Economics and Mathematical Systems; Springer: Berlin/Heidelberg, Germany, 1980; Volume 177, pp. 109–127.
22. Martins, E.Q.V. On a multicriteria shortest path problem. *Eur. J. Oper. Res.* **1984**, *16*, 236–245. [CrossRef]
23. Tarapata, Z. Selected multicriteria shortest path problems: An analysis of complexity, models and adaptation of standard algorithms. *Int. J. Appl. Math. Comput. Sci.* **2007**, *17*, 269–287. [CrossRef]
24. Suurballe, J.W. Disjoint paths in a network. *Networks* **1974**, *4*, 125–145. [CrossRef]
25. Bhandari, R. *Survivable Networks: Algorithms for Diverse Routing*; Kluwer Academic Publishers: Boston, MA, USA, 1999.
26. Ahuja, R.K.; Magnanti, T.L.; Orlin, J.B. *Network Flows: Theory, Algorithms, and Applications*; Prentice Hall: Englewood Cliffs, NJ, USA, 1993.
27. Szcześniak, I.; Woźna-Szcześniak, B. Adapted and constrained Dijkstra for elastic optical networks. In Proceedings of the 2016 International Conference on Optical Network Design and Modeling (ONDM), Cartagena, Spain, 9–12 May 2016; pp. 1–6. [CrossRef]

28. Cetinkaya, E.; Alenazi, M.; Cheng, Y.; Peck, A.; Sterbenz, J. On the fitness of geographic graph generators for modelling physical level topologies. In Proceedings of the 2013 5th International Congress on Ultra Modern Telecommunications and Control Systems and Workshops (ICUMT), Almaty, Kazakhstan, 10–13 September 2013; pp. 38–45. [CrossRef]
29. Wan, X.; Hua, N.; Zheng, X. Dynamic routing and spectrum assignment in spectrum-flexible transparent optical networks. *J. Opt. Commun. Netw.* **2012**, *4*, 603–613. [CrossRef]

entropy

MDPI

Article

ECG Signal Classification Using Deep Learning Techniques Based on the PTB-XL Dataset

Sandra Śmigiel [1,*], Krzysztof Pałczyński [2] and Damian Ledziński [2]

[1] Faculty of Mechanical Engineering, UTP University of Science and Technology in Bydgoszcz, 85-796 Bydgoszcz, Poland
[2] Faculty of Telecommunications, Computer Science and Electrical Engineering, UTP University of Science and Technology in Bydgoszcz, 85-796 Bydgoszcz, Poland; krzysztof@palczynski.com.pl (K.P.); damian.ledzinski@utp.edu.pl (D.L.)
* Correspondence: sandra.smigiel@utp.edu.pl; Tel.: +48-52-340-8346

Abstract: The analysis and processing of ECG signals are a key approach in the diagnosis of cardiovascular diseases. The main field of work in this area is classification, which is increasingly supported by machine learning-based algorithms. In this work, a deep neural network was developed for the automatic classification of primary ECG signals. The research was carried out on the data contained in a PTB-XL database. Three neural network architectures were proposed: the first based on the convolutional network, the second on SincNet, and the third on the convolutional network, but with additional entropy-based features. The dataset was divided into training, validation, and test sets in proportions of 70%, 15%, and 15%, respectively. The studies were conducted for 2, 5, and 20 classes of disease entities. The convolutional network with entropy features obtained the best classification result. The convolutional network without entropy-based features obtained a slightly less successful result, but had the highest computational efficiency, due to the significantly lower number of neurons.

Keywords: ECG signal; classification; PTB-XL; deep learning

Citation: Śmigiel, S.; Pałczyński, K.; Ledziński, D. ECG Signal Classification Using Deep Learning Techniques Based on the PTB-XL Dataset. *Entropy* **2021**, *23*, 1121. https://doi.org/10.3390/e23091121

Academic Editor: Ernestina Menasalvas

Received: 5 July 2021
Accepted: 25 August 2021
Published: 28 August 2021

Publisher's Note: MDPI stays neutral with regard to jurisdictional claims in published maps and institutional affiliations.

1. Introduction

According to publicly available reports, cardiovascular disease remains the leading cause of mortality worldwide [1]. One of the main causes of cardiovascular diseases is cardiac arrhythmia, in which the heartbeat deviates from typical beating patterns [2]. However, there are many types of irregular heartbeat. Accurate classification of heart disease types can aid in diagnosis and treatment [3].

An electrocardiogram (ECG) is a widely used, reliable, noninvasive approach to diagnosing cardiovascular disease. The standard ECG consists of 12 leads [4]. Traditionally, ECG results are manually interpreted by cardiologists based on a set of diagnosis criteria and experience. However, manual interpretation is time consuming and requires skill. Incorrectly interpreted ECG results may give rise to incorrect clinical decisions and lead to a threat to human life and health. With the rapid development of ECG and, at the same time, an insufficient number of cardiologists, the accurate and automatic diagnosis of ECG signals has become an interesting research topic for many scientists.

Over the past decade, numerous attempts have been made to identify a 12-lead clinical ECG, largely on the basis of the availability of large, public, open-source ECG data collections. Previous literature on ECG databases has shown a methodological division: signal processing and machine learning [5,6]. On the one hand, digital signal processing methods mainly include low- or high-pass filters, fast Fourier transform, and wavelet transform [7]. In this area, many algorithms are based on three processes: feature extraction, feature selection, and classification [8]. On the other hand, an alternative method is the application of machine learning methods. Such an application would primarily focus on the automatic recognition of patterns that classify various disease entities, a method that is gaining greater importance in medical practice.

Algorithms known as deep neural networks have become particularly important in the last five years. Deep learning models have proven to be useful in increasing the effectiveness of diagnoses of cardiovascular diseases using ECG signals. By using the cascade of heterogeneous layers of neural networks to gradually extract increasingly high-level features, they lead to ever-improving neural networks built on their basis. Deep neural networks are reaching their zenith in various areas where artificial intelligence algorithms are applied.

In recent years, machine learning models have given rise to huge innovations in many areas, including image processing, natural language processing, computer games, and medical applications [9]. To date, however, the lack of adequate databases, well-defined assessment procedures, and unambiguous labels identifying signals has limited the possibilities for creating an automatic interpretation algorithm for the ECG signal. Known databases provided by PhysioNet, such as the MIT-BIH Arrhythmia Database and the PTB Diagnostic ECG Database, were deemed insufficient [10,11]. Data from single, small, or relatively homogeneous datasets, further limited by a small number of patients and rhythm episodes, prevented the creation of algorithms in machine learning models.

The work of the PhysioNet/Computing in Cardiology Challenge 2020 project to develop an automated ECG classifier provided an opportunity to address this problem by adding data from a wide variety of sources. Among these, there are numerous works, including the development of a comprehensive deep neural network model for the classification of up to 27 clinical diagnoses from the electrocardiogram. The authors of one of these achieved results, using the ResNet model, at the level of AUC = 0.967 and ACC = 0.43 in their study [12]. A similar approach was proposed [13], using the SE_ResNet model to improve the efficiency of the classification of various ECG abnormalities. Others, focusing on the comparative analysis of the recently published PTB-XL dataset, assessed the possibility of using convolutional neural networks, in particular those based on the ResNet and Inception architectures [14]. A different approach in the classification of cardiovascular diseases was demonstrated by the authors of a work [15] related to the detection of QRS complexes and T & P waves, together with the detection of their boundaries. The ECG classification algorithm was based on 19 classes. Features were extracted from the averaged QRS and from the intervals between the detected points.

The 12-lead ECG deep learning model found its reference mainly to ECG diagnosis in the automatic classification of cardiac arrhythmias. A deep learning model trained on a large ECG dataset was used with a deep neural network [16] based on 1D CNN for automatic multilabel arrhythmia classification with a score of ACC = 0.94 − 0.97. The authors of this study also conducted experiments on single-lead ECG with an analysis of the operation of every single lead. The subject of arrhythmia classification is also of interest to other authors [17], where, with the use of long-short term memory (LSTM), a model with an LSTM score of 0.6 was proposed. The choice of ECG for arrhythmia detection was undertaken by the authors of the paper [18], where they designed a computer-aided diagnosis system for the automatic diagnosis of four types of serious arrhythmias. In this approach, the ECG was analyzed using thirteen nonlinear features, known as entropy. The features extracted in this way were classified using ANOVA and subjected to automated classification using the K-nearest neighbor and decision tree classifiers. The obtained results were for KNN − ACC = 93.3% and DT − ACC = 96.3%. Various deep learning models for the examination of the ECG signal have also been proposed for atrial fibrillation, obtaining the result of ACC = 0.992 [19]. It is worth noting that the presented model successfully detected atrial fibrillation, and the tests were carried out with the use of various ECG signals. Attempts to investigate cardiac arrhythmias and cardiovascular diseases were also carried out in a new convolutional neural network [9] with a nonlocal convolutional block attention module (NCBAM), which focused on representative features along space, time, and channels. For the classification problem of ECG arrhythmia detection, the authors obtained AUC = 0.93. The approach to convolutional neural networks, the possibilities and usability of tools, and the analysis of biomedical signals were also proposed by the authors of other papers [20]. The research included the implementation of a multilabel classification algorithm with the use of machine learning methods based on a CNN. The work described the details of the algorithm necessary for reconstruction and presented limitations and suggestions for improvement. A different approach to the ECG signal was presented by the authors of [21], where the focus was instead placed on processing the ECG signal, data sampling, feature extraction, and classification. They used a deep learning class model with gated recursive complex (GRU) and extreme learning machine (ELM) to recognize the ECG signal.

The aim of the study was to check the effectiveness of multiclass classification of ECG signals with the use of various neural network architectures. An additional aim was to test the effectiveness of very light nets for classification. A novelty in the article is the combination of a neural network with entropy-based features.

2. Materials and Methods

2.1. PTB-XL Dataset

In this article, data from the PTB-XL ECG database were used [11]. The PTB-XL database is a clinical ECG dataset of unprecedented size, with changes applied to evaluate machine learning algorithms. The PTB-XL ECG dataset contains 21,837 clinical 12-lead ECGs from 18,885 patients of 10 s in length, sampled at 500 Hz and 100 Hz with 16 bit resolution. Figure 1 shows examples of rhythms, consistent with the data contained in Table 1, which were used in the work. Among them there are examples of the following ECG signals: NORM—normal ECG, CD—myocardial infarction, STTC—ST/T change, MI—conduction disturbance, HYP—hypertrophy.

Table 1. The numbers of individual classes.

Number of Records	Class	Description
7185	NORM	Normal ECG
3232	CD	Myocardial Infarction
3064	STTC	ST/T Change
2936	MI	Conduction Disturbance
815	HYP	Hypertrophy

The PTB-XL database is gender balanced. The data included were derived from 52% males and 48% females, ranging in age from 2 to 95 years (median 62). The data were enriched with additional information about the patient (age, sex, height, weight). Each ECG by the authors of the dataset was classified into one or more of 23 diagnostic subclasses in 5 diagnostic classes, or into classes that are not diagnostic classes. Each class was assigned a probability. Classes are marked according to the standard with the codes SCP_ECG.

The research methodology included classification studies carried out in 3 categories of binary classifications, where the classes were NORM (healthy patient) and all other classes (sick patient), where 5 diagnostic classes were used and where 20 diagnostic subclasses were used.

The research methodology was as follows (Figure 2): Data from the PTB-XL database were filtered and then divided into training, validation, and test groups. These data were then normalized and used as inputs for the neural networks that were examined. The network performed a classification. The signal class was obtained as an output, and this was then subjected to evaluation.

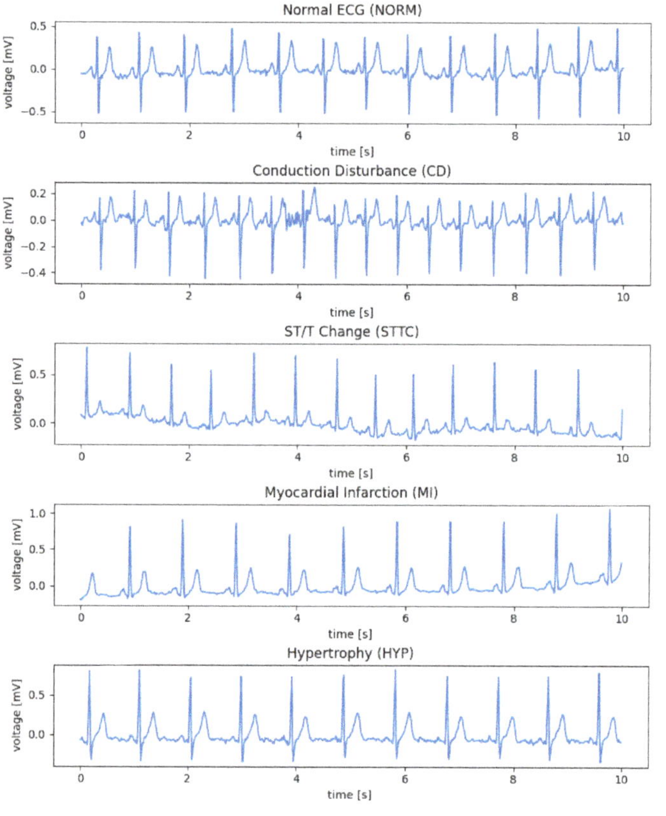

Figure 1. Examples of rhythm ECG signals.

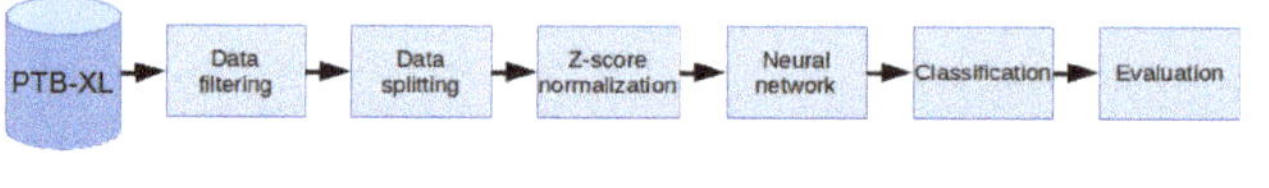

Figure 2. General overview diagram of the method.

During the filtering stage, a set of 21,837 ECG records from the PTB-XL database was included in the simulation. ECGs not classified into diagnostic classes were filtered from the dataset. Subsequently, the ECGs in which the probability of classification was less than 100% were filtered out. In the next stage, ECGs were filtered out of those subclasses whose presence in the dataset was less than 20. A sampling frequency of 100 Hz was selected for the study, with 10 s as the length.

The dataset was divided into training, validation, and test sets in proportions of 70%, 15%, and 15%, respectively. The training set was used to train the network; the validation set was used to select the model; the test set was used to test the network's effectiveness.

As a result of the above activities, a total of 17,232 ECG records were used for the experimental analysis (Figure 3).

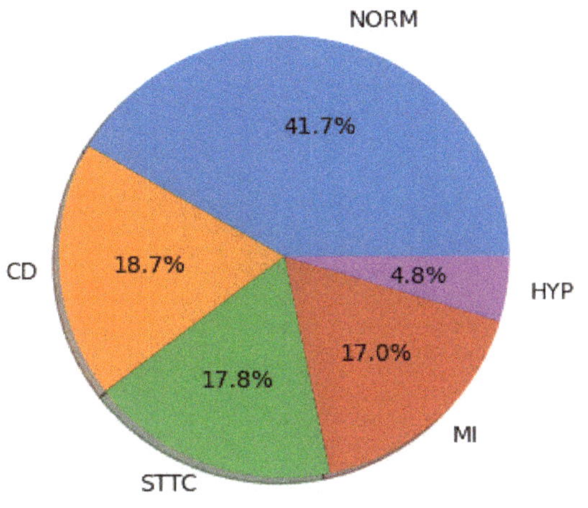

Figure 3. Diagnostic classes used in the study.

A detailed summary of the size of the individual classes used in the study and resulting from the above-described activities on the basis of PTB-XL is presented in Tables 1 and 2. The tables show the number of individual records used in the study, assigned to the appropriate diagnostic classes and subclasses defining cardiovascular diseases sorted by number of records.

Table 2. Numbers of individual subclasses.

Number of Records	Subclass	Class	Description
7185	NORM	NORM	Normal ECG
1713	STTC	STTC	Non-diagnostic T abnormalities, suggests digitalis effect, long QT interval, ST-T changes compatible with ventricular aneurysm, compatible with electrolyte abnormalities
1636	AMI	MI	Anterior myocardial infarction, anterolateral myocardial infarction, in anteroseptal leads, in anterolateral leads, in lateral leads
1272	IMI	MI	Inferior myocardial infarction, inferolateral myocardial infarction, inferoposterolateral myocardial infarction, inferoposterior myocardial infarction, in inferior leads, in inferolateral leads
881	LAFB/LPFB	CD	Left anterior fascicular block, left posterior fascicular block
798	IRBBB	CD	Incomplete right bundle branch block
733	LVH	HYP	Left ventricular hypertrophy
527	CLBBB	CD	(Complete) left bundle branch block
478	NST_	STTC	Nonspecific ST changes
429	ISCA	STTC	In anterolateral leads, in anteroseptal leads, in lateral leads, in anterior leads
385	CRBBB	CD	(Complete) right bundle branch block
326	IVCD	CD	Nonspecific intraventricular conduction disturbance
297	ISC_	STTC	Ischemic ST-T changes
204	_AVB	CD	First-degree AV block, second-degree AV block, third-degree AV block
147	ISCI	STTC	In inferior leads, in inferolateral leads
67	WPW	CD	Wolff–Parkinson–White syndrome
49	LAO/LAE	HYP	Left atrial overload/enlargement
44	ILBBB	CD	Incomplete left bundle branch block
33	RAO/RAE	HYP	Right atrial overload/enlargement
28	LMI	MI	Lateral myocardial infarction

2.2. Designed Network Architectures

This research compared three neural networks (convolutional network, SincNet, convolutional network with entropy features) in terms of the correct classification of the ECG signal. The research consisted of the implementation and testing of the proposed models of the neural networks. Cross-entropy loss as a loss function was applied to all networks.

The artificial neural networks proposed in this article were based on layers performing one-dimensional convolutions. This is a state-of-the-art solution in signal processing using deep learning due to its ability to extract features based on changes in consecutive samples, while simultaneously being faster and easier to train than recurrent layers such as LSTMs. The convolutional networks described in this article also contain residual connections between convolutional layers as described in [22]. These shortcut connections eliminate the so-called vanishing gradient problem and increase the capacity of models for better representation learning.

The networks were trained using the Adam optimizer as described in [23]. The optimizer trained the neural network using mini-batches of 128 examples in one pass. The learning rate was set at 0.001 at the beginning of the training and was later adjusted to 0.0001 to perform final corrections before ending the training. To prevent overfitting, early stopping was employed as described in [24]. The training of the neural network was stopped as soon as the network was unable to obtain better results on the validation dataset. This was to prevent overfitting. Following testing, the neural network was trained on the test dataset.

The tests were carried out using hardware configurations on a dual-Intel Xeon Silver 4210R, 192 GB RAM, and Nvidia Tesla A100 GPU. In this research, PyTorch and Jupyter Lab programming solutions were used for the implementation of the neural networks.

2.2.1. Convolutional Network

The first network examined is presented in Figure 4. It consists of five layers of one-dimensional convolutions with LeakyReLU activation functions and one fully connected layer with a softmax activation function. The network accepts ECG signals consisting of 12 channels containing 1000 samples each as inputs and outputs a class distribution vector normalized by application of the softmax function. The network determines the class to which an input signal belongs by determining the index of the vector maximum value. The class represented by this index is considered as a class of the input signal.

LeakyReLU was used instead of basic ReLU to preserve gradient loss in neurons outputting negative values. The coefficient describing a negative slope was set to 0.01; thus, the activation function can be described by the equation below:

$$f(x) = \begin{cases} 0.01x & \text{for } x < 0 \\ x & \text{for } x \geq 0 \end{cases} \tag{1}$$

This configuration was used in every network proposed in this article.

This architecture was tested on both the normalized signal taken from the dataset without any other transformations and a spectrogram, and the results obtained from the former were better than from the latter. The network computing the spectrogram interpreted each spectrogram as a multichannel one-dimensional signal. Each of the twelve signals' spectrograms was processed by five one-dimensional convolutional blocks with the LeakyReLU activation function. The results of the convolutions where aggregated by performing adaptive average pooling. Afterwards, the results of pooling were flattened to the format of a one-dimensional vector and processed by a fully connected layer with a softmax activation function, and the output was used as a vector describing the probability distribution of the input signals belonging to each of the defined classes.

This is a simplified architecture designed to achieve both better computation time and memory storage efficiency. This network design has only 6 layers and, depending on the number of classes in classification, has just 8882 weights for binary classification and

11,957 weights for detecting 5 different classes of signal. The last segment of the network is a fully connected layer, which has a number of neurons equal to the quantity of possible classes to which the signal may belong. As a result, the more granular the classification process is, the more neurons are required, which increases the number of total weights in the network. The addition of residual connections did not increase the performance of the network significantly, but enlarged the quantity of parameters and computational steps required to process the signal.

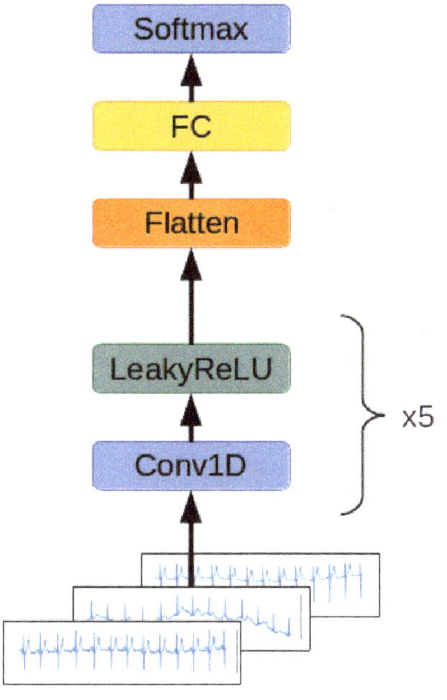

Figure 4. Convolutional network architecture. A twelve-channel ECG signal is passed through five subsequent one-dimensional convolutional layers with the LeakyReLU activation function. The results of the computation are flattened to the format of a one-dimensional vector. The results of the calculation are processed by a fully connected layer with a softmax activation function. The output value is a one-dimensional vector describing the probability distribution of the input signal belonging to each of the defined classes.

2.2.2. SincNet

The second examined network uses the SincNet layers described in [25]. SincNet layers are designed for the extraction of low-level features from a raw signal's data samples. SincNet layers train "wavelets" for feature extraction by performing convolution on the input signal:

$$y[n] = x[n] \cdot g[n, \theta] \quad (2)$$

where n is the index of the probe and θ are the parameters of the wavelets determined during training. The wavelet function g is described with the equation:

$$g[n, f_1, f_2] = 2f_2 sinc(2\pi f_2 n) - 2f_1 sinc(2\pi f_1 n) \quad (3)$$

where *sinc* function is defined as:

$$sinc(x) = \frac{sin(x)}{x} \quad (4)$$

f_1 and f_2 are the cutoff frequencies determined by the SincNet layer during the training phase and form a set of trainable parameters θ:

$$\theta = \{(f_{i,1}, f_{i,2}) | i \in C^+ \cap i \leq l\} \quad (5)$$

where l is the number of wavelets in the SincNet layer.

The pair of filters (f_1,f_2) are initialized using the frequencies used for calculation of Mel-frequency cepstral coefficients [26].

SincNet layers are designed to interpret only the signal's singular channel at once, so the second network's architecture consists of a subnetwork using a SincNet layer, which encodes each signal's channel separately. The features extracted by the subnetwork are concatenated into one feature vector, which is fed to a block of fully connected layers. The softmax layer serves the role of the output classification layer, while the SincNet subnetwork consists of the SincNet layer adjusting the wavelets to the raw signal, two convolutional layers with LeakyReLU activation functions and layer normalizations, and three fully connected layers with batch normalization and LeakyReLU activation functions (Figure 5).

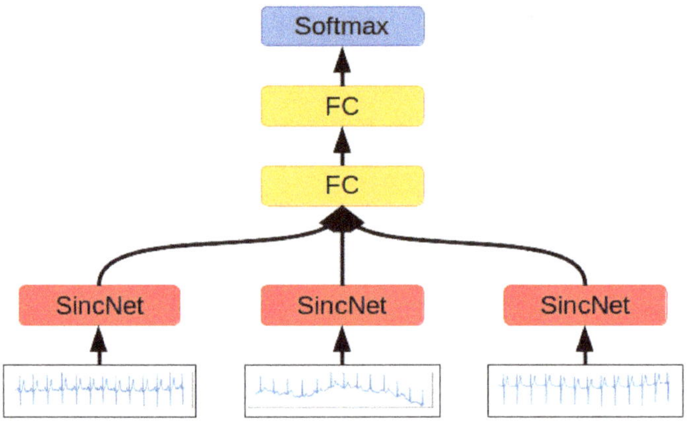

Figure 5. SincNet-based network architecture. Each channel of the 12-channel ECG signal is processed by a dedicated SincNet block. The results of each block are concatenated, flattened to the format of a one-dimensional vector, and used as an input for two subsequent fully connected layers, with LeakyReLU and softmax activation functions, respectively. The output value is a one-dimensional vector describing the probability distribution of the input signal belonging to each of the defined classes.

2.2.3. Convolutional Network with Entropy Features

The third network examined is presented in Figure 6. This network is an extended variant of the convolutional network. The network processes the ECG signal, and the values of the entropies are calculated for every channel of the signal. These entropies are:

- Shannon entropy—the summation of the informativeness of every possible state in the signal by measuring its probability. As a result, Shannon entropy is the measurement of the spread of the data [27];
- Approximate entropy—the measurement of series regularity. It provides information on how much the ECG fluctuates and its predictability [28];
- Sample entropy—an improvement on approximate entropy due to the lack of the signal length's impact on the entropy computations [28];
- Permutation entropy—the measurement of the order relations between ECG samples. This quantifies how regular and deterministic the signal is [29];

- Spectral entropy—the quantification of the energy spread uniformness across the frequency spectrum [30];
- SVD entropy—the measurement of how possible the dimensionality reduction of time series matrix is through factorization using the eigenvector approach;
- Rényi entropy—the generalization of the Shannon entropy by introducing the fractal order of the subsequent informativeness of each signal's state [31];
- Tsallis entropy—the generalization of the Boltzmann–Gibbs entropy, able to detect long-term memory effects on the signal [32];
- Extropy—the measurement of the amount of uncertainty represented by the distribution of the values in the observed ECG signal [33].

Granelo-Belinchon et al., in their article [34], stated that the tools of information theory can be straightforwardly applied to any nonstationary time process when considering small chunks of data spanning a short enough time range, allowing a slow evolution of higher-order moments to be neglected. The augmented Dickey–Fuller test has been conducted on ten-second-long training chunks of signals to determine the momentary stationarity of ECG signals. It turned out that 89.5% of tested signals were deemed stationarity in this small period of time, allowing the use of entropy methods for their interpretation.

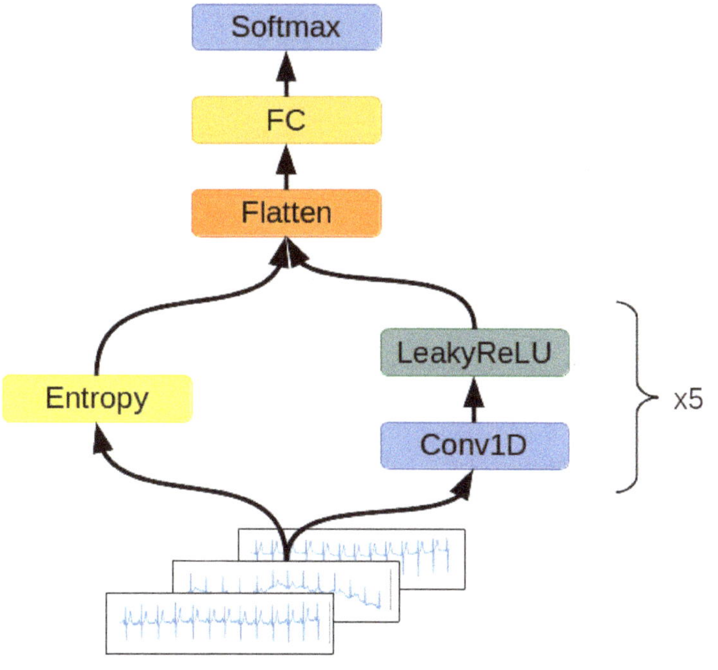

Figure 6. Convolutional network with the entropy features' block architecture. The computational graph of the network is made up of two branches. In the first branch, a twelve-channel ECG signal is passed through five subsequent one-dimensional convolutional layers with the LeakyReLU activation function. In the second branch, the input signal is used to compute the vector of entropies for every channel of the signal. The results of the computations from both branches are concatenated and flattened to the format of a one-dimensional vector. The results of the calculation are processed by a fully connected layer with softmax activation function. The output value is a one-dimensional vector describing the probability distribution of the input signal belonging to each of the defined classes.

The artificial neural network consists of two blocks: convolutional and fully connected. In the first step, a raw ECG signal is encoded by a convolutional block formed by five one-dimensional convolutional layers with the LeakyReLU activation function. Each layer has a stride parameter equal to 2 to reduce the number of samples representing the time vector. Each layer also has a residual connection with the original, raw signal. Because of the signal's sample reduction due to the applied stride parameter, the ECG signal for each step of the residual connection is shrunk by average pooling with a window size of 2.

The encoded raw ECG signal is concatenated with the values of the entropies of every channel. Such a feature vector is fed to three fully connected layers with LeakyReLU activation functions in the first two and a softmax function in the last layer. The result of the softmax function is the output vector of the network and is used in order to classify the signal. For regularization purposes, there was a dropout with a chance of zeroing the input equal to 20% applied before each layer. The dropout was turned off during the network's evaluation.

2.3. Metrics

The neural networks were evaluated using the metrics described below. For the purpose of the simplicity of the equations, certain acronyms were created, as follows: TP—true positive, TN—true negative, FP—false positive, FN—false negative. The metrics used for the network evaluation are:

- Accuracy: Acc = (TP + TN)/(TP + FP + TN + FN);
- Precision = TP/(TP + FP);
- Recall = TP/(TP + FN);
- F1 = 2 * precision * recall/(precision + recall);
- AUC—area under the curve, ROC—area under the receiver operating characteristic curve. The ROC is a curve determined by calculating TFP = true positive rate = TP/(TP + FN) and FPR = false positive rate = FP/(TN + FP). The false positive rate describes the x-axis and the true positive rate the y-axis of a coordinate system. By changing the threshold value responsible for the classification of an example as belonging to either the positive or negative class, pairs of TFP-FPR are generated, resulting in the creation of the ROC curve. The AUC is a measurement of the area below the ROC curve;
- Total Params—number of neurons in the network. The smaller this number, the better, as less computation is required in order to perform classification.

3. Results

The results of the networks based on the convolutional network, SincNet, and the convolutional network with entropy features are summarized in Tables 3–5. With the recognition of two classes, the network based on the convolutional network achieved 88.2% ACC and with five classes 72.0% ACC. Similarly, the network based on SincNet achieved 85.8% ACC with the recognition of two classes and 73.0% with the recognition of five classes. The network based on the convolutional network with entropy features achieved 89.82% ACC with the recognition of two classes and 76.5% with the recognition of five classes. The network based on the convolutional network turned out to be slightly better than that based on SincNet. The situation changed with the recognition of 20 classes, where SincNet turned out to be slightly more effective. However, the network based on the convolutional network with entropy features turned out to be the best in all cases. It is worth noting that, depending on the number of recognized classes, the convolutional network had 200–600-times less weight than the SincNet-based network, which means it is much lighter. Adding entropy-based features to the convolutional network increases its weight two- to seven-fold. The convolutional neural network with entropy features achieved the highest accuracy in every classification task, scoring 89.2%, 76.5%, and 69.8% for 2, 5, and 20 classes, respectively. The basic convolutional network achieved better accuracy than SincNet during the classification of two classes (healthy/sick), but SincNet performed better on the classification of five and twenty classes. As described by Ravanelli et al. in [25], the neural network was designed to process the human voice without any data preprocessing and did so successfully according to the authors. However, the results of its usage on ECG signals are far from ideal, as presented in Tables 3–5.

Table 3. The results of the convolutional network.

Number of Classes	Acc	Avg Precision	Avg Recall	Avg F1	Avg AUC	Total Params
2	0.882	0.879	0.882	0.88	0.953	8882
5	0.72	0.636	0.602	0.611	0.877	11,957
20	0.589	0.259	0.228	0.238	0.856	27,332

Table 4. The results of SincNet.

Number of Classes	Acc	Avg Precision	Avg Recall	Avg F1	Avg AUC	Total Params
2	0.858	0.855	0.854	0.855	0.93	6,109,922
5	0.73	0.666	0.589	0.6	0.884	6,109,922
20	0.593	0.287	0.269	0.262	0.807	6,269,204

Table 5. The results of the convolutional network with entropy features.

Number of Classes	Acc	Avg Precision	Avg Recall	Avg F1	Avg AUC	Total Params
2	0.892	0.889	0.893	0.891	0.96	58,178
5	0.765	0.714	0.662	0.68	0.910	58,259
20	0.698	0.355	0.339	0.332	0.815	58,664

Figures 7–15 show the confusion matrices of the results of the evaluated networks.

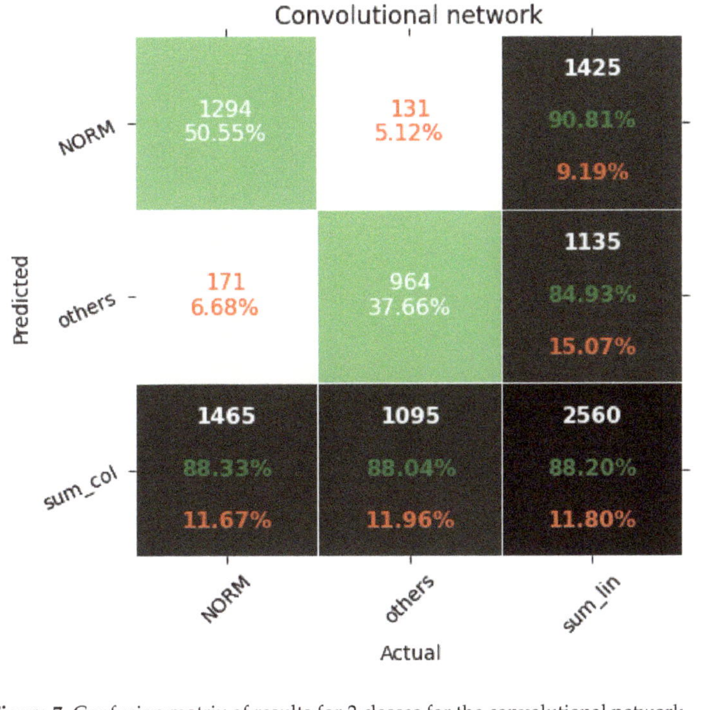

Figure 7. Confusion matrix of results for 2 classes for the convolutional network.

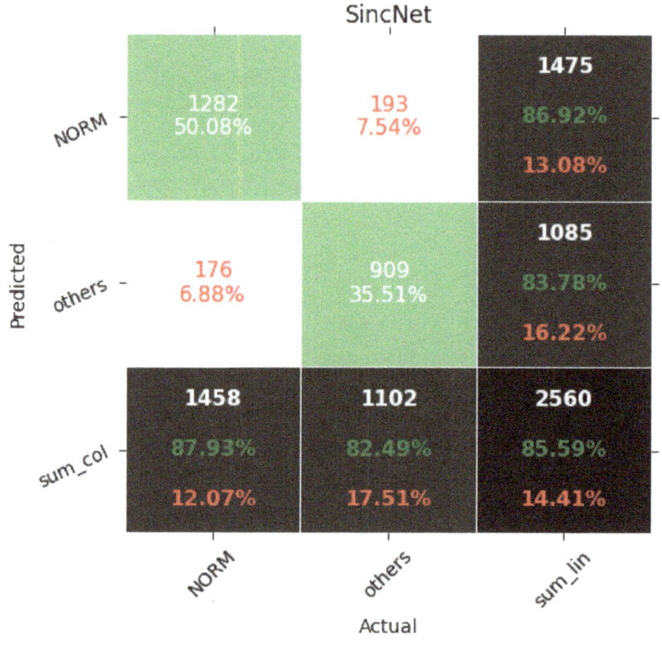

Figure 8. Confusion matrix of results for 2 classes for SincNet.

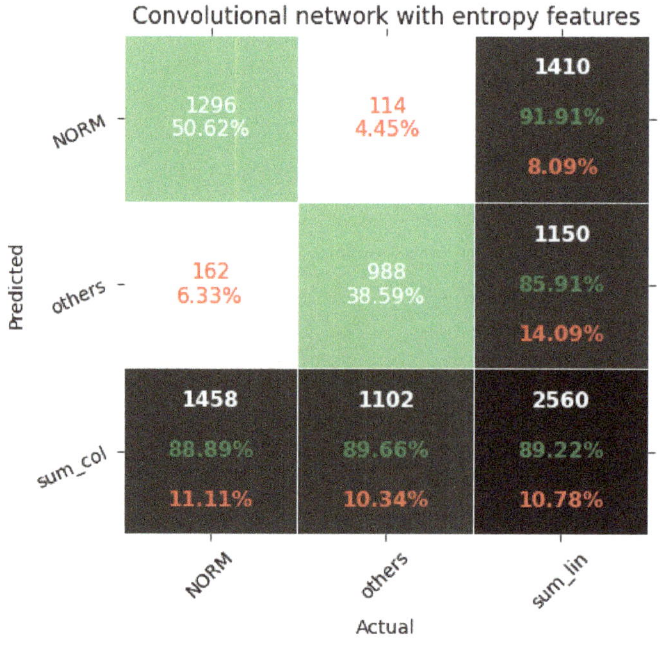

Figure 9. Confusion matrix of results for 2 classes for the convolutional network with entropy features.

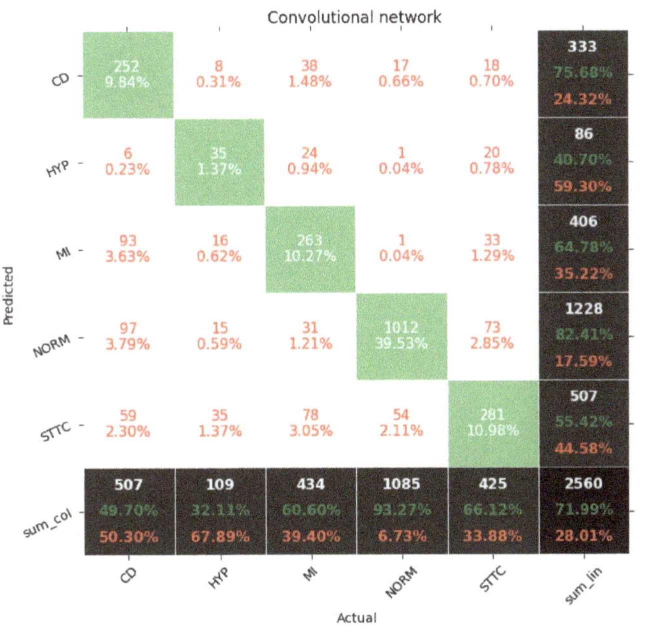

Figure 10. Confusion matrix of results for 5 classes for the convolutional network.

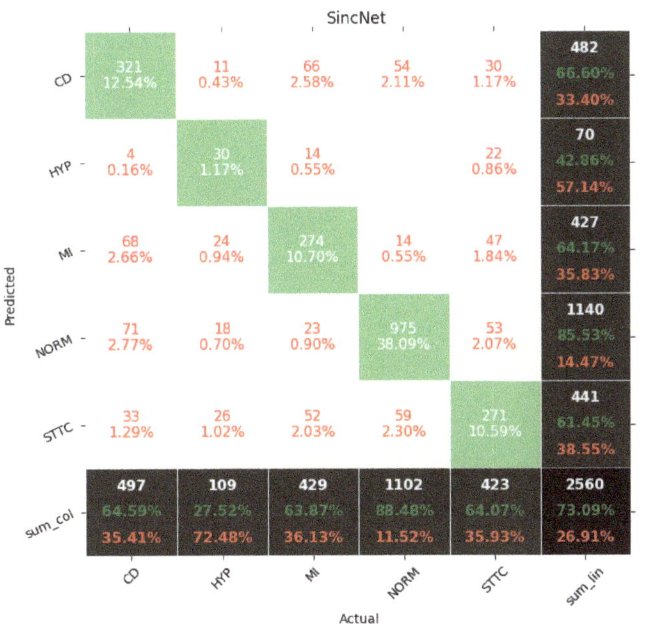

Figure 11. Confusion matrix of results for 5 classes for SincNet.

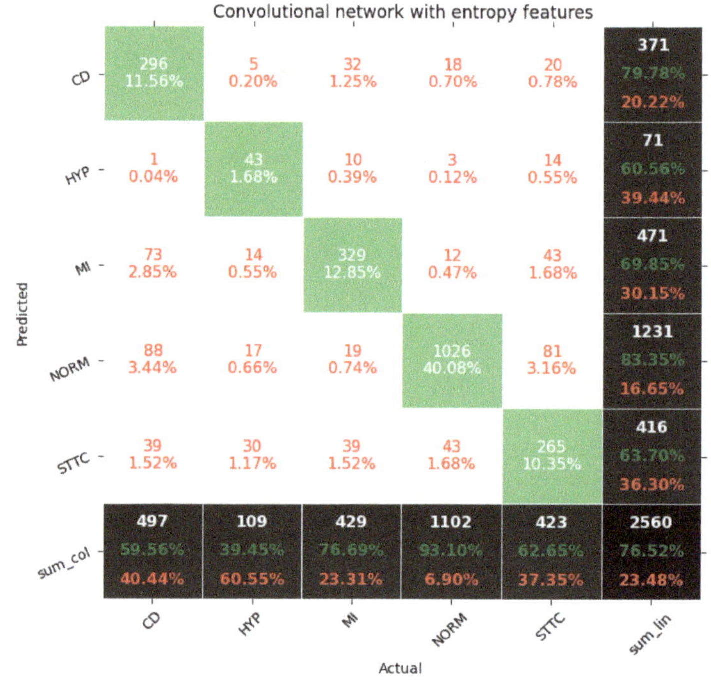

Figure 12. Confusion matrix of results for 5 classes for the convolutional network with entropy features.

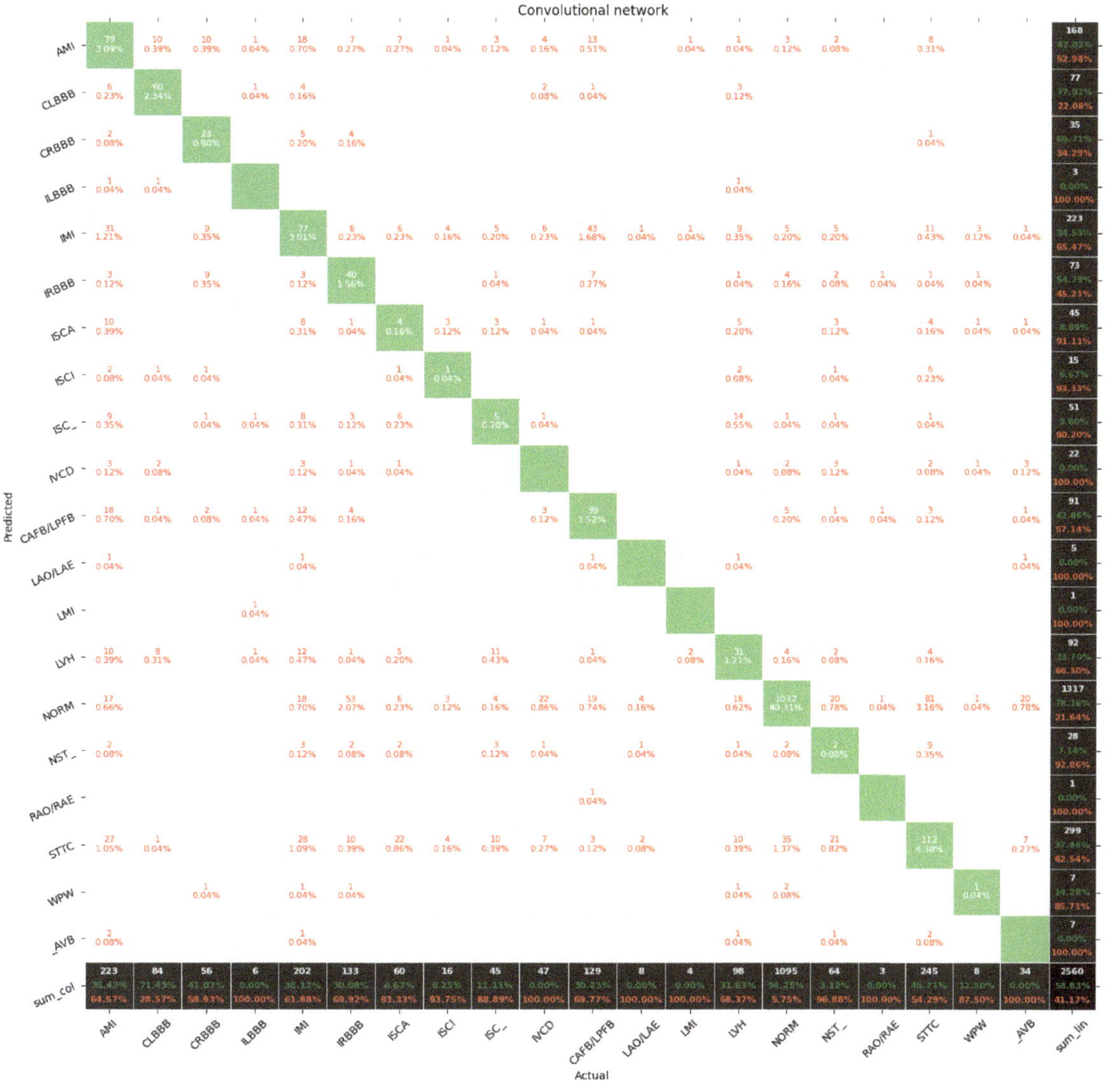

Figure 13. Confusion matrix of results for 5 classes for the convolutional network.

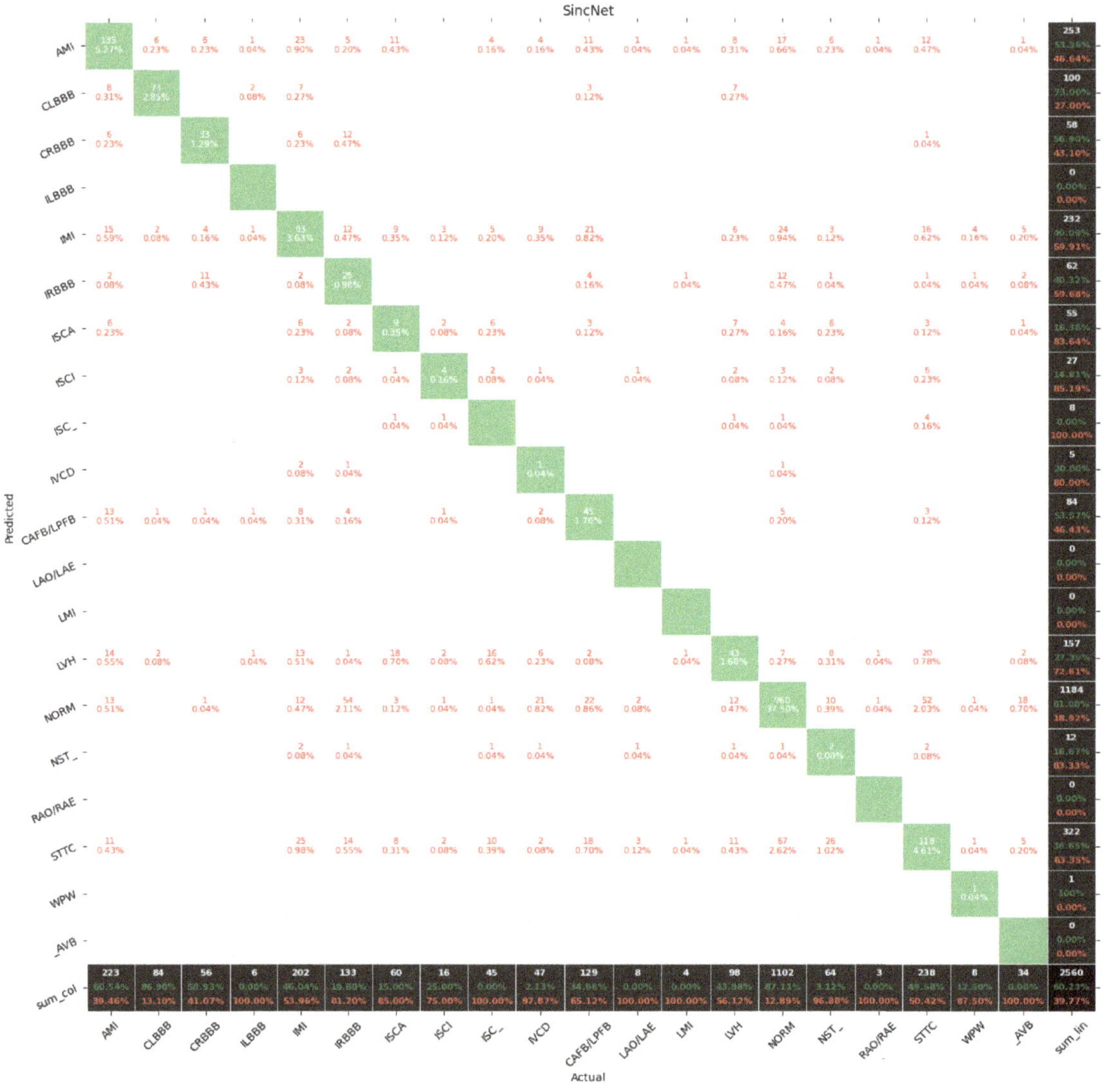

Figure 14. Confusion matrix of results for 5 classes for SincNet.

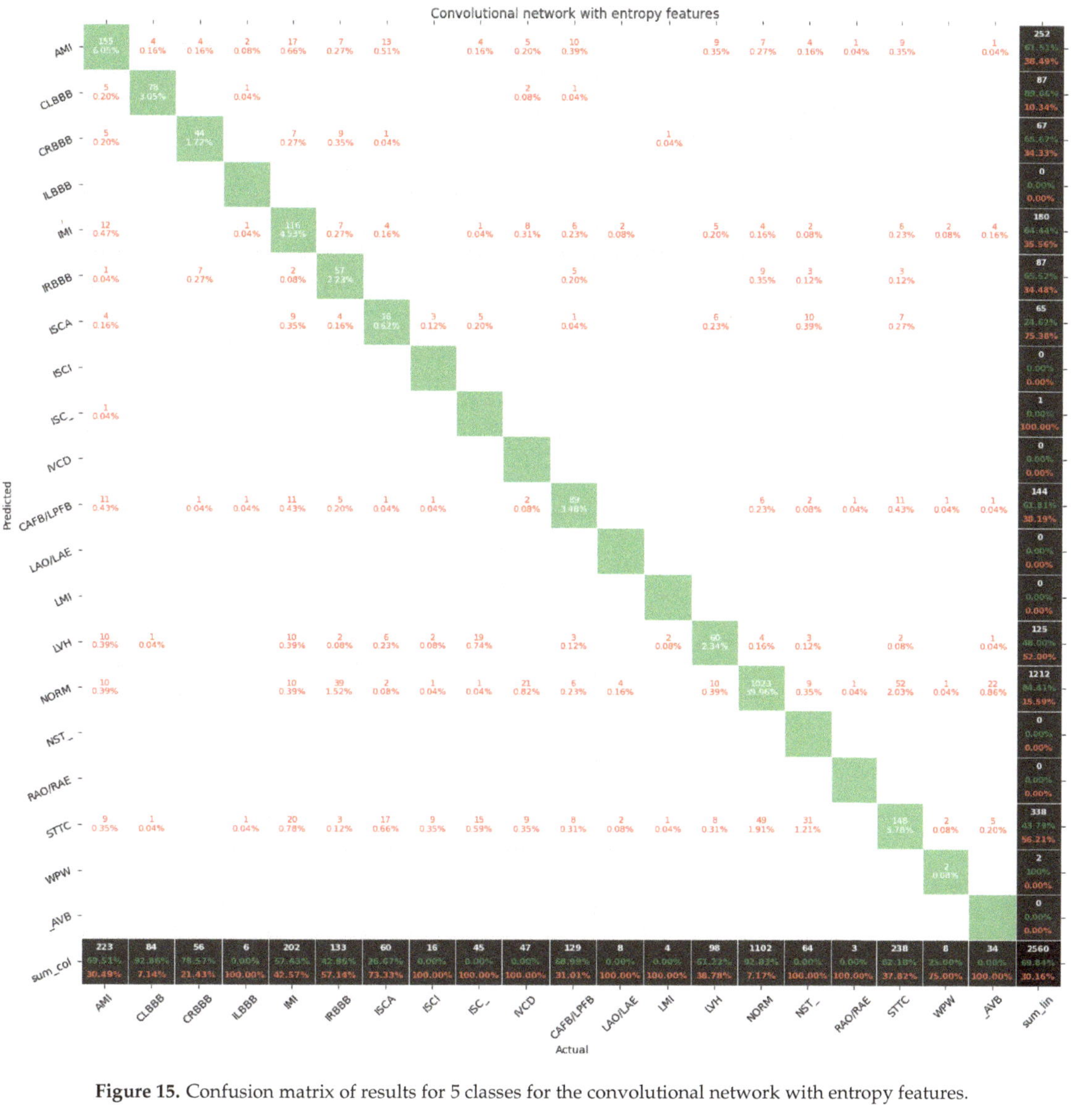

Figure 15. Confusion matrix of results for 5 classes for the convolutional network with entropy features.

In all cases of the evaluated networks, the NORM class obtained the highest value, which resulted from the large number of ECG recordings in this class.

4. Discussion

This paper presented a new model of convolutional neural networks, optimized to limit the computational and memory complexity for ECG recognition and classification of cardiovascular diseases. The research was carried out using a CNN network based on the convolutional network, which is relatively light and yields good results. The advantage of this approach is the possibility of using it on mobile and embedded devices, such as a Raspberry Pi or smartphone graphics cards.

The application of additional entropy-based features significantly improved the results. Such a solution also increased the weight of the network several times, however. As a result, in applications where a very light network is needed, a compromise between weight and accuracy should be sought.

SincNet is a promising solution, but due to being designed to work with the human voice, it does not cope well with ECG signals in its original format. This results from the use of a set of initialization frequencies used in the computation of the Mel-frequency cepstral coefficients that are adapted to the spectral characteristics of the human voice. In the future, it would be worth considering the possibility of adapting SincNet to work with ECG.

The authors were unable to obtain better results due to the issue of overfitting on the training dataset. It was presumed that the addition of customized features may further boost the performance. The authors plan to investigate this claim in their next work.

Sampling determines the amount of measurements used to describe the signal. By changing the sampling, the signal is described by either more or fewer samples, whereas a stack of convolutional layers processes a fixed number of measurements in one context window. As a result, through a modification of the signal sampling, the network may either come to focus on more global features by reducing the amount of samples describing the signal or increase its attention to the details by increasing the measurements per signal.

Interpreting signals with different samplings may prove beneficial. In this work, we used only signals encoding 10 s of experiment on 1000 samples. It may well be the case that a network simultaneously interpreting a signal sampled with frequencies of 500 samples per second, 100 samples per second, and 50 samples per second will return better results. This is because signals sampled at lower frequencies can have entire ECG waves interpreted by one convolutional block, while signals sampled more frequently provide more detailed series for the extraction of features encoded by a small part of an ECG wave.

The proposed network based on a convolutional network is relatively uncomplicated. It is likely that better results could be obtained with the use of Inception models. This model uses heterogeneous subnets to improve the result. It is comparable to the case of wavelet transform, which may prove to be more advantageous than the use of fast Fourier transform. According to the authors, the proposed solution could be used in small devices for continuous monitoring of ECG signals, for example to alert about anomalies and make an initial diagnosis or support a doctor in this.

The authors assumed that a network's performance may be improved with a manageable cost increase by expanding its architecture with Inception-style heterogeneous subnetworks with varying kernels and poolings. The authors intend to investigate this assumption in their future work.

The authors further assumed that the integration of SincNet layers for low-level feature extraction in the first step of signal processing with the successful implementation of the first network based on convolutional layers may prove a benefit. The authors intend to investigate this assumption in their future work.

5. Conclusions

This study presented the capability of convolutional neural networks in the classification of heart diseases by the examination of ECG signals. The network proposed by the authors is both accurate and efficient as it is lightweight, allowing it to be computed on nonspecialized devices. The application of entropy-based features proved beneficial due to the improvements in the accuracy of heart disease classification. Entropy-based features are promising additions to data preprocessing that may prove beneficial in other signal-processing-related tasks.

Author Contributions: Conceptualization, S.Ś., K.P. and D.L.; methodology, S.Ś., K.P. and D.L.; software, S.Ś., K.P., and D.L.; validation, S.Ś., K.P. and D.L.; formal analysis, S.Ś., K.P., and D.L.; investigation, S.Ś., K.P. and D.L.; resources, S.Ś., K.P., and D.L.; data curation, S.Ś., K.P. and D.L.; writing—original draft preparation, S.Ś., K.P., and D.L.; writing—review and editing, S.Ś., K.P. and D.L.; visualization, S.Ś., K.P. and D.L. All authors read and agreed to the published version of the manuscript.

Funding: This research received no external funding.

Institutional Review Board Statement: Not applicable.

Informed Consent Statement: Not applicable.

Data Availability Statement: The data presented in this study are available upon request from the corresponding author.

Conflicts of Interest: The authors declare no conflict of interest.

References

1. Benjamin, E.J.; Virani, S.S.; Callaway, C.W.; Chamberlain, A.M.; Chang, A.R.; Cheng, S.; Chiuve, S.E.; Cushman, M.; Delling, F.N.; Deo, R.; et al. Heart disease and stroke statistics—2018 update: A report from the American Heart Association. *Circulation* **2018**, *137*, e67–e492. [CrossRef] [PubMed]
2. Gupta, D.; Bajpai, B.; Dhiman, G.; Soni, M.; Gomathi, S.; Mane, D. Review of ECG arrhythmia classification using deep neural network. *Mater. Today Proc.* **2021**, In Press. [CrossRef]
3. World Health Organization. *Global Status Report on Noncommunicable Diseases*; WHO: Geneva, Switzerland, 2014.
4. Bogun, F.; Anh, D.; Kalahasty, G.; Wissner, E.; Serhal, C.B.; Bazzi, R.; Weaver, W.D.; Schuger, C. Misdiagnosis of atrial fibrillation and its clinical consequences. *Am. J. Med.* **2004**, *117*, 636–642. [CrossRef] [PubMed]
5. Schläpfer, J.; Wellens, H.J. Computer-interpreted electrocardiograms: Benefits and limitations. *J. Am. Coll. Cardiol.* **2017**, *70*, 1183–1192. [CrossRef]
6. Houssein, E.H.; Kilany, M.; Hassanien, A.E. ECG signals classification: A review. *Int. J. Intell. Eng. Informatics* **2017**, *5*, 376–396. [CrossRef]
7. Jambukia, S.H.; Vipul, K.D.; Harshadkumar, B.P. Classification of ECG signals using machine learning techniques: A survey. In Proceedings of the 2015 International Conference on Advances in Computer Engineering and Applications, Ghaziabad, India, 19–20 March 2015.
8. Macfarlane, P.W.; Devine, B.; Clark, E. The university of Glasgow (Uni-G) ECG analysis program. In Proceedings of the Computers in Cardiology, Lyon, France, 25–28 September 2005.
9. Wang, J.; Qiao, X.; Liu, C.; Wang, X.; Liu, Y.; Yao, L.; Zhang, H. Automated ECG classification using a non-local convolutional block attention module. *Comput. Methods Programs Biomed.* **2021**, *203*, 106006. [CrossRef] [PubMed]
10. Goldberger, A.; Amaral, L.A.; Glass, L.; Hausdorff, J.M.; Ivanov, P.C.; Mark, R.G.; Mietus, J.E.; Moody, G.B.; Peng, C.K.; Stanley, H.E.; et al. PhysioBank, PhysioToolkit, and PhysioNet: Components of a new research resource for complex physiologic signals. *Circulation* **2000**, *101*, e215–e220. [CrossRef]
11. Wagner, P.; Strodthoff, N.; Bousseljot, R.; Samek, W.; Schaeffter, T. PTB-XL, a large publicly available electrocardiography dataset (version 1.0.1). *Sci. Data* **2020**, *7*, 1–5. [CrossRef]
12. Jia, W.; Xu, X.; Xu, X.; Sun, Y.; Liu, X. Automatic Detection and Classification of 12-lead ECGs Using a Deep Neural Network. In Proceedings of the Computing in Cardiology, Rimini, Italy, 13–16 September 2020; pp. 1–4.
13. Zhu, Z.; Lan, X.; Zhao, T.; Guo, Y.; Kojodjojo, P.; Xu, Z.; Liu, Z.; Liu, S.; Wang, H.; Sun, X.; et al. Identification of 27 abnormalities from multi-lead ECG signals: An ensembled SE_ResNet framework with sign loss function. *Physiol. Meas.* **2021**, *42*, 065008. [CrossRef]
14. Strodthoff, N.; Wagner, P.; Schaeffter, T.; Samek, W. Deep learning for ECG analysis: Benchmarks and insights from PTB-XL. *arXiv* **2020**, arXiv:2004.13701.
15. Smisek, R.; Nemcova, A.; Marsanova, L.; Smital, L.; Vitek, M.; Kozumplik, J. Cardiac Pathologies Detection and Classification in 12-lead ECG. In Proceedings of the Computing in Cardiology, Rimini, Italy, 13–16 September 2020; pp. 1–4.
16. Zhang, D.; Yang, S.; Yuan, X.; Zhang, P. Interpretable deep learning for automatic diagnosis of 12-lead electrocardiogram. *Iscience* **2021**, *4*, 102373. [CrossRef]
17. Warrick, P.A.; Lostanlen, V.; Eickenberg, M.; Andén, J.; Homsi, M.N. Arrhythmia Classification of 12-lead Electrocardiograms by Hybrid Scattering-LSTM Networks. In Proceedings of the Computing in Cardiology, Rimini, Italy, 13–16 September 2020; pp. 1–4.
18. Acharya, U.R.; Fujita, H.; Adam, M.; Lih, O.S.; Hong, T.J.; Sudarshan, V.K.; Koh, J.E. Automated characterization of arrhythmias using nonlinear features from tachycardia ECG beats. In Proceedings of the 2016 IEEE International Conference on Systems, Man, and Cybernetics (SMC), Budapest, Hungary, 9–12 October 2016.
19. Jo, Y.Y.; Cho, Y.; Lee, S.Y.; Kwon, J.M.; Kim, K.H.; Jeon, K.H.; Cho, S.; Park, J.; Oh, B.H. Explainable artificial intelligence to detect atrial fibrillation using electrocardiogram. *Int. J. Cardiol.* **2021**, *328*, 104–110. [CrossRef] [PubMed]
20. Lepek, M.; Pater, A.; Muter, K.; Wiszniewski, P.; Kokosińska, D.; Salamon, J.; Puzio, Z. 12-lead ECG Arrythmia Classification Using Convolutional Neural Network for Mutually Non-Exclusive Classes. In Proceedings of the Computing in Cardiology, Rimini, Italy, 13–16 September 2020; pp. 1–4.
21. Ramaraj, E.; Virgeniya, S.C. A Novel Deep Learning based Gated Recurrent Unit with Extreme Learning Machine for Electrocardiogram (ECG) Signal Recognition. *Biomed. Signal Process. Control* **2021**, *68*, 102779.
22. He, K.; Zhang, X.; Ren, S.; Sun, J. Deep residual learning for image recognition. In Proceedings of the IEEE Conference on Computer Vision and Pattern Recognition, Las Vegas, NV, USA, 27–30 June 2016; pp. 770–778.

23. Kingma, D.P.; Ba, J. Adam: A method for stochastic optimization. *arXiv* **2014**, arXiv:1412.6980.
24. Caruana, R.; Lawrence, S.; Giles, L. Overfitting in neural nets: Backpropagation, conjugate gradient, and early stopping. In Proceedings of the 14th Annual Neural Information Processing Systems Conference, Denver, CO, USA, 27 November–2 December 2020, pp. 402–408.
25. Ravanelli, M.; Yoshua, B. Speaker recognition from raw waveform with sincnet. In Proceedings of the 2018 IEEE Spoken Language Technology Workshop (SLT), Athens, Greece, 18–21 December 2018.
26. Molau, S.; Pitz, M.; Schluter, R.; Ney, H. Computing Mel-frequency cepstral coefficients on the power spectrum. In Proceedings of the 2001 IEEE International Conference on Acoustics, Speech, and Signal Processing, Salt Lake City, UT, USA, 7–11 May 2001.
27. Shannon, C. A mathematical theory of communication. *Bell Syst. Tech. J.* **1948**, *27*, 379–423. [CrossRef]
28. Richman, J.S.; Moorman, J.R. Physiological time-series analysis using approximate entropy and sample entropy. *Am. J. Physiol.-Heart Circ. Physiol.* **2000**, *278*, H2039–H2049. [CrossRef]
29. Bandt, C.H.; Bernd, P. Permutation entropy: A natural complexity measure for time series. *Phys. Rev. Lett.* **2002**, *88*, 174102. [CrossRef]
30. Inouye, T.; Shinosaki, K.; Sakamoto, H.; Toi, S.; Ukai, S.; Iyama, A.; Katsuda, Y.; Hirano, M. Quantification of EEG irregularity by use of the entropy of the power spectrum. *Electroencephalogr. Clin. Neurophysiol.* **1991**, *79*, 204–210. [CrossRef]
31. Renyi, A. On measures of entropy and information. In *Proceedings of the Fourth Berkeley Symposium on Mathematical Statistics and Probability, Volume 1: Contributions to the Theory of Statistics*; University of California Press: Oakland, CA, USA, 1961; pp. 547–561.
32. Bezerianos, A.; Tong, S.; Thakor, N. Time dependent entropy of EEG rhythm changes following brain ischemia. *Ann. Biomed. Eng.* **2003**, *31*, 221–232. [CrossRef]
33. Lad, F.; Sanfilippo, G.; Agrò, G. Extropy: A complementary dual of entropy. *arXiv* **2011**, arXiv:1109.6440.
34. Granero-Belinchón, C.; Roux, S.G.; Garnier, N.B. Information Theory for Non-Stationary Processes with Stationary Increments. *Entropy* **2019**, *21*, 1223. [CrossRef]

entropy

Article

Selection of the Optimal Smart Meter to Act as a Data Concentrator with the Use of Graph Theory

Piotr Kiedrowski

Institute of Telecommunications and Computer Science, UTP University of Science and Technology in Bydgoszcz, Al. Prof. S. Kaliskiego 7, 85-796 Bydgoszcz, Poland; pkied@utp.edu.pl

Abstract: Changing the construction of mart Meter (SM) devices, more specifically equipping them with more than one communication module, facilitates the elimination of a Transformer Station Data Concentrator (TSC) module, moving its function to one of the SMs. The opportunity to equip a chosen device in an additional communication module makes it possible to use it as an acquisition node. The introduction of this solution creates a problem with the optimum selection of the above-mentioned node out of all the nodes of the analyzed network. This paper suggests the criterion of its location and, as per the criterion, the way of conduct using the elements of the graph theory. The discussion is illustrated with the examples of the possibility to use the method for the optimization of the architecture of the network. The described method makes it possible to choose the location of a backup acquisition node as well as locate intermediary nodes (signal repeaters) in case of a failure (removal) of some SM devices. In the era of the common introduction of dispersed telemetric systems requiring an adequate level of performance and reliability of information transmission, the offered method can be used for the optimization of the structures of Smart Grids.

Keywords: graph theory; smart meter; smart metering; wireless sensor network

Citation: Kiedrowski, P. Selection of the Optimal Smart Meter to Act as a Data Concentrator with the Use of Graph Theory. *Entropy* **2021**, *23*, 658. https://doi.org/10.3390/e23060658

Academic Editors: Michal Choras, Robert Burduk, Agata Giełczyk, Rafal Kozik and Tomasz Marciniak

Received: 29 April 2021
Accepted: 21 May 2021
Published: 24 May 2021

Publisher's Note: MDPI stays neutral with regard to jurisdictional claims in published maps and institutional affiliations.

1. Introduction

Currently, a rapid growth of technologies based on computers with increasingly good parameters, a progressing miniaturization of electronic systems and the development of professional software and methods of information transfer are being observed. Wireless transmission technologies are becoming more and more important thanks to their low cost and quick deployment. Amongst them are wireless sensor networks (WSNs) [1]. They feature low transmitting power, very low energy consumption, are operated with the most recent monolithic transceivers, and their nodes cooperate directly with sensors of various measurable parameters.

The networks are widely used, for example, in municipal buildings as early warning systems for terrorist attacks or environmental contamination [2], for monitoring the environment (temperature, humidity, pollution level) [3], for industrial surveillance (providing information about the overall condition of machines used in the production process, e.g., vibration, fluid level, temperature of components, etc., received from sensors placed in hard-to-reach or inaccessible places) [4], for the control of traffic [5], and in medicine (allowing higher effectiveness of diagnosing and monitoring patients' health without the need to connect them to medical devices with limited mobility, e.g.,wireless body sensor networks (WBSNs)) [6,7]. WSNs areused for the creation of smart cities, professionally known as Cyberville, Digital City, or Electronic Communities. Their most important features are digital and computerized management of energy production and distribution;, automatic smart security systems; lighting and heating control; management of car parks; fast data transmission using the latest wireless technologies in the 5G network; municipal serviceless meters and usage sensors with digital water, electricity, and gas distribution management systems; smart medical care systems remote life quality management and improvement such as systems of e-participation in various forms of cultural, sporting, leisure, and social activity; and surveillance systems, biometric systems, and safety systems for terrorist and criminal threats, etc. [8–10].

An example of the utilization of the WSN technology is the wireless scattered telemetry network intended for the last-mile service of the Smart Metering system. The term Smart Metering refers to smart power grids which provide the communication between producers and recipients of energy as well as with power repositories. The basic component of these networks is an expanded metering systemthat gives access to information about the power consumption. It consists of ICT systems transferring the measurements to decision points, as well as smart information, forecasting, and decision algorithms. The Smart Metering system makes it possible to transfer and process information important for a power grid, such as the power consumption by receivers and the production of energy from conventional and renewable sources. It ensures a high flexibility level of the power grid which in turn allows it to control the demand and supply of power quickly and optimally [11].

A data concentrator collecting the information from Smart Meter (SM) devices is one of the basic components of ICT systems [9]. Usually, it is installed next to an MV/LV transformer, hence its name Transformer Station Data Concentrator (TSC) [12]. TSC has the same role as an acquisition node in sensor networks. Similarly to sensor networks, in order to increase the range of the last-mile network operation, a multi-hop technique is used [13].

In the communication between the two above-mentioned components of the system, two basic transmission methods are used, wired and wireless. Amongst the wireless technologies, the most frequently used are radio transmission in the Industrial-Scientific-Medical (ISM) band (known as RF) and General Packet Radio Service (GPRS), available as one of the services offered by the Global System for Mobile Communications (GSM). SM devices equipped with GPRS communication modules are used only in countries with low population density such as Denmark or Sweden [13,14]. If the density is higher, the GPRS modules are installed only in TSCs, thus giving them access to IP networks.

Amongst the wired technologies, another two main techniques can be mentioned. These are Power Line Communication (PLC) and M-bus. The M-bus technology, designed for the reading of various kinds of metering devices, is used in Smart Metering only in last-mile network with direct hubs—Local Metering Concentrators (LMCs)—to which multiple SM devices can be connected through the use of M-bus [9,15,16]. LMC communicates with TSC by the use of the RF or the PLC technology. Modern SMs are equipped with M-buses and/or Modbus interfaces, which facilitate the installation of a few communication modules in one meter [17] thanks to the ongoing miniaturization of components, and equips the SMs in more than one LMC unit whereby each of them can use a different communication technique. Adopting such solutions makes it possible to realize and utilize last-mile networks more flexibly while increasing their security, reliability, quality, productivity, and the ease of migration process, e.g., from RF to PLC [18–21].

This article touches upon a new approach in the last-mile network structural solutions, i.e.,the elimination of TSC and moving its functions to an SM equipped with two LMC units, namely RF or PLC and GSM. The benefits of introducing this solution are lower network creation costs, the elimination of additional TC–SM connections, which are often unreliable due to the long distance between a transformer and an edge of a last-mile network, and there is no need for the maintenance of TSCs involving complex procedures, assuring the safety of transmission and service. However, moving functions from TSC to SM, apart from the above-mentioned benefits, also causes a problem—a solution for which the author offers in this paper. The solution consists of choosing a parameter of the optimal SM localization, which additionally works as a data concentrator.

2. Research Method

For the analysis of telemetric networks, parts of the graph theory were used [22]. The topology of such networks can be defined by a formula describing geometrical random graphs [9]:

$$G = (V, E, R) \tag{1}$$

where V is a set of vertices, E is a set of edges, and R is set of radii determining the transmission range.

Each v_i vertex of the G graph represents a single network node. Each node is surrounded by a circle of radius r_i depicting the range of the signal emitted by a node's transmitter. All nodes in the circle surrounding the node (assuming omnidirectional antennas are used) can freely communicate with the surrounded node bidirectionally and are considered directly adjacent, as it is shown in Figure 1.

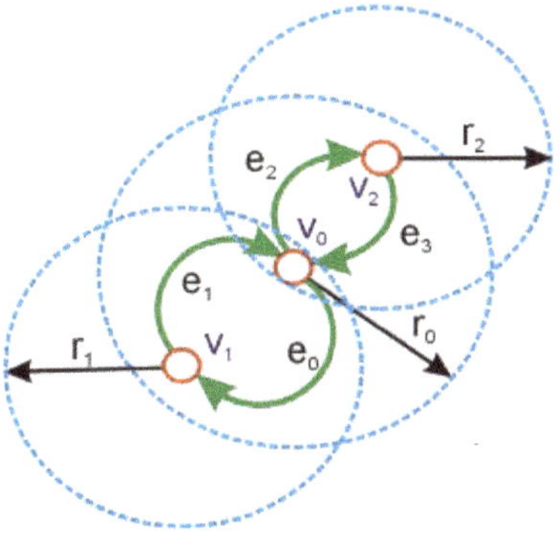

Figure 1. An example of a graph describing the topology of a network.

The below formula [23] defines a set of nodes in the transmission range of a given node:

$$Br(x) = \{y : |x - y| < r\}, \quad x, y \in V \tag{2}$$

where $Br(x)$ is a set of nodes within the transmission range, x and y are the localization of the nodes, and r_i is the radius of the signal range.

The edge set E contains all edges of the G graph and represents direct paths connecting any chosen vertices and all their adjacent items within the circle, representing the communication range of each node.

It is assumed that the edges do not have to represent bidirectional connections, which means that the presence of edges (x,y) does not mean that edges (y,x) exist. The assigned measurable value between two vertices of a graph, such as distance, angle, and amount of energy in the node or probability, can be assigned to the edge as its weight, provided that it can have different values for edges (x,y) and (y,x). The definition of asymmetric edges is expressed with the Formula (3):

$$E = \{(x, y, d) : y \in Br(x) \wedge d = |x - y|\}, \quad x, y \in V \tag{3}$$

where E is a set of asymmetric edges and d is the measure of the asymmetry of weights.

A graph obtained this way, describing the whole connection topology of the network, is called a maximum power graph (in real-life conditions when first determining the connection topology, the transmitters work with maximum power). The structure of this graph has to be reduced because of the presence of many redundant connections, which are undesirable due to their transmitting of identical information, leading to collisions, unnecessary power consumption, and the increase in the final network emissivity. The reduction in these connections leads to the creation of a minimum spanning tree whose root is an acquisition node in which all the information consistent with the network's function are collected.

The configuration of the discussed network can be presented with a graph with an adjacency matrix [24]. To explain the way of conduct, a graph describing an example of a virtual network is shown in Figure 2. The edges of the graph correspond to the links connecting its nodes—that is, the nodes intermediary to the transmission and acting commutatively.

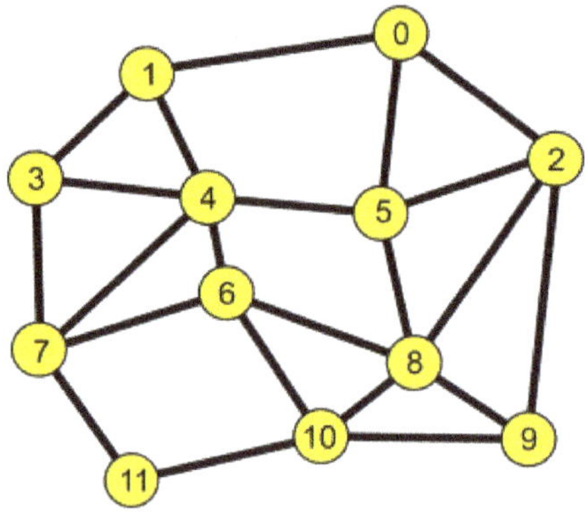

Figure 2. An example of a telemetry network.

The network shown in Figure 2 is described by an adjacency matrix (*Ms*):

$$[Ms] = \begin{bmatrix} 0 & 1 & 1 & 0 & 0 & 1 & 0 & 0 & 0 & 0 & 0 & 0 \\ 1 & 0 & 0 & 1 & 1 & 0 & 0 & 0 & 0 & 0 & 0 & 0 \\ 1 & 0 & 0 & 0 & 0 & 1 & 0 & 0 & 1 & 1 & 0 & 0 \\ 0 & 1 & 0 & 0 & 1 & 0 & 0 & 1 & 0 & 0 & 0 & 0 \\ 0 & 1 & 0 & 1 & 0 & 1 & 1 & 1 & 0 & 0 & 0 & 0 \\ 1 & 0 & 1 & 0 & 1 & 0 & 0 & 0 & 1 & 0 & 0 & 0 \\ 0 & 0 & 0 & 0 & 1 & 0 & 0 & 1 & 1 & 0 & 1 & 0 \\ 0 & 0 & 0 & 1 & 1 & 0 & 1 & 0 & 0 & 0 & 0 & 1 \\ 0 & 0 & 1 & 0 & 0 & 1 & 1 & 0 & 0 & 1 & 1 & 0 \\ 0 & 0 & 1 & 0 & 0 & 0 & 0 & 0 & 1 & 0 & 1 & 0 \\ 0 & 0 & 0 & 0 & 0 & 0 & 1 & 0 & 1 & 1 & 0 & 1 \\ 0 & 0 & 0 & 0 & 0 & 0 & 0 & 1 & 0 & 0 & 1 & 0 \end{bmatrix}$$

By exponentiating the (*Ms*) matrix, we obtain a set of paths created by the edges connecting the graph nodes, whose length corresponds to the power of the matrix. The process is carried on until it is confirmed that all the nodes are interconnected and the maximum power of the (*Ms*) matrix is the diameter of the analyzed graph. By the analysis of the obtained components of the matrix, minimum lengths of the paths connecting chosen nodes (created by the smallest number of edges) are determined, and their number is calculated, as well.

Table 1 includes the calculated lengths of minimum paths connecting the nodes of the analyzed graph, and Table 2 shows the numbers of these paths.

For example, there are four minimum paths connecting node 4 with node 9, which consist of three edges.

However, on the basis of the obtained results, it is not possible to determine which edges of the set create individual paths. To define a configuration of every path, each edge is given a name, presented in Table 3 and in Figure 3.

Table 1. The lengths of the minimum paths (the number of edges).

1	**1**										
2	1	2									
3	2	1	3								
4	2	1	2	1							
5	1	2	1	2	1						
6	3	2	2	2	1	2					
7	3	2	3	1	1	2	1				
8	2	3	1	3	2	1	1	2			
9	2	3	1	4	3	2	2	3	1		
10	3	3	2	3	2	2	1	2	1	1	
11	4	3	3	2	2	3	2	1	2	2	1
Node	0	1	2	3	4	5	6	7	8	9	10

Table 2. The numbers of the minimum paths.

1	**1**										
2	1	1									
3	1	1	2								
4	2	1	1	1							
5	1	2	1	1	1						
6	4	1	1	2	1	2					
7	3	2	2	1	1	1	1				
8	2	4	1	3	2	1	1	1			
9	1	1	1	8	4	2	2	3	1		
10	3	1	2	3	1	1	1	2	1	1	
11	6	2	2	1	1	2	2	1	1	1	1
Node	0	1	2	3	4	5	6	7	8	9	10

Table 3. Adopted edge marking.

Edge	**a**	**b**	**c**	**d**	**e**	**f**	**g**	**h**	**i**	**j**	**k**
Nodes	0–1	0–2	0–5	1–3	1–4	2–5	2–8	2–9	3–4	3–7	4–5
Edge	l	m	n	o	p	r	s	t	u	v	w
Nodes	4–6	4–7	5–8	6–7	6–8	6–10	7–11	8–9	8–10	9–10	10–11

By performing the exponentiation of the sign matrix according to the rule of matrix exponentiation, sets of minimum paths are obtained together with their structure. It is shown in Table 4.

While reviewing the components in Table 4, it is clearly visible that the numbers and the lengths of the minimum paths connecting the nodes are consistent with the calculation results presented in Tables 1 and 2.

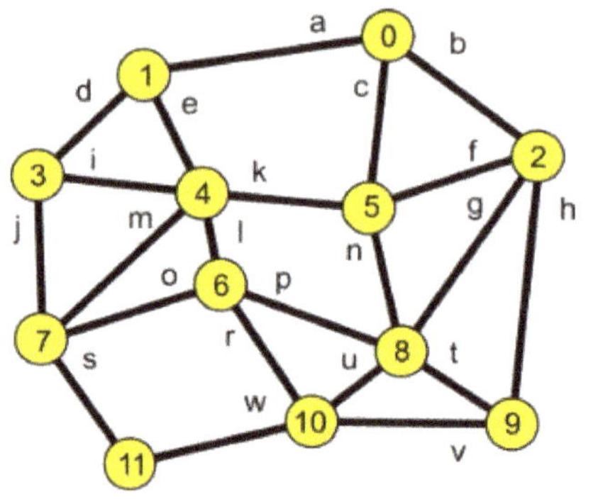

Figure 3. Edge marking.

Table 4. Minimum paths configurations.

Node	0	1	2	3	4	5	6	7	8	9	10	11
0	-	a	b	ad	ae ck	c	ael ckl cnp bgp	adj ckm aem	cn bg	bh	bhv cnu bgu	adjs aems cnuw bguw bhvw ckms
1	a	-	ab	d	e	ek ac	el	dj em	elp ekn acn abg	abh	elr	djs ems
2	b	ab	-	bad fki	fk	f	gp	fkm gpo	g	h	hv gu	hvw guw
3	ad	d	bad fki	-	i	ik	il jo	j	ilp jop ikn	ilpt iknt ikfh ikrv jorv jopt jswv dabh	ilr jor jsw	js
4	ae ck	e	fk	i	-	k	l	m	Lp kn	lpt lrv knt kfh	lr	ms
5	c	ek ac	f	ik	k	-	kl np	km	n	nt fh	nu	nuw kms
6	ael ckl cnp bgp	el	gp	il jo	l	kl np	-	o	p	pt rv	r	rw os
7	adj ckm aem	dj em	fkm gpo	j	m	km	o	-	op	opt swv	or sw	s
8	cn bg	elp ekn acn abg	g	ilp jop ikn	lp kn	n	p	op	-	t	u	uw

Table 4. *Cont.*

Node	0	1	2	3	4	5	6	7	8	9	10	11
9	bh	abh	h	ilpt iknt ikfh ikrv jorv jopt jswv dabh	lpt lrv knt kfh	nt fh	pt rv	opt swv	t	-	v	vw
10	bhv cnu bgu	elr	hv gu	ilr jor jsw	lr	nu	r	or sw	u	v	-	w
11	adjs aems cnuw bguw bhvw ckms	djs ems	hvw guw	Js	ms	nuw kms	rw os	s	uw	vw	w	-

3. Determining the Optimal Position for the Acquisition Node of a Telemetry Network

So far, the analysis was made with the assumption that the probability of realization of a correct transmission by each node is 1, which is in contradiction to reality, and the parameter is especially important for radio networks.

It is known that radio links are less resistant to external interference than fiber-optic or cable links, and a radio wave carrying information is subject to suppression depending on the distance between a transmitter and a receiver. The dependency is partially described by the free space signal suppression value formula *FSL* (Free Space Loss):

$$FSL = 32.44\ \text{dBm} + 20\log(f) + 20\log(d) \tag{4}$$

where f is the transmission frequency in MHz and d is the distance between a transmitter and a receiver in km [25].

By assuming the same level of radio signal emission, e.g., 0 dBm, the same sensitivity of the receivers and the use of omnidirectional antennas (which causes the increase in the suppression value by 40 dBm of the signal reaching the receiver), the parameter RSSI (Received Signal Strength Indicator) was determined on the basis of the dependency shown in Figure 4, taken from paper [26], and PER (Packet Error Ratio) parameter values were found (it was assumed that 433 MHz radio frequency was used).

In the analyzed case, assuming the distances between the nodes are as shown in Table 5 (they are correlated with the data presented in Figure 1), the RSSI and PER values were determined.

Table 5 includes the RSSI values resulting from the distance between the network nodes linearly. The possibility that suppressing signals for a given link can vary for individual directions of information transmission was not taken into consideration.

To check the construction of the structure after including different signal suppression depending on the information transfer direction, it was assumed that the values of the probability of a return transmission from the target nodes to the source nodes reaches the values given in Table 6 (the '$\leftarrow$' sign means the change of the information transfer direction).

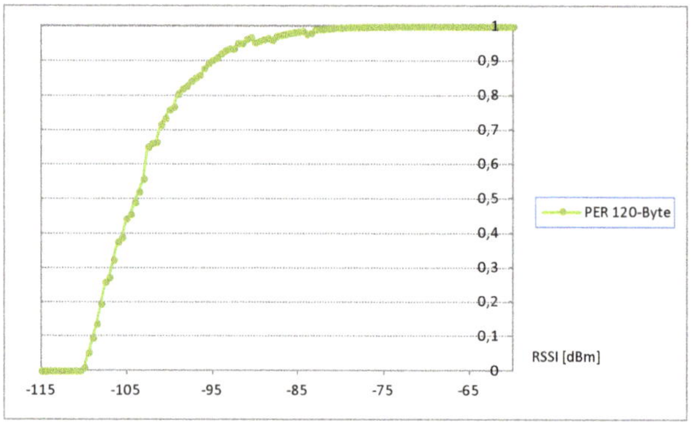

Figure 4. The chart of the dependency of link quality in the function of the level of the signal reaching the receiver.

Table 5. The determined RSSI and PER values.

Edge	a	b	c	d	e	f	g	h	i	j	k
d(m)	48.6	36.45	34.02	29.16	26.73	36.45	51.03	55.89	34.02	33.04	34.02
RSSI (dBm)	−95.38	−92.88	−92.28	−90.94	−90.19	−92.88	−95.80	−96.59	−92.28	−92.03	−92.28
PER	0.8922	0.9347	0.9354	0.9579	0.9620	0.9347	0.8592	0.8592	0.9354	0.9505	0.9354
Edge	l	m	n	o	p	r	s	t	u	v	w
d(m)	19.44	43.74	31.59	37.91	38.88	33.05	31.59	26.73	23.33	38.88	36.45
RSSI (dBm)	−87.42	−94.47	−91.64	−93.22	−93.44	−92.03	−91.64	−90.19	−89.01	−93.44	−92.88
PER	0.9723	0.9092	0.9513	0.9347	0.9299	0.9505	0.9513	0.9620	0.9657	0.9347	0.9347

Table 6. The assumed RSSI and PER values for the second transmission direction.

Edge	$\leftarrow a$	$\leftarrow b$	$\leftarrow c$	$\leftarrow d$	$\leftarrow e$	$\leftarrow f$	$\leftarrow g$	$\leftarrow h$	$\leftarrow i$	$\leftarrow j$	$\leftarrow k$
PER	0.9723	0.9092	0.9513	0.9347	0.9299	0.9505	0.9513	0.9620	0.9657	0.9347	0.9347
Edge	$\leftarrow l$	$\leftarrow m$	$\leftarrow n$	$\leftarrow o$	$\leftarrow p$	$\leftarrow r$	$\leftarrow s$	$\leftarrow t$	$\leftarrow u$	$\leftarrow v$	$\leftarrow w$
PER	0.8922	0.9347	0.9354	0.9579	0.9620	0.9347	0.8592	0.8592	0.9354	0.9505	0.9354

By using the data from Table 6 and assuming that the data of the information transfer direction from the source nodes to the target nodes are above the diagonal of the table, and from the target nodes to the source nodes are below the diagonal, the probability values were calculated and are presented in Figure 5.

The way of conduct described in the previous part of the paper was used for the determination of the optimal position of the acquisition node of the network, i.e., the node whose average probability of the realization of a correct transmission is the highest in comparison with the other nodes. For this purpose, the probability values for both transmission values were multiplied. They corresponded to individual paths connecting the nodes of the graph (for example, nodes 0 and 5) and the obtained values were averaged. It is shown in Table 7.

Node	0	1	2	3	4	5	6	7	8	9	10	11
0		0.892	0.935	0.855	0.867	0.935	0.810	0.796	0.846	0.803	0.795	0.900
1	0.972		0.834	0.958	0.962	0.867	0.924	0.893	0.806	0.716	0.878	0.849
2	0.909	0.884		0.808	0.874	0.935	0.799	0.771	0.859	0.859	0.816	0.763
3	0.909	0.935	0.842		0.935	0.875	0.893	0.951	0.812	0.870	0.828	0.904
4	0.897	0.930	0.888	0.966		0.935	0.960	0.909	0.860	0.799	0.848	0.865
5	0.951	0.897	0.951	0.903	0.935		0.891	0.850	0.951	0.859	0.919	0.845
6	0.826	0.830	0.915	0.878	0.892	0.867		0.935	0.930	0.891	0.951	0.889
7	0.842	0.871	0.854	0.935	0.935	0.874	0.958		0.869	0.834	0.889	0.951
8	0.877	0.829	0.951	0.845	0.902	0.935	0.962	0.921		0.962	0.966	0.903
9	0.875	0.850	0.962	0.873	0.784	0.859	0.857	0.778	0.859		0.935	0.874
10	0.824	0.775	0.902F	0.798	0.834	0.875	0.935	0.850	0.935	0.951		0.935
11	0.887	0.749	0.844	0.803	0.803	0.767	0.849	0.859	0.875	0.889	0.935	

Figure 5. The calculated probability values of obtaining a correct transmission.

Table 7. The resulting probability of obtaining a correct transmission.

Node	0	1	2	3	4	5	6	7	8	9	10	11	P_{awr}
0		0.867	0.850	0.777	0.777	0.890	0.669	0.670	0.743	0.702	0.656	0.562	0.7421
1	0.867		0.737	0.895	0.895	0.777	0.776	0.778	0.670	0.783	0.689	0.636	0.7931
2	0.850	0.737		0.681	0.777	0.888	0.731	0.657	0.817	0.827	0.736	0.644	0.7587
3	0.777	0.895	0.681		0.903	0.790	0.790	0.888	0.705	0.586	0.679	0.726	0.7655
4	0.777	0.895	0.777	0.903		0.874	0.868	0.850	0.777	0.653	0.771	0.695	0.8035
5	0.890	0.777	0.888	0.790	0.874		0.777	0.743	0.890	0.735	0.804	0.655	0.8021
6	0.669	0.776	0.731	0.790	0.868	0.777		0.895	0.895	0.764	0.888	0.754	0.8001
7	0.670	0.778	0.657	0.888	0.850	0.743	0.895		0.801	0.648	0.755	0.817	0.7731
8	0.743	0.670	0.817	0.705	0.777	0.890	0.895	0.801		0.827	0.903	0.790	0.8016
9	0.702	0.783	0.827	0.586	0.653	0.735	0.764	0.648	0.827		0.888	0.777	0.7446
10	0.656	0.689	0.736	0.679	0.771	0.804	0.888	0.755	0.903	0.888		0.874	0.7859
11	0.562	0.636	0.644	0.726	0.695	0.655	0.754	0.817	0.790	0.777	0.874		0.7209

On the basis of the data presented in the table above, it was concluded that the optimal localization for an acquisition node was node 4.

Table 8 includes the set of minimum length paths, taking into consideration the presence of parallel, internodal paths (the parallel paths are the paths consisting of the same number of edges connecting the same nodes—source and target).

Table 8. The set of minimum length paths obtained with the assumption that the acquisition node is node 4.

Node	0	1	2	3	5	6	7	8	9	10	11
Set of Paths	ae ck	E	fk	I	K	l	m	lp kn	lpt lrv knt kfh	lr	*ms*
P_{path}	0.777	0.895	0.777	0.903	0.874	0.868	0.850	0.777	0.653	0.771	0.695

Where: Set of Paths, the set of all minimum length paths; P_{path}, the probability of the realization of transmission between nodes.

In Figure 6, the image of the obtained graph describing the network created by the minimum length paths is shown.

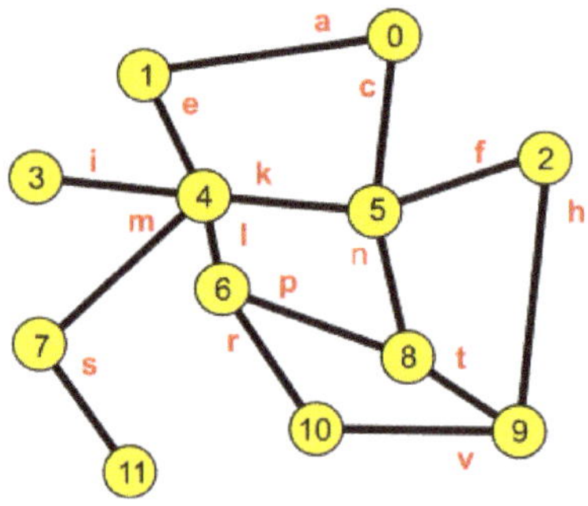

Figure 6. The image of the obtained graph.

Thanks to the analysis of the elements in Table 8, a minimum spanning tree was created; its root is node 4, so by removing the redundant connections, paths were chosen that ensured the highest probability of a correct transmission. The set of paths creating the minimum spanning tree is included in Table 9.

Table 9. The set of paths creating the minimum spanning tree.

Node	0	1	2	3	5	6	7	8	9	10	11
C_{min}	ck	E	fk	i	K	l	M	nk	vrl	rl	sm
$P_{pathtree}$	0.778	0.895	0.777	0.903	0.874	0.868	0.850	0.778	0.685	0.771	0.695

Where: C_{min}, the configuration of a minimum path; $P_{pathtree}$, the probability of the realization of transmission.

In Figure 7,the obtained minimum spanning tree is shown.

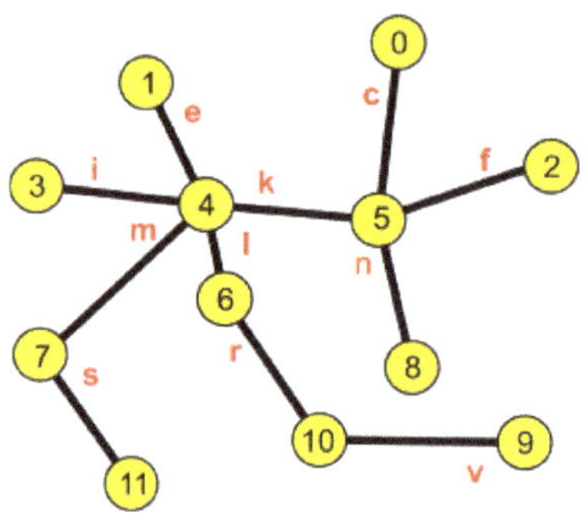

Figure 7. The image of a minimum spanning tree.

The described way of conduct can be used for the analysis of a situation caused by a failure of individual links due to which the graph describing the obtained tree becomes a disconnected graph resulting in isolated nodes. When there are two parallel paths connecting chosen nodes created by various combinations of edges (in the examples 4—0, 4—8, and 4—9), the graph stays connected. Thanks to the proposed method, that is, the analysis of the matrix powers (*Ms*), it is possible to determine a set of emergency connections and avoid a connection shortage. The set of paths used in case of a link failure is included in Table 10.

Table 10. The set of paths used in case of a link failure.

Link Failure	Node	0	1	2	3	5	6	7	8	9	10	11
E	Sets of Paths	ck	Di	fk	I	K	l	m	nk	vrl	rl	sm
F		ck	E	bck	I	K	l	m	nk	vrl	rl	sm
K		ae	E	bae	I	Cae	l	m	lp	vrl	rl	sm
I		ck	E	fk	Jm	K	l	m	nk	vrl	rl	sm
L		ck	E	fk	I	K	om	m	nk	knt	unk	Sm
M		ck	E	fk	I	K	l	ol	nk	vrl	unk	Wrl
R		ck	E	fk	I	K	l	m	nk	knt	unk	Sm
S		ck	E	fk	I	K	l	m	nk	vrl	rl	Wrl

The determined set of paths can be saved in routing memories of individual nodes and used to support transmission in case of a failure of certain links.

4. The Analysis of the Possibility to Ensure the Operation of a Network after Node Removal

In practice, due to changes in networks (for example, by replacing the energy meters with devices working in a different technology), 'holes' appear, which causes the graphs describing these networks to be disconnected. In such cases, a signal repeater is often installed in place of one of the nodes. Its task is only to mediate in the transmission of signals between sensors and an acquisition node. Using such a solution is led by the use of an already existing infrastructure enabling the supply of power to the repeater. There is also a case of choosing a location for the repeater. In this case, the proposed method of network analysis can work as well.

To explain the accepted way of conduct, the analysis of a more complex network was used. The network is shown in Figure 8.

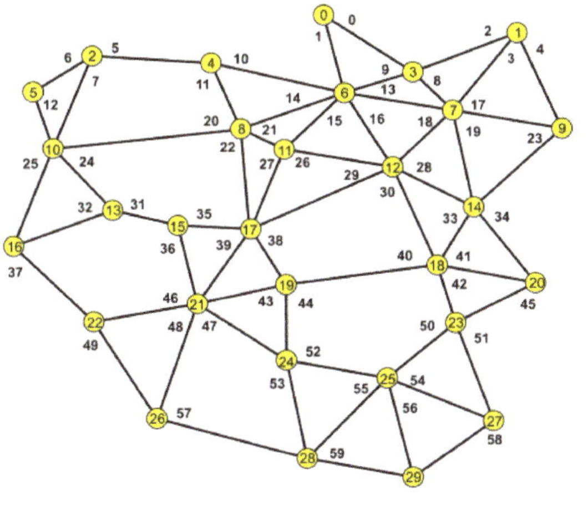

Figure 8. The analyzed network before nodes failure.

By following the rules described above, the optimum location of the acquisition node was determined to be node 17.

It was assumed that for some reasons, nodes 2, 6, 7, and 10 had been excluded which caused nodes 0,1, 3, and 5 to lose the possibility to contact the acquisition node. This is shown in Figure 9.

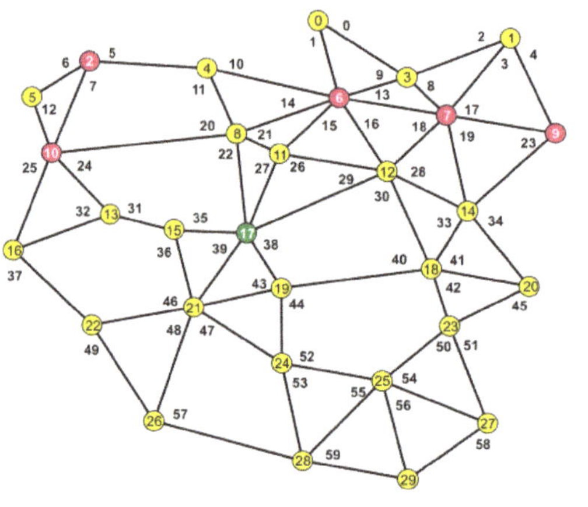

Figure 9. An image of the network after the nodes failure.

By analyzing the changed structure of the network, it was divided into two parts. One related to the fifth node (Figure 10a) and one associated with the remaining nodes 0, 1, and 3 (Figure 10b).

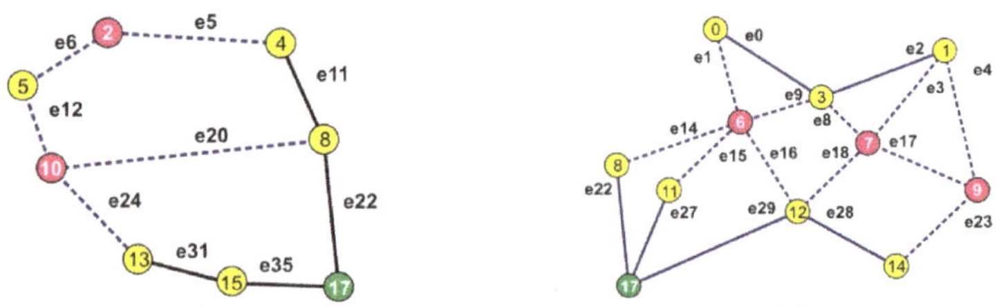

(a) (b)

Figure 10. The structures after the exclusion of the chosen nodes. (**a**) related to the fifth node; (**b**) associated with the remaining nodes i.e., 0, 1 and 3.

The sets of virtual shortest paths connecting the isolated nodes are given in Table 11.

Table 11. The sets of the shortest paths (for the node 5).

Node	Paths	P_{path}
5	e12, e20, e11	0.6570
	e6, e5, e11, e22	0.6706
	e12, e24, e31, e35	0.7900

The data in Table 11 show that when it comes to node 5, it is clear for a signal repeater to be installed in node 10.

The situation presented in Figure 10b is more difficult to analyze. In that case, it should be verified which of the edges creating the paths ensure obtaining the maximum probability of a correct transmission.

Even a superficial scan of the network scheme shows that placing a repeater in node 9 is futile, so the focus was moved towards node 6 or 7. The set of minimum length paths for nodes 0, 1, and 3 is presented in Table 12.

Table 12. The set of minimum length paths (nodes 0, 1, 3).

Node	Paths	P_{path}
0	e1, e15, e27	0.78192
	e1, e16, e29	0.68325
	e1, e14, e22	0.72990
1	e3, e18, e29	0.66908
	e2, e9, e15, e27	0.71395
	e2, e9, e14, e22	0.66645
3	e9, e15,e27	0.81298
	e9, e14,e22	0.75889
	e9, e16,e29	0.71039
	e8, e18,e29	0.71623

The configurations of paths connecting individual nodes with the acquisition node ensuring the highest probability of a successful transmission were analyzed: from node 0—e1, e15, e27; from node 1—e2, e9, e15, e27; from node 3—e9, e15, e27. Each of the above-mentioned paths is connected with node 6, so it is therefore the location of an intermediary node.

5. Discussion

The proposed method allows for the determination of the best location ofan SM device for an additional role as a concentrator, thanks to which it is possible to eliminate a TSC module as well as redundant connections between the TSC and its adjacent SM modules. The absence of need for the installation of TSC lowers the cost of the network construction and its location maintenance (that is a transformer station MV/LV). It leads to the avoidance of time-consuming, complex, and expensive procedures. The removal of unreliable links between the TSC and the SM modules, resulting from a long distance between them, causes the improvement of the reliability and traffic parameters of the network. Improvement of the traffic parameters allows for more frequent reading, which is crucial in energy consumption forecasting. One additional benefit is the possibility to perform more frequent reading of the meters, which is particularly crucial, for example, in Smart Grid systems, with distributed generation, increase in the number of SMs in a last-mile network, and therefore the extension of the operating area of the network.

Sometimes, in order to improve the traffic parameters of a network, it is necessary to divide it into several subnetworks. The division of networks within the structure using TSC was possible only if the main network was extended with a new transformer. The proposed method makes it possible to determine the best location for the SMs, which are also supposed to work as concentrators in particular subnetworks. The paper also proves that the proposed method can be used for finding a place for backup concentrators and repeaters.

6. Conclusions

All the presented solutions referred to last-mile networks using the RF technology. Thanks to the introduction of the ITU-T recommendations defining the PLC interfaces for Smart Metering [27–29], the technology can be used in many areas of a Smart Grid [30]. Currently, the author is working on the adaptation of the proposed method for networks based on the G3 PLC [28] and PRIME [29] interfaces. In this case, it is necessary to develop a method for the calculation of the PER values for the links using PLC (not only RSSI, but also Signal to Noise Ratio (SNR) and a way to describe the topology of a network (in case of a three-phase mains, the installed SM devices for transmission can use three common media or only one of them).

Funding: This research received no external funding.

Institutional Review Board Statement: Not applicable.

Informed Consent Statement: Not applicable.

Data Availability Statement: The data presented in this study are available on request from the author.

Conflicts of Interest: The author declares no conflict of interest.

References

1. Raghavendra, C.S.; Sivalingam, K.M.; Znati, T. *Wireless Sensor Network*; Springer: New York, NY, USA, 2004; p. 426.
2. Bello, L.L.; Lombardo, A.; Milardo, S.; Patti, G.; Reno, M. Experimental Assessments and Analysis of an SDN Framework to Integrate Mobility Management in Industrial Wireless Sensor Networks. *IEEE Trans. Ind. Inform.* **2020**, *16*, 5586–5595. [CrossRef]
3. Steere, D.; Baptista, A.; McNamee, D.; Pu, C.; Walpole, J. Research Challenges in Environmental Observation and Forecasting Systems. In Proceedings of the 6th Annual International Conference on Mobile Computing and Networking, Boston, MA, USA, 6–11 August 2000; Association for Computing Machinery: New York, NY, USA, 2000; pp. 292–299. [CrossRef]
4. Zabłudowski, Ł.; Marciniak, B.; Lutowski, Z.; Bujnowski, S. Quality Parameters in Wireless Distributed Control Systems. *Solid State Phenom.* **2014**, *223*, 231–237. [CrossRef]
5. Leow, W.; Ni, D.; Pishro-Nik, H. A Sampling Theorem Approach to Traffic Sensor Optimization. *IEEE Trans. Intell. Transp. Syst.* **2008**, *9*, 369–374. [CrossRef]
6. Chiang, S.-Y.; Kan, Y.-C.; Chen, Y.-S.; Tu, Y.-C.; Lin, H.-C. Fuzzy Computing Model of Activity Recognition on WSN Movement Data for Ubiquitous Healthcare Measurement. *Sensors* **2016**, *16*, 2053. [CrossRef] [PubMed]
7. Laurijssen, D.; Saeys, W.; Truijen, S.; Daems, W.; Steckel, J. Synchronous Wireless Body Sensor Network Enabling Human Body Pose Estimation. *IEEE Accesss* **2019**, *7*, 49341–49351. [CrossRef]
8. Kirimtat, A.; Krejcar, O.; Kertesz, A.; Tasgetiren, M.F. Future Trends and Current State of Smart City Concepts: A Survey. *IEEE Access* **2020**, *8*, 86448–86467. [CrossRef]
9. Guo, J.; Ding, X.; Wu, W. A Blockchain-Enabled Ecosystem for Distributed Electricity Trading in Smart City. *IEEE Internet Things J.* **2021**, *8*, 2040–2050. [CrossRef]
10. Garakani, H.G.; Moshiri, B.; Safavi-Naeini, S. Cyber Security Challenges in Autonomous Vehicle: Their Impact on RF Sensor and Wireless Technologies. In Proceedings of the 2018 18th International Symposium on Antenna Technology and Applied Electromagnetics (ANTEM), Waterloo, ON, Canada, 19–22 August 2018; pp. 1–3.
11. Morello, R.; De Capua, C.; Fulco, G.; Mukhopadhyay, S.C. A Smart Power Meter to Monitor Energy Flow in Smart Grids: The Role of Advanced Sensing and IoT in the Electric Grid of the Future. *IEEE Sens. J.* **2017**, *17*, 7828–7837. [CrossRef]
12. Petrusevski, I.; Živanović, M.; Rakić, A.; Popović, I. Novel AMI architecture for real-time Smart Metering. In Proceedings of the 2014 22nd Telecommunications Forum Telfor (TELFOR), Belgrade, Serbia, 25–27 November 2014; pp. 664–667. [CrossRef]
13. Miao, H.; Chen, G.; Zhao, Z.; Zhang, F. Evolutionary Aggregation Approach for Multihop Energy Metering in Smart Grid for Residential Energy Management. *IEEE Trans. Ind. Inform.* **2021**, *17*, 1058–1068. [CrossRef]
14. Stefan, M. Automation of Smart Grid Operations Through Spatio-Temporal Data-Driven Systems. Ph.D. Thesis, Technical Faculty of IT and Design, Aalborg University, Aalborg, Denmark, 2019; p. 134.
15. Polak, M. Experiences of Smart Grid Implementation in Sweden. *RynekEnergii* **2014**, *91*, 252–256.
16. Rinaldi, S.; Ferrari, P.; Flammini, A.; Sisinni, E.; Vezzoli, A. Uncertainty Analysis in Time Distribution Mechanisms for OMS Smart Meters: The Last-Mile Time Synchronization Issue. *IEEE Trans. Instrum. Meas.* **2018**, *68*, 693–703. [CrossRef]
17. Zhou, J.; Hu, R.Q.; Qian, Y. Scalable Distributed Communication Architectures to Support Advanced Metering Infrastructure in Smart Grid. *IEEE Trans. Parallel Distrib. Syst.* **2012**, *23*, 1632–1642. [CrossRef]
18. Bouhafs, F.; Mackay, M.; Merabti, M. *Towards a Unified Smart Grid ICT Infrastructure*; Springer: New York, NY, USA, 2014; p. 95. [CrossRef]
19. Hyun, D.; Lee, Y.; Moon, Y.A. Study on the Composite Power Line Communication Network. In *Communications in Computer and Information Science*; Slezak, D., Kim, T., Chang, A.C., Vasilakos, T., Li, M., Sakurai, K., Eds.; Springer: Berlin/Heidelberg, Germany, 2009; Volume 56, pp. 547–554. [CrossRef]
20. Sensus and GE Combine Forces to Create Residential Meter with FlexNet Communications. Available online: https://sensus.com/news-events/news-releases/sensus-and-ge-combine-forces-to-create-residential-meter-with-flexnet-communications/ (accessed on 20 April 2021).
21. Aloui, A.; Ben Rhouma, O.; Rebai, C. PLC and RF Communications Evaluation For Smart Grid Application In Rural Configuration. In Proceedings of the 2019 IEEE 19th Mediterranean Microwave Symposium (MMS), Hammamet, Tunisia, 31 October–2 November 2019; pp. 1–4. [CrossRef]
22. Deo, N. *Graph Theory with Applications to Engineering and Computer Science*; Dover Publications: Dover, UK, 2016; p. 496.
23. Penrose, M. *Random Geometric Graphs*; Oxford University Press: Oxford, UK, 2003; p. 344. [CrossRef]
24. Graham, R.L.; Knuth, D.E.; Patashnik, O. *Concrete Mathematics: A Foundation for Computer Science*, 2nd ed.; Addison-Wesley: Boston, MA, USA, 1994; p. 657.
25. Faruque, S. *Radio Frequency Propagation Made Easy*; Springer: Cham, Switzerland, 2015; p. 103. [CrossRef]

26. Boryna, B.; Dubalski, B.; Kiedrowski, P.; Zabłudowski, A. Errors Nature in Indoors Low Power 433 MHz Wireless Network. In *Image Processing and Communications Challenges 2. Advances in Intelligent and Soft Computing*; Choraś, R.S., Ed.; Springer: Berlin/Heidelberg, Germany, 2010; Volume 84, pp. 373–378. [CrossRef]
27. International Telecommunication Union, Telecommunication Standardization Sector of ITU, ITU-T Recommendation. *G.9902: Narrowband Orthogonal Frequency Division, Multiplexing Power Line Communication Transceivers for ITU-T G.hnem Networks*; International Telecommunication Union: Geneva, Switzerland, 2014.
28. International Telecommunication Union, Telecommunication Standardization Sector of ITU, ITU-T Recommendation. *G.9903: Narrowband Orthogonal Frequency Division, Multiplexing Power Line Communication Transceivers for G3-PLC Networks*; International Telecommunication Union: Geneva, Switzerland, 2014.
29. International Telecommunication Union, Telecommunication Standardization Sector of ITU, ITU-T Recommendation. *G.9904: Narrowband Orthogonal Frequency Division, Multiplexing Power Line Communication Transceivers for PRIME Networks*; International Telecommunication Union: Geneva, Switzerland, 2013.
30. Hoffmann, S.G. Layer-2 security for PLC—A comparison between ITU-T G.9903 and IEEE 1901.2. In Proceedings of the 2016 International Symposium on Power Line Communications and Its Applications (ISPLC), Bottrop, Germany, 20–23 March 2016; pp. 173–178. [CrossRef]

entropy

MDPI

Article

Hfinger: Malware HTTP Request Fingerprinting

Piotr Białczak [1,2,*] and Wojciech Mazurczyk [2]

[1] CERT Polska/Research and Academic Computer Network (NASK), Kolska 12, 01-045 Warsaw, Poland
[2] Institute of Computer Science, Warsaw University of Technology, Nowowiejska 15/19, 00-665 Warsaw, Poland; wojciech.mazurczyk@pw.edu.pl
* Correspondence: piotr.bialczak@cert.pl

Abstract: Malicious software utilizes HTTP protocol for communication purposes, creating network traffic that is hard to identify as it blends into the traffic generated by benign applications. To this aim, fingerprinting tools have been developed to help track and identify such traffic by providing a short representation of malicious HTTP requests. However, currently existing tools do not analyze all information included in the HTTP message or analyze it insufficiently. To address these issues, we propose Hfinger, a novel malware HTTP request fingerprinting tool. It extracts information from the parts of the request such as URI, protocol information, headers, and payload, providing a concise request representation that preserves the extracted information in a form interpretable by a human analyst. For the developed solution, we have performed an extensive experimental evaluation using real-world data sets and we also compared Hfinger with the most related and popular existing tools such as FATT, Mercury, and p0f. The conducted effectiveness analysis reveals that on average only 1.85% of requests fingerprinted by Hfinger collide between malware families, what is 8–34 times lower than existing tools. Moreover, unlike these tools, in default mode, Hfinger does not introduce collisions between malware and benign applications and achieves it by increasing the number of fingerprints by at most 3 times. As a result, Hfinger can effectively track and hunt malware by providing more unique fingerprints than other standard tools.

Keywords: fingerprinting; malware analysis; malicious network traffic analysis; HTTP protocol analysis; pcap file analysis; malware tracking; malware identification

Citation: Białczak, P.; Mazurczyk, W. Hfinger: Malware HTTP Request Fingerprinting. *Entropy* **2021**, *23*, 507. https://doi.org/10.3390/e23050507

Academic Editor: Amelia Carolina Sparavigna

Received: 26 March 2021
Accepted: 19 April 2021
Published: 23 April 2021

1. Introduction

Currently, malicious software (malware) developers use Hypertext Transfer Protocol (HTTP) as one of the primary carriers for malicious communication. According to Miller et al. [1], HTTP is the most common protocol used in the Command and Control (C&C) traffic, more popular than Hypertext Transfer Protocol Secure (HTTPS). It is utilized by malware, e.g., to connect to the C&C server to register or obtain commands, check the infected machine's IP address, or download additional modules. Moreover, it can be used to perform DDoS (Distributed Denial of Service) attacks or to click on referral links, thus creating revenue.

To identify and discern different malware communication activities, network traffic fingerprinting methods can be applied. The notions of a fingerprint and fingerprinting as the act of creating a fingerprint are similar to the notions of classic forensic work, where the fingerprint is an impression of human fingers' friction ridges. In the field of computer science, a working definition of a fingerprint is a short representation of a larger object [2]. The most crucial property of fingerprinting is that two different objects have different fingerprints, and the probability of a collision, i.e., an event when two different objects have the same fingerprint, is low. File fingerprinting is one of the application examples, where cryptographic hash functions, e.g., such as SHA-256, are used to create identification tags for the fingerprinted files.

However, it is not only files that can be fingerprinted. Network traffic can also be used for this purpose. Various network protocols can be analyzed to represent the exchanged data, which then, in turn, can be used for identification purposes. The process of fingerprinting can be conducted in an active or passive mode. The former is performed with a modification of the standard network traffic, for example, by sending carefully crafted messages. In the latter, no changes are introduced, and network traffic is only monitored. The most popular practical usage examples are passive Operating System fingerprinting (as realized, e.g., in p0f https://lcamtuf.coredump.cx/p0f3, accessed on 26 March 2021), web browser fingerprinting (like in, e.g., privacy research and advocacy service at https://panopticlick.eff.org/, accessed on 26 March 2021) or network service discovery (as performed by, e.g., Nmap https://nmap.org/, accessed on 26 March 2021). Network protocols can also be fingerprinted to identify, track, or detect malware. Until now, several such examples of network traffic fingerprinting methods and tools exist, and they are discussed in detail in Section 2.

From the variety of network protocols, HTTP fingerprinting is a promising approach to provide identification and tracking of malware communications, which is crucial for malware analysts in their daily work while defending networks. Currently, several tools have been proposed that help fingerprinting HTTP traffic, such as FATT (https://github.com/0x4D31/fatt, accessed on 26 March 2021), p0f (https://lcamtuf.coredump.cx/p0f3, accessed on 26 March 2021), or Mercury (https://github.com/cisco/mercury, accessed on 26 March 2021). However, they all share the same limitation. In our opinion, they do not analyze all information included in the HTTP messages or analyze it insufficiently. For example, the above-mentioned tools do not analyze the payload of the request, and the URI analysis is at most limited to value encoding. Note that both these features have already been proven to be useful for malware detection and identification purposes, for example, as described by Li et al. [3] or Perdisci et al. [4]. Fingerprints created with such an approach omit information that can potentially identify and discern various malware families' requests.

To address these issues, we propose the Hfinger tool that aims to fingerprint malware HTTP requests more comprehensively. In more detail, Hfinger processes HTTP requests and generates a fingerprint based on the URI, protocol version, the request method, headers and their values, and the request's payload. The tool's main goal is to produce unique representations of malware HTTP requests, thus providing a mechanism for the identification of such requests in network traffic of various applications. The fingerprint created by Hfinger represents a malware request in a short and concise form that can still be interpretable by a human analyst. Hfinger was designed to be used with exact match searching mechanisms, which provide means for direct fingerprint searching without using wildcard techniques. Exact match searching is supported by, for example, many security monitoring and logging tools. In this vein, it is intended to provide similar search functionality as malware sample hashes such as SHA-256. Nevertheless, its overt nature can still help analysts by giving basic information about the request or by finding patterns in the network traffic.

Hfinger does not provide direct identification of particular malware families or directly detect malware per se. However, it can identify requests that can be labeled as malicious using other information sources, such as Intrusion Detection Systems. The tool can also be used in threat hunting to uncover unknown requests that were omitted by other security solutions but that share fingerprint with those identified as malicious. The tool is open source and has been published at https://github.com/CERT-Polska/hfinger, accessed on 26 March 2021. The research presented in this paper is focused on HTTP requests of Windows-based malware. The utilization of only HTTP requests is related to the fact that the server responses may be unavailable in some of the fingerprinting tool usage scenarios. For example, when analyzing an old malware sample for which C&C servers no longer work, or server responses are unreliable because the malicious infrastructure was sinkholed and the received messages are different from the original. From this perspective, using only requests for fingerprinting purposes can be more reliable in analyzing malware's actual behavior. Moreover, focusing on Windows-based malware is related to the fact that it is still the most frequently attacked platform despite the constant increase in the number of threats on mobile platforms. According to AV-TEST Institute in 2019, more than 75% of malware targeted Windows operating system [5].

To prove the effectiveness of the proposed Hfinger tool we have conducted an extensive experimental study. To this aim, we based the performed evaluation on four main metrics that measure fingerprint collision probability for various malware families (separately and including benign software), the number of created fingerprints, and their entropy. Moreover, to determine the performance of the proposed solution for malware HTTP request fingerprinting, we use real-world malware and benign applications data sets containing HTTP traffic. In more detail, the former consists of 121 popular malware families represented using 401,566 HTTP requests, while the latter incorporates 248,657 HTTP requests generated by popular Windows applications, including web browsers. Additionally, the effectiveness of the developed tool has been compared with the three existing, previously mentioned community-proven HTTP fingerprinting tools, i.e., FATT, p0f, and Mercury.

Considering the above, the main contributions of this paper are as follows:

- Proposing Hfinger—a new malware HTTP request fingerprinting tool;
- Performing a review and analysis of popular HTTP fingerprinting tools;
- Providing an extensive experimental evaluation of the proposed approach and its comparison with the popular, existing HTTP fingerprinting tools.

The rest of the paper is structured as follows. First, Section 2 describes the most notable related work. Then, in Section 3, we present the proposed HTTP fingerprinting tool. Next, Section 4 contains details on the chosen experimental methodology, while in Section 5 obtained results are included and discussed. Section 6 showcases potential practical usage scenarios for Hfinger and pinpoints its main limitations. Finally, Section 7 concludes our work and indicates potential future research directions.

2. Related Work

In this section, first we review the most important work related to the topic of this paper and compare it to Hfinger. Then we describe existing popular tools used for HTTP traffic fingerprinting and we discuss their limitations. Finally, we compare them to the solution proposed in this paper.

An important distinction between the presented research solutions and tools must be drawn. The former were created to provide an extensive research analysis of a particular problem which, in some cases, resulted in creation of a tool or a system that solves the stated problem. On the other hand, the latter were primarily focused on creation of a tool that solves a specific technical (rather than a research) problem and the tool's analysis is typically quite limited.

2.1. Proposed Research Solutions

Research on web browser fingerprinting is directly related to HTTP malware fingerprinting, and this topic has been extensively covered in the literature (cf. Laperdrix et al. [6]). This type of fingerprinting is based on active and passive techniques in which information about different features of the environment, web browser, and OS are extracted. While for active fingerprinting different techniques are used, such as JavaScript to query information about the canvas, a list of browser plugins, or screen resolution, passive fingerprinting techniques analyze requests sent by web browsers. Common techniques involve checking the values of popular headers such as *User-Agent*, *Accept*, or *Content-Encoding* but also headers' order. Hfinger utilizes these passive fingerprinting techniques; however, they are extended, for example, with URI and payload analysis.

Fingerprinting of HTTP network traffic can be used to create models of applications present in a monitored network and used as a baseline for detecting unknown applications that can be malicious. Bortolameotti et al. presented in [7] DECANTeR a system for detection of HTTP network traffic that is anomalous for analyzed host. It passively extracts fingerprints of benign applications running on the host. This process involves extracting information from clustered *POST* and *GET* requests in the form of *Host* header value, constant header fields, average request size, *User-Agent* header value, *Accept-Language* header value, and the size of outgoing information in the cluster.

Bortolameotti et al. presented in [8] a system for the detection of anomalous traffic. Their system uses two models of header-value entropy and header sequence extracted from multiple requests to create known applications' fingerprints. After the training phase, the system can evaluate if an unknown request is similar to already known applications or it originates from a new application. Comparing these two approaches to Hfinger shows they were designed with different objectives than the latter. Their goal is to provide a model of application behavior based on multiple HTTP requests to create a baseline for detecting outlying applications in a particular network, thus providing malware detection. On the other hand, Hfinger is focused on the unique representation of malware HTTP requests, providing a fingerprint for each separate request. Such an approach allows analyzing network traffic without the baseline model creation stage and analyzing network traffic with single requests, for example, when the infrastructure of analyzed malware is not working anymore. Furthermore, while all these systems analyze similar parts of the requests, Hfinger utilizes a broader set of features for fingerprint generation and analyzes all requests, regardless of their method. In contrast, for example, DECANTeR fingerprints clustered *GET* and *POST* requests only.

Various approaches have been proposed for fingerprinting other popular network protocols. Many studies focus on HTTPS protocol, where the primary research objectives are HTTPS network traffic presence identification or identification of services utilizing HTTPS for communication (cf. [9]). SMTP network traffic fingerprinting can be used to identify malware families as presented in [10,11]. SMTP messages, SMTP extensions, and IMF fields are used to create different e-mail clients' dialects, thus providing a method for their identification. DNS protocol fingerprinting can be used as a method to detect DNS amplification DDoS attacks [12], identify DNS servers [13], or for the detection of bots [14]. Segal et al. in [15] presented a white paper on fingerprinting of HTTP/2 protocol clients. When used for malware network traffic fingerprinting, the presented approaches can be applied to identify malware families but also to detect some specific operations, such as sending spam messages or performing DDoS attacks.

Other approaches for network traffic fingerprinting with more generic methods exist, too. For example, Holland et al. in [16] proposed nPrint—a system for standard representation of network traffic. For every analyzed packet, its representation is created that maps all packet bytes to a feature vector representing all possible headers of a particular protocol. The authors claim that their system can generate data suitable as input for machine learning algorithms in classification problems. Unfortunately, when fingerprinting HTTP network traffic, nPrint needs to be configured with appropriate rules, defining which parts of the messages should be extracted. Therefore, it does not provide ready-made methods for HTTP fingerprinting.

Table 1 introduces the classification of the existing fingerprinting research based on its application scenario. It also shows that these approaches utilize fingerprinting for various purposes—some provide identification of benign services or applications, while the other uncover malicious activities and software.

Although the research solutions described above provide fingerprinting mechanisms of various network protocols, they differ from Hfinger in several aspects. Browser fingerprinting methods use, mainly, active analysis techniques and are focused on one type of HTTP clients. Hfinger utilizes only passive analysis techniques, and, despite the focus on malware requests, its design and performance analysis considered the presence of benign HTTP clients. DECANTeR and HeadPrint systems utilize multiple HTTP requests to create a baseline model of the observed network to identify requests that do not fit the created model, thus detecting unknown applications. Hfinger is focused on the unique representation of single malware HTTP requests, which is a different research goal. nPrint provides generic fingerprints of various network protocols and requires the creation of configuration to produce HTTP request fingerprints. The configuration is not provided by the authors, so research into optimal HTTP representation is needed. Conversely, Hfinger is focused only on HTTP, for which a complete optimization has been performed. Segal et al. analyze HTTP in version 2 that is different than previous protocol versions in data transfer and representation techniques. Thus their analysis techniques cannot be directly compared to Hfinger. Other reviewed research solutions analyze protocols different than HTTP.

Table 1. Classification of the reviewed fingerprinting research solutions based on their application scenario.

Usage of Fingerprinting	Examples of Research Solution
Client application or user identification	Laperdrix et al. [6], Segal et al. [15]
Detection of unknown applications	Bortolameotti et al. [7,8]
Service or server identification	Shbair et al. [9], Kim et al. [13]
Malware family identification	Stringhini et al. [10], Bazydło et al. [11]
Malware detection	Blaise et al. [14]
Attack detection	Fachkha et al. [12]
Generic protocol fingerprinting	Holland et al. [16]

2.2. *Existing HTTP Fingerprinting Tools*

In this subsection, HTTP fingerprinting tools similar to the Hfinger are described. Three tools (FATT, Mercury, and p0f) have been selected based on capability of passive, pcap file based analysis of HTTP requests without any major code modification. Source code and documentation of these three applications are public, and they are well known in professional network security community.

Other tools fingerprinting HTTP do exist, but they use active fingerprinting for web browser identification (e.g., FingerprintJS—https://github.com/fingerprintjs/fingerprintjs, accessed on 26 March 2021) or they perform only server fingerprinting (httprecon—https://www.computec.ch/projekte/httprecon/, accessed on 26 March 2021, httprint—https://net-square.com/httprint.html, accessed on 26 March 2021, or nmap—https://nmap.org/, accessed on 26 March 2021). As such, they cannot be compared to the same extent to Hfinger as FATT, Mercury, or p0f.

When using the classification of fingerprinting research solutions presented in Table 1 the three reviewed tools, with some extensions, can be classified into groups providing identification of client applications, unknown applications, or malware families. All these tools produce fingerprints that, after the labeling process, can be used for searching the application defined by the labeled fingerprint. Conversely, if the created fingerprint does not fit a list of known applications, it can be used to identify an unknown application.

The three presented tools will be further analyzed and compared with Hfinger in Section 5. Details about their source code and configuration used in the analysis are presented in Section 4.3.

2.2.1. FATT

FATT—fingerprint all the things (https://github.com/0x4D31/fatt/, accessed on 26 March 2021)—is a tool for fingerprinting protocols such as SSL/TLS, SSH, RDP, HTTP, gQUIC. For HTTP, it provides means for fingerprinting headers of requests and responses by extracting header names into a list and computing MD5 hash from it. Depending on the chosen reporting format, the tool outputs additional contextual information, such as request URI, request full URI, request method and protocol version, and *User-Agent* value. However, these components are not used as a part of the fingerprint.

2.2.2. p0f

p0f—passive OS fingerprinting (https://lcamtuf.coredump.cx/p0f3/, accessed on 26 March 2021)—is a tool mainly known for its capabilities of OS fingerprinting. In version 3 of the tool, additional functionality of the HTTP fingerprinting was added. It can fingerprint client requests but also server responses. However, the request support is limited only to *GET* and *HEAD* methods, which in the authors' opinion is a huge drawback. The request fingerprint provides information about protocol version, order-preserving list of headers present in the request, absent headers, and *User-Agent* header value. When creating a list of headers, headers defined as optional are marked with a question mark "?", values of Host and User-Agent headers are skipped. For other headers, their values are provided, creating a list of header name and value pairs. If any of the headers *User-Agent*, *Host*, *Connection*, *Accept*, *Accept-Encoding*, *Accept-Language*, *Accept-Charset* or *Keep-Alive* is absent, such information is provided by the fingerprinter. Note that Hfinger, proposed in this paper, also provides an order-preserving list of headers, but header values are provided in a separate part of the fingerprint and only for popular headers.

p0f provides information about automatic fingerprint generation. The tool can also handle user-provided fingerprints and search in pcap files for such fingerprints. The search can identify requests even when some other headers occur between those present in the fingerprint or when some of the headers are missing—those that are marked as optional, providing that p0f can detect mismatches between the identified fingerprint and the declared *User-Agent* value. The Hfinger does not offer this functionality.

2.2.3. Mercury

Mercury is a network metadata capture and analysis framework (https://github.com/cisco/mercury, accessed on 26 March 2021). It provides fingerprinting capabilities for protocols such as TLS, DTLS, SSH, HTTP, and TCP. Additionally, it can perform application identification using the created fingerprints. The HTTP fingerprinting can be performed on both requests and responses. Note that the description presented below covers only request fingerprinting.

The tool analyzes the HTTP request to extract information about request method, protocol version, and a list of present headers, if they are on a predefined list of popular headers, including *Accept-Charset*, *Accept-Language*, *Cache-Control*, *Host*, and *User-Agent*. Some of the headers are presented with their values, for example *Accept*, *Accept-Encoding*, or *Connection*. All these features are represented using their hexadecimal values, forming the actual fingerprint. Beside the fingerprint, the tool provides contextual information that presents in a clear form URI and values of headers such as: *User-Agent*, *Host*, *X-Forwarded-For*, and *Via*.

2.2.4. Limitations of Current Tools

The described tools use a limited set of features for HTTP request fingerprinting, and the performed analysis is limited. FATT neglects URI, method, protocol version, payload, and headers' values during the fingerprint generation process. p0f does not analyze URI, payload, and method of request. Mercury does not process the payload, and URI analysis is limited to simple encoding that is even not added to the fingerprint. In both p0f and Mercury, the list of popular headers included in fingerprint creation can be improved as well as the list of headers whose values are added to the fingerprint.

The described tools' analysis is insufficient to achieve a satisfactory level of malware HTTP request fingerprinting uniqueness and can be improved. This paper will try to prove this statement by comparing the results of these three tools with the proposed approach. Moreover, to the authors' best knowledge, there is no extensive academic study that systematically analyzes the effectiveness of FATT, p0f, or Mercury for malware HTTP fingerprinting.

3. Hfinger

In this section, Hfinger functioning is presented in detail, along with request features that the tool investigates. Moreover, the process of the fingerprint generation is thoroughly explained.

3.1. Hfinger's Workflow

Hfinger has been created using Python3 language and additionally it utilizes TShark (https://www.wireshark.org, accessed on 26 March 2021, minimum version 2.2.0) to reassemble TCP segments into HTTP requests. An overview of the tool's workflow is presented in Figure 1.

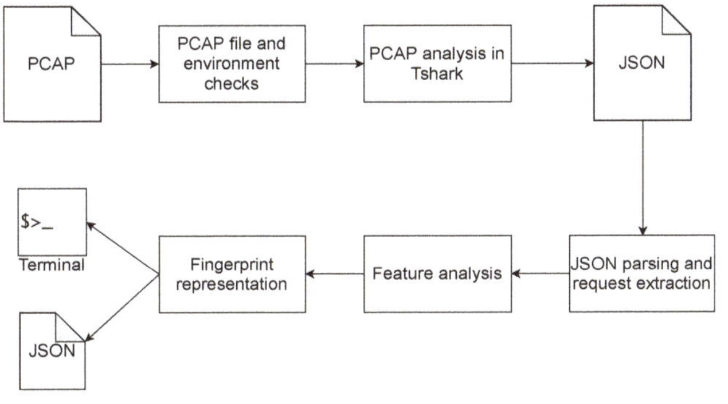

Figure 1. Hfinger's data workflow.

The tool firstly checks the execution environment to determine whether minimal criteria for running are met (e.g., Tshark binary is present) and if the input file is a valid PCAP file. If successful, the tool calls Tshark binary and feeds the analyzed PCAP file into it. TShark is configured to output a JSON file covering only HTTP requests. Then, the output JSON file is parsed by Hfinger and the requests are extracted. In the next step, the extracted requests are analyzed to generate the feature values. In the final step the feature values are joined together with a " | " (pipe) in a particular order, forming the HTTP request fingerprint. Depending on the users' choice, the results in the JSON format are either printed to the terminal or written to a file. The JSON output consists of the basic network information about each request: request timestamp, IP addresses, utilized ports, and the actual request fingerprint. Features analyzed by Hfinger are described in detail in the next section.

3.2. Analyzed Features

In this section, the features analyzed by Hfinger are presented and discussed. The chosen feature set utilized by the developed tool relies on the authors' previous work [17], previously published research (see [18] for URI features), and the authors' own malware analysis experience. In general, extracted features can be divided into three groups depending on the part of the request that they refer to: URI, headers, and payload.

3.2.1. URI Features

These features are used to extract information from the URI part of a request. They include:

- *Length of the URI*, represented as a logarithm with base 10 of the actual URI length (provided as a floating-point number rounded to one decimal place or rounded to an integer);
- *Number of directory levels in the URI*, represented as an integer;

- *Average length of the directory*, represented as a logarithm with base 10 of the actual average length of the directory (provided as a floating-point number rounded to one decimal place or rounded to an integer);
- *Extension of the file requested in the URI*, if applicable. The extension is extracted only if it is present on a defined list of popular extensions to prevent extracting nonsensical values;
- *Length of the variable part of the URI*, where the URI parameters are stored, represented as a logarithm with base 10 of the length (provided as a floating-point number to rounded one decimal place or rounded to an integer);
- *Number of variables in the URI*, represented as an integer;
- *Average value length*, represented as a logarithm with base 10 of the actual average value length (provided as a floating-point number rounded to one decimal place or rounded to an integer).

3.2.2. Header Structure Features

They provide information about headers, their values, extended with information about the request method, and HTTP version. The analyzed features consist of (in the order used in the fingerprint):

- *Request method*, presented as the first two characters of the method name;
- *HTTP version*, expressed as a single number, depending on the first digit after the dot in the protocol definition, for example, "1" for "1.1" version and "9" if no protocol version is defined;
- *Representation of header order in the analyzed request*, where the headers are expressed by the chosen encoding scheme. This scheme provides a list of popular headers for which encoding is provided to shorten the fingerprint length. However, if the header is not on the list, its name is hashed using the 32-bit Fowler–Noll–Vo hash function in version 1a (FNV1a) [19], and the hexadecimal representation of the hash is used as the name. If the header name does not begin with an upper case letter (or any first letter of the parts of a compound header name, for example, *Accept-Encoding*), an exclamation mark *!* is prepended to the header representation;
- *Representation of popular header's values*—the following headers are analyzed to extract their value:
  - *Connection*,
  - *Accept-Encoding*,
  - *Content-Encoding*,
  - *Cache-Control*,
  - *TE*,
  - *Accept-Charset*,
  - *Content-Type*,
  - *Accept*,
  - *Accept-Language*,
  - *User-Agent*.

  If the value is present on a list of popular values, it is encoded with a chosen short encoding representation. If it is not on the list, the values are hashed using FNV1a. The representation is provided as an encoded header name and its encoded value, separated by ":" (colon), and such pairs are separated using "/" (forward slash). If the header can have multiple values, their representation is separated by "," (comma). The order of the headers is preserved. Additionally, the value of the *User-Agent* header is always represented as the FNV1a hash.

3.2.3. Payload Features

They are extracted if the payload of a request is not empty. Payload features consist of three features (in the order used in the fingerprint):

- *Presence of non-ASCII characters*, represented as a single letter "N" if non-ASCII characters are present, and "A" otherwise,
- *Shannon entropy of the payload*, represented as a floating-point number rounded to one decimal place or rounded to an integer,
- *Payload length*, represented as a logarithm with base 10 of the actual payload length (provided as a floating-point number rounded to one decimal place or rounded to an integer).

3.2.4. Numerical Features' Representation

As presented above, some of the numerical features are inherently real numbers and have to be represented as a float type. During the design phase, a decision was made to round such values to one decimal place or round them to an integer. The rounding mode can have a significant impact on Hfinger evaluation; thus, in Section 5.1 an analysis of which version of the representation should be chosen for each of these features is discussed.

3.3. Fingerprint Generation

Features described above are used to create the HTTP request fingerprint. Figure 2 illustrates an overview of an exemplary Hfinger fingerprint generation. All analyzed features are presented, including floating-point representation, what may vary from the final feature set selection presented in Section 5.1.

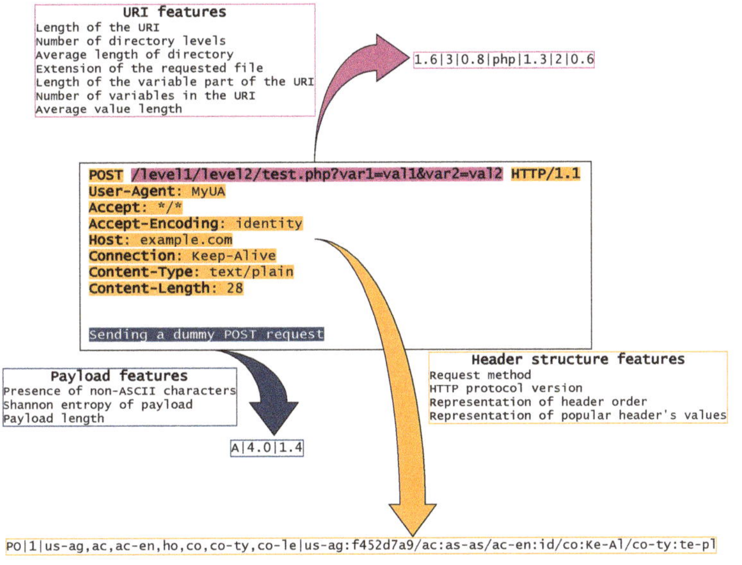

Figure 2. An example of a HTTP POST request fingerprint produced by Hfinger.

As presented in Figure 2, Hfinger analyzes three parts of the HTTP request to generate a fingerprint. Firstly, the URI part is analyzed and the feature values are generated. For example, in Figure 2, the URI length is 43 characters, there are 3 directory levels, and a *PHP* file is requested. These features are represented in the generated fingerprint part as *1.6*, *3*, and *php* respectively.

Secondly, header structure features are extracted to generate the second part of the final fingerprint. For instance, using the example in Figure 2, the method is *POST*, protocol is in version *1.1*, *User-Agent* header has value of *MyUA*, and *Connection* header has value of *Keep-Alive*. These values are transformed into corresponding fingerprint parts: *PO*, *1*, *us-ag:f452d7a9*, and *co:Ke-Al*, respectively. Header names on their own and in pair with values are parts of broader structure features, representing order of all headers in the request, or representing popular header's values, also order wise. The encoding of header names and values is provided by Hfinger's configuration file.

The third part of the fingerprint is generated on the basis of request's payload data. In the example in Figure 2, the request contains payload of *Sending a dummy POST request*. This string is built only from ASCII characters and is 28 characters long. The corresponding features generated by Hfinger are *A* and *1.4*.

Final fingerprint is created by combining the three generated parts in predefined manner: URI features, header structure features, and payload features. The fingerprint length is variable as it is dependent on the request's structure and data, for example, payload features are provided only if payload data is present. The final selection of features in particular fingerprint parts and their rounding mode also affects the format and length of the fingerprint. This will be described in Section 5.1.

4. Experimental Methodology

In this section, we first present details related to the malware and benign application data sets that are later used during experimental evaluation. Next, we describe existing fingerprinting tools utilized while conducting comparison analysis. Finally, we outline and define performance measures and experimental methodology.

4.1. Malware Data Set

Malware data set is compiled from two pcap data sets: one was used in the authors' previous work [17] and originates from CERT Polska's sandboxing environment and Malware Capture Facility Project (https://www.stratosphereips.org/datasets-malware, accessed on 26 March 2021). The second data set was derived from a newer version of CERT Polska's malware analysis platform.

The first data set contains 26,133 pcap files analyzed and labeled in the previous work. To this end, Snort IDS with Emerging Threats Pro (ET Pro—https://www.proofpoint.com/us/threat-insight/et-pro-ruleset, accessed on 26 March 2021) and Snort Registered (https://www.snort.org/downloads#rules, accessed on 26 March 2021) rulesets were used. More information about this data set can be found in [17].

The second pcap data set consists of 8674 files and it was created specifically for the purpose of this research. The analyzed pcap files originate from the CERT Polska's malware analysis platform, where Windows-based malware samples are analyzed. The malware samples are obtained from various open-source feeds, for example, Abuse.ch (https://abuse.ch/, accessed on 26 March 2021), from external user uploads via mwdb.cert.pl, accessed on 26 March 2021 malware service, and from the CERT Polska's internal malware hunting systems. The analyzed pcap files were labeled using Suricata IDS and ET Pro ruleset. The labeling process was performed in multiple steps. Firstly, all pcap files were analyzed using Suricata IDS and these logs were saved. Secondly, alert messages from the IDS logs were reviewed semimanually to include only those related to HTTP requests and the malware family's name. Based on the SID rule identification number, the alert messages were labeled with the corresponding malware family name using the information from the corresponding IDS rule. Thirdly, all HTTP requests belonging to a particular network flow, for which the reviewed IDS alert existed, were labeled with corresponding alerts. This step was performed with the assumption that all requests within such a network flow should be treated as malicious. Network flows were identified by source and destination IP addresses and ports. Note that in many cases, HTTP requests were labeled with multiple IDS alerts. As the last step, malware requests were labeled with the malware family name. Again this process was performed semiautomatically by reviewing the names of families corresponding to alerts of particular requests. Requests with multiple different family names were analyzed manually. In most cases, it involved aliases of malware, when names were merged to one, or forks of malware families. These were merged to one name or a specific fork name was chosen. For example, all Ursnif family forks were merged, because alert messages were written by different rule authors, thus incorporating inconsistencies in naming, and the provided fingerprints were identical.

The two data sets mentioned above were merged based on the labeled malware family name. They originate nearly from the same source of malware traffic and use the same intelligence source for labeling, mainly ET Pro rulesets. The final malware data set used in further analyses covers 121 popular malware families with 401,566 HTTP requests. The complete data set provides more data; however, only those malware families were chosen that have at least 20 requests. The top 10 malware families sorted by the number of HTTP requests are presented in Table 2, while the complete list is presented in Appendix C.

Table 2. Top 10 malware families by the number of HTTP requests in the final data set.

Malware Family Name	Number of Requests	Percentage of All Requests [%]
Upatre	62,257	15.50
Simda	57,730	14.38
Locky	44,498	11.08
Dridex	30,070	7.49
Arkei	22,057	5.49
DirtJumper	18,486	4.60
Chthonic	14,410	3.59
Vflooder	14,252	3.55
Ursnif	11,756	2.93
Arid Viper APT	10,063	2.51

4.2. Data Set of Benign Application

Apart from the malicious data set, the benign one was also necessary. To obtain it, network traffic of benign applications was collected from two sources: (i) popular web browsers, including background traffic, and (ii) popular benign applications running on Windows 10.

4.2.1. Popular Web Browsers

The data set of popular web browsers' network traffic was generated by the authors in their previous research [17], where it is described in detail. Various web browsers under the control of different versions of the Windows OS were used to visit websites from the list of 500 most popular websites worldwide, extracted from Alexa top 1 million websites worldwide (http://s3.amazonaws.com/alexa-static/top-1m.csv.zip, accessed on 9 February 2017). The websites were accessed between 9 and 15 February 2017 and between 13 and 18 October 2017, depending on the browser. Table 3 contains information about the networking environment and the number of requests observed in each web browser traffic. Including background traffic, this part of the data set contains 194,940 HTTP requests.

Table 3. Networking environments in which web browser HTTP traffic was analyzed.

Browser Name	Operating System	Number of Requests
Microsoft Edge	Windows 10	17,659
Google Chrome	Windows 7	30,281
Mozilla Firefox (Adobe Flash Player installed)	Windows 7	19,523
Mozilla Firefox	Windows 7	26,131
Microsoft Internet Explorer 11	Windows 7	29,216
Google Chrome	Windows 8.1	22,133
Mozilla Firefox	Windows 8.1	19,082
Microsoft Internet Explorer 11	Windows 8.1	19,807

4.2.2. Network Traffic of Popular Benign Applications Running on Windows 10

Network traffic of popular benign applications running on Windows 10 was obtained using an experimental environment equipped to perform a man-in-the-middle (MitM) attack on HTTPS network traffic. The main objective was to create a data set of network traffic that would be highly similar to the traffic observed in a home or a small business network.

The experimental environment consisted of two virtualized hosts: one, in the remainder of this section called analysis host, was running Windows 10, and the second one, called *MitM host*, was used to provide Internet connectivity and network traffic dumping. Windows 10 OS was obtained from https://modern.ie, accessed on 26 March 2021 in version 1809. Additional root X.509 certificate was installed in the system to provide means for the MitM mechanism. *MitM host* was based on Ubuntu 20.04 LTS OS, equipped with sslsplit tool to perform a man-in-the-middle attack on HTTPS traffic. All network traffic was routed through the *MitM host*. The experiment was divided into three parts that were executed during six consecutive days. The network traffic was not deciphered during the first part, mainly giving unmangled situation and normal traffic. All OS updates were performed during this period.

In the second part of the experiment, the MitM mechanism was enabled and network traffic on ports 80 and 443 was forwarded through sslsplit. sslsplit was configured to work with the least offensive mode to minimize its impact on the network traffic. The traffic was dumped to pcap files for later analysis. During this period, popular benign applications were installed and run. This includes VLC media player, Adobe Acrobat Reader, Steam, Spotify, Discord, Libre Office, and Microsoft Office. The complete list is available in Appendix B. The applications were used to mimic the behavior of a standard user: creating files with Microsoft Office/LibreOffice suites, saving them to OneDrive cloud repositories, opening some saved files, using e-mail clients to download and send messages, listening to music, or downloading files. In all applications, update modules were used to download any available updates. Additionally, some well-known websites were visited using Google Chrome and Microsoft Edge based on Chromium, including registering and logging on popular social media sites such as Facebook, Instagram, and Outlook.com, accessed on 26 March 2021. Internal Windows applications were also used, including weather, calendar, and movie services. Usage of the MitM during this part of the experiment caused some essential OS services to stop working, including Windows Update and Windows App Store. According to multiple sources [20,21], these services send telemetry data using HTTPS with internal, additional X.509 certificate repository and certificate pinning mechanism. When sslsplit was enabled, these applications encountered the error 80245006.

In the third part, the sslsplit was disabled and traffic was dumped in the same manner as during the first part. This phase provided an environment not impacted by the MitM mechanism, with all OS services operating normally and background services of previously installed applications.

The data set contains 53,717 HTTP requests. The top 10 values of the *User-Agent* header value ordered by the number of requests are presented in Table 4. Note that 2.26% of requests do not contain *User-Agent* header or its value is empty.

Table 4. The top 10 values of *User-Agent* header value ordered by the number of requests in the data set of network traffic of popular benign applications running on Windows 10.

User-Agent Header Value	Percentage of All Requests in the Data Set [%]
Mozilla/5.0 (Windows NT 10.0; Win64; x64) AppleWebKit/537.36 (KHTML, like Gecko) Chrome/83.0.4103.116 Safari/537.36 Edg/83.0.478.58	35.87
Microsoft-Delivery-Optimization/10.0	10.57
Mozilla/5.0 (Windows NT 10.0; Win64; x64) AppleWebKit/537.36 (KHTML, like Gecko) Chrome/83.0.4103.116 Safari/537.36 Edg/83.0.478.61	9.46

Table 4. *Cont.*

User-Agent Header Value	Percentage of All Requests in the Data Set [%]
Mozilla/5.0 (Windows NT 10.0; Win64; x64) AppleWebKit/537.36 (KHTML, like Gecko) Chrome/83.0.4103.116 Safari/537.36	6.50
Mozilla/5.0 (Windows NT 10.0; Win64; x64) AppleWebKit/537.36 (KHTML, like Gecko) Chrome/64.0.3282.140 Safari/537.36 Edge/18.17763	5.27
Mozilla/5.0 (Windows; U; Windows NT 10.0; en-US; Valve Steam Client/default/1591251555;) AppleWebKit/537.36 (KHTML, like Gecko) Chrome/79.0.3945.117 Safari/537.36	2.82
Mozilla/5.0 (Windows NT 10.0.17763; Win64; x64) AppleWebKit/537.36 (KHTML, like Gecko) Slack/4.7.0 Chrome/83.0.4103.119 Electron/9.0.5 Safari/537.36 Sonic Slack_SSB/4.7.0	2.52
Valve/Steam HTTP Client 1.0 (0)	1.65
microsoft.windowscommunicationsapps	1.51
Microsoft Office/16.0 (Windows NT 10.0; Microsoft Outlook 16.0.13001; Pro)	1.45

4.3. Fingerprinting Tools Used for Comparison

Three HTTP fingerprinting tools were used for comparison with Hfinger: FATT, p0f, and Mercury. Their overview is presented in Section 2.2. Their source code versions are presented in Appendix A. Code changes and configuration of the tools used in the analysis are described below.

FATT is used in the version provided by its GitHub repository (https://github.com/0x4D31/fatt/, accessed on 26 March 2021). Additionally, to provide similar test conditions between all tested tools, two types of FATT output are further analyzed. The first one is the header hash as provided by the tool. The second one is the header hash with the value of the *User-Agent* header that is the output of the default reporting mode when used with the command-line interface.

p0f was analyzed using source code parts of its Python port (https://github.com/FlUxIuS/p0f3plus, accessed on 26 March 2021). As the tool fingerprints only *GET* requests, the code was patched to analyze all request types to provide the same base for comparison with Hfinger. The code was also patched to support requests with a nonstandard end of line tag: *LF* instead of *CRLF*.

For analysis of Mercury its Python version *pmercury* was used. Mercury's analysis process can be modified using a configuration file to manipulate, for example, the list of analyzed headers or the list of headers that should be represented with their values. Thus, for comparison with other tools, two configurations were used: (i) the default, provided by the authors of the tool and (ii) the same as the default but extended with representing the value of the *User-Agent* header in the fingerprint. The source code was patched to support the analysis of nonstandard requests. These were present when analyzing requests with the nonstandard end of line tag: *LF* instead of *CRLF* and those without protocol version definition. Even though such requests are rarely observed in malware traffic, they should be properly handled.

4.4. Comparison Measures and Methodology

In this research, the performed analyses and comparisons are based on four measures that, in our opinion, provide useful insights into real-life applications of malware HTTP traffic fingerprinting tools. This includes, for example, the uniqueness of the fingerprint across malware families.

Please note that to minimize the effect of different sizes of request sets of analyzed malware families, the measures are computed as averages of each family's partial value, not a global value. Firstly, the analyzed phenomena occurrences are counted separately for each family and then the average value is computed and provided as the final measure. With such an approach, all requests in each malware family set can be analyzed, which could not be achieved if, for example, undersampling methods of data set balancing were used. If the measures were counted with a global approach, the families with a significantly higher number of requests (e.g., Dridex, Upatre, or Chthonic) would bias measure's value impacting the whole analysis.

The comparison measures utilized in this research include: malware collision level, fingerprint generation level, level of collision with benign applications, and entropy. All of them are explained in detail below.

Malware collision level provides information on whether any collisions of request fingerprints between malware families occur, that is, whether request fingerprints are seen across multiple families and are not unique to one family. This measure should be as low as possible to provide exclusive and reliable fingerprints. Malware collision level is computed by firstly counting the ratio of requests with fingerprint collision to all requests for each family, then counting the mean value of these ratios across all families. Malware collision level is expressed by Equation (1), where N is the number of malware families, r_i^c denotes the number of requests with fingerprint collision for malware family i, and r_i expresses the number of all requests for malware family i.

$$C_M = \frac{\sum_{i=1}^{N} \frac{r_i^c}{r_i}}{N} \tag{1}$$

Fingerprint generation level provides information about the number of fingerprints generated for a particular malware family set of requests. It can be interpreted as a measure of an average number of fingerprints generated per analyzed request set (for example, in a single pcap file) but also a measure of the degree to which requests are grouped together. Thus indirectly informing about the degree of a fingerprinting tool's request information generalization. This measure should be as low as possible but still capable of discerning requests that are actually different. It results from requirements that fingerprinter should extract only necessary information from requests and minimize the number of produced fingerprints, not to overwhelm logging and analytic systems. The measure is computed by counting the average ratio of request fingerprints to all malware families' requests. Fingerprint generation level is calculated using Equation (2), where N is the number of malware families, f_i is the number of fingerprints for malware family i, and r_i is the number of all requests for a malware family i.

$$G = \frac{\sum_{i=1}^{N} \frac{f_i}{r_i}}{N} \tag{2}$$

Note that a trade-off between malware collision and fingerprint generation levels exists. When the fingerprinter extracts more information from requests to provide a more unique set of fingerprints, it decreases the collision level. However, it also provides a larger number of fingerprints, as a result increasing the fingerprint generation level. This trade-off is further analyzed in Section 5.1, where optimization of these measures is performed.

The third measure is the level of *collision with benign applications* that provides information about the number of malware fingerprint collisions with some popular, benign applications. In real-life environments, malware operates along with standard, nonmalicious applications. A good fingerprinter should be capable of producing unique fingerprints both to malware and benign applications, thus providing means for discerning these types of applications. This measure is computed similarly to *malware collision level*, i.e., an average value of the ratio between malware requests with fingerprint collision with benign applications and the number of all requests. The level of collision with benign applications is expressed by Equation (3), where N is the number of malware families, r_i^{bc} expresses the number of requests with fingerprint collision with benign applications for malware family i, and r_i denotes the number of all requests for the malware family i.

$$C_B = \frac{\sum_{i=1}^{N} \frac{r_i^{bc}}{r_i}}{N} \tag{3}$$

The final, fourth measure provides information about *entropy* of the tool. It is represented as an average Shannon entropy of fingerprints across analyzed malware families represented in bits. Equation (4) provides the formula of this measure, where N is the number of malware families and H_i Shannon entropy for malware family i.

$$E = \frac{\sum_{i=1}^{N} H_i}{N} \tag{4}$$

Shannon entropy H_i of fingerprints of a particular malware family i is defined by Equation (5), where M_i denotes the number of fingerprints produced by the tool for the malware family i, p(f_j) represents the occurrence probability of fingerprint f_j (computed as the number of occurrences of requests with a particular fingerprint j divided by the number of all requests of the particular family), and log_2 is a logarithm with base 2.

$$H_i = \sum_{j=1}^{M_i} p(f_j) log_2(p(f_j)) \tag{5}$$

Fingerprint entropy E can be interpreted as a measure of the average amount of information provided by malware fingerprints of a particular tool. The higher the value, the better, as in this case, fingerprints are more informative.

5. Experimental Results

Below we present the experimental evaluation of the proposed Hfinger tool. First of all, we demonstrate how the optimal feature set selection has been performed. Then, we outline the results of the comparison of Hfinger with other existing HTTP fingerprinting tools. Note that all tools were analyzed using fingerprint exact match search, and no fuzzy search mechanisms were utilized, even if the tool under evaluation supports it.

Data sets presented in Section 4 were divided randomly into two equal parts based on the malware family (malware data set) or the application name present in the *User-Agent* string (benign data set). For each malware family/application, 50% of the requests were assigned to the first part used to select the optimal feature set, while the rest of the requests were assigned to the part used for the final evaluation of fingerprinting tools.

5.1. Selecting the Optimal Feature Set

The goal of the selection of the optimal feature set is to provide a list of features from those presented in Section 3.2 that will provide the optimal results of the four measures defined in Section 4.4 (i.e., malware collision level, fingerprint generation level, level of collisions with benign applications, and fingerprint entropy). Additionally, some numerical features can be presented with different rounding: with or without fractional component, in the remainder of the text described as a float or as an integer, respectively. Thus, this process will provide information on which rounding would be best for each feature.

The process of feature set selection is based on two steps. Firstly, the defined measures are computed for all 186,623 subsets of features. Secondly, the actual selection was performed using different methods described further in the text, including multiobjective optimization techniques, with the results from the first step.

Figure 3 presents the relationships between all pairs of defined measures: malware collision level, fingerprint generation level, level of collisions with benign applications, and fingerprint entropy for all possible combinations of feature sets.

Analysis of the relationship diagrams in Figure 3 suggests that for many feature sets, with the increase of malware collision levels, the level of collisions with benign applications also increases, while the fingerprint generation level and fingerprint entropy decrease. This follows the intuition that with the increase of fingerprint information, fewer malware requests are incorrectly tagged with the exact representation but for the price of an elevated number of fingerprints.

Based on Figure 3 an interesting observation can also be made. Two distinct result groups can be seen for all diagrams except for fingerprint generation level and fingerprint entropy relationship. One of the groups represents results better suited for optimization. Analysis of feature sets showed that this group contains sets with the order of headers or popular headers and their values, thus indicating a significant impact of those two features on results.

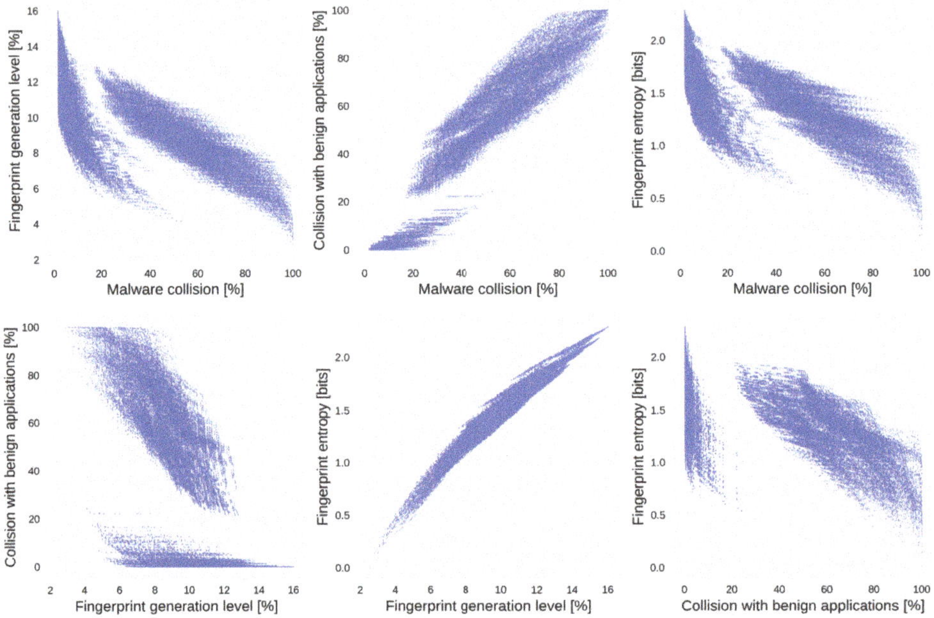

Figure 3. Relationships between defined measures for all possible combinations of feature subsets using training data set. From left to right upper row: (**a**) fingerprint generation level in function of malware collision level, (**b**) level of collision with benign applications in function of malware collision level, (**c**) fingerprint entropy in function of malware collision level. From left to right lower row: (**d**) Level of collision with benign applications in function of fingerprint generation level, (**e**) Fingerprint entropy in function of fingerprint generation level, (**f**) Fingerprint entropy in function of collision level with benign applications.

Using results described above, five feature sets (A–E) are selected, and they are presented in Table 5.

The feature sets (A–E) were selected using the following methods. The descriptions include a short explanation of the main motive behind each selection method:

A lexicographic method from multiobjective optimization techniques (see [22]), where firstly malware collision level was minimized, then using this minimal value, the minimal value of fingerprint generation level was selected. Obtained feature sets had equal values of collision level with benign applications and fingerprint entropy. Finally, a feature set with the lowest number of features was chosen. This set has been chosen using a proven method of multiobjective optimization.

B from all feature sets with maximal feature number, the set with the minimal level of fingerprint generation level was selected. One such set existed. This set has been chosen to provide information about all proposed request features but also to minimize number of generated fingerprints. In some analysis scenarios, e.g., highly similar malware, such complete information might be crucial for discerning malware families.

C firstly, feature sets were limited to those containing features from the defined list. The list was compiled based on the authors' experience with malware analysis and how commonly such features are used in their operational work. The list is formed by features: length of the URI, the extension of the file requested in the URI, representation of header order or representation of popular header's values (at least one feature from this pair), request method, protocol version, payload length, and Shannon entropy of payload. Secondly, the feature sets were filtered to provide only those with a null level of collisions with benign applications and a minimal level of collisions with malware. The feature set with a higher number of features was chosen from two sets with an identical value of fingerprint generation level and fingerprint entropy.

D firstly, fingerprint generation level was limited to 6% (approximately half of the value observed for the sets chosen with the lexicographic method). Then, feature sets with the lowest number of malware collisions were chosen. From four such sets, one with the lowest number of features was chosen. This set has been chosen to provide significantly lower fingerprint generation levels than other Hfinger's feature sets that also are comparable to other tools.

E firstly feature sets with the maximum level of fingerprint entropy were chosen, and then feature sets with minimal fingerprint generation level were chosen. Four feature sets were obtained with this method, where malware collision level and level of collision with benign applications were equal. Hence, the set with the highest number of features was chosen. This set has been chosen to provide the highest entropy level but with the minimal possible number of generated fingerprints, thus giving the most informative fingerprints from all feature sets.

Table 5. Selected feature sets.

Feature Set Name	Feature List
A	average directory length represented as an integer average value length represented as a float number of directories extension of requested file order of headers popular headers and their values payload length represented as a float
B	average directory length represented as an integer average value length represented as an integer number of directories extension of requested file URI length represented as an integer variable length represented as an integer number of variables request method version of protocol order of headers popular headers and their values presence of non-ASCII characters payload entropy represented as an integer payload length represented as an integer

Table 5. *Cont.*

Feature Set Name	Feature List
C	average directory length represented as an integer average value length represented as a float number of directories extension of requested file URI length represented as an integer request method version of protocol order of headers popular headers and their values presence of non-ASCII characters payload entropy represented as an integer payload length represented as a float
D	average directory length represented as an integer average value length represented as an integer extension of requested file URI length represented as an integer order of headers
E	average directory length represented as a float average value length represented as a float number of directories extension of requested file URI length represented as a float variable length represented as a float request method version of protocol order of headers popular headers and their values presence of non-ASCII characters payload entropy represented as a float payload length represented as a float

The results for these five feature sets, along with the results for other analyzed, existing tools, are presented in Table 6.

Table 6. Optimization results for five selected feature sets compared to other analyzed tools. The *UA* suffix marks nondefault configuration of tools supporting *User-Agent* header value as a part of the fingerprint.

Tool	Malware Collision Level [%]	Fingerprint Generation Level [%]	Level of Collisions with Benign Applications [%]	Fingerprint Entropy [bits]
Hfinger (A)	1.76	11.76	0.00	1.72
Hfinger (B)	3.49	11.19	0.00	1.57
Hfinger (C)	1.76	12.09	0.00	1.77
Hfinger (D)	16.85	5.95	1.11	0.87
Hfinger (E)	1.76	15.96	0.00	2.29
FATT	53.04	4.16	24.45	0.54
FATT UA	22.11	6.63	11.87	0.88
Mercury	64.15	4.11	31.33	0.49
Mercury UA	27.13	6.58	15.26	0.85
p0f	15.70	16.71	11.25	1.99

5.2. Comparison of Hfinger to Other Existing Tools

A final comparison of the results was performed using the remaining 50% of the data set, as described in Section 5. The results for the four defined measures are presented in Table 7.

Based on the results presented in Table 7, a general observation can be made that with the default configuration, Mercury provides the worst levels of collision (both malware and benign applications) and fingerprint entropy for all analyzed tools, 63.34%, 31.95%, and 0.46 bits, respectively. It is followed by FATT with the default configuration with malware collision level at 53.45%, benign application collision level at 25.11%, and fingerprint entropy at 0.51 bits. Nevertheless, these two tools provide the lowest fingerprint generation levels: 3.79% for Mercury and 3.83% for FATT. p0f compared to these two tools results in a lower level of malware and benign applications collisions (15.25% and 10.96% respectively) and a higher level of fingerprint entropy (1.98 bits) but at the cost of a higher fingerprint generation level, i.e., 16.41%.

Table 7. Final evaluation of Hfinger's five selected feature sets compared to other analyzed tools. The *UA* suffix marks nondefault configuration of tools supporting *User-Agent* header value as a part of fingerprint.

Tool	Malware Collision Level [%]	Fingerprint Generation Level [%]	Level of Collisions with Benign Applications [%]	Fingerprint Entropy [bits]
Hfinger (A)	1.85	11.76	0.00	1.72
Hfinger (B)	3.58	11.01	0.00	1.58
Hfinger (C)	1.85	12.15	0.00	1.78
Hfinger (D)	16.78	5.78	1.51	0.85
Hfinger (E)	1.78	15.96	0.00	2.30
FATT	53.45	3.83	25.11	0.51
FATT UA	21.77	6.32	12.22	0.87
Mercury	63.34	3.79	31.95	0.46
Mercury UA	26.46	6.27	15.76	0.84
p0f	15.25	16.41	10.96	1.98

When the *User-Agent* header value is used as a part of a fingerprint for FATT or Mercury, the tools provide lower levels of collisions, both for malware and benign applications but also higher fingerprint entropy. For FATT, the malware collision level decreases by nearly 32 pp (percentage points), from 53.45% to 21.77%, while for Mercury, it decreases by almost 37 pp, from 63.34% to 26.46%. Collisions with benign applications decrease by nearly 13 pp from 25.11% to 12.22% for FATT and by 16 pp from 31.95% to 15.76% for Mercury. An increase in fingerprint entropy value is observed from 0.51 to 0.87 bits for FATT and from 0.46 to 0.84 bits for Mercury. These improvements of the three measures' values come with the worsening of the fingerprint generation level that nearly doubles both for FATT (from 3.83% to 6.32%) and Mercury (from 3.79% to 6.27%). These results support intuition of the relationship between the collision level and the fingerprint generation level, i.e., if the tool better discerns applications, the number of fingerprints it provides also elevates.

Further analysis of Table 7 shows that, except feature set D, all other feature sets of Hfinger provide significantly lower levels of malware collision than other tools. Feature sets A, C, and E achieve levels lower than 2% and feature set B lower than 4%. These levels are lower by nearly 60 pp (30 times), compared to the worst value for Mercury, and at least 11 pp (4 times) when compared to the best value of the p0f tool. These feature sets also provide a null value of collisions with benign applications that is lower by 10 to 30 pp when compared to other tools. Hfinger's feature sets A, B, and C achieve higher fingerprint entropy levels than Mercury and FATT, i.e., 1.72, 1.58, and 1.78 bits, respectively. It is about one bit higher than these two tools. However, only feature set E that was chosen to provide the maximum fingerprint entropy of 2.29 bits achieves a higher level than p0f. As observed before with other tools, Hfinger's measure improvements increase fingerprint generation level. For feature sets A, B, and C, they are nearly 12% (11.76%, 11.01%, and 12.15% respectively), and for feature set E, it is 15.85%. These values are higher by 5 to 12 pp (2–3 times) compared to FATT and Mercury but lower by 5 pp or nearly equal compared to p0f.

Hfinger's feature set D was chosen with a focus on providing a lower fingerprint generation level than other feature sets, more comparable to FATT and Mercury; thus, it will be analyzed separately for the sake of brevity of the argument. As Table 7 presents, feature set D achieves lower collision levels for both malware and benign applications when compared to FATT and Mercury. Differences appear in comparing feature set D with the default and nondefault configurations of these two tools. Default configurations achieve lower fingerprint generation levels by nearly 2 pp compared to feature set D but almost the same level for nondefault configurations. The same is observed for fingerprint entropy: default versions achieve lower levels, while nondefault achieve nearly the same values as Hfinger with feature set D. When compared to p0f, feature set D achieves 1.5 pp higher malware collision level, but 10 pp lower fingerprint generation level. The level of collisions with benign software is also lower for feature set D by 9 pp. However, p0f achieves a higher fingerprint entropy level: 1.98 bits when the feature set D: 0.85 bits. The additional perspective of the results for feature set D is provided by the fact that this feature set does not contain information about the values of *User-Agent* header, unlike other Hfinger's feature sets and unlike p0f, and FATT's and Mercury's nondefault configurations. It can be used as a starting point for future work on analyzing how much the *User-Agent* header's value can impact the fingerprint and its capabilities to identify applications.

Overall, the analysis results show that in the majority, Hfinger achieves significantly lower levels of malware and benign applications collisions than other analyzed tools. It results in higher fingerprint generation levels compared to FATT and Mercury but still lower than those of p0f. Fingerprint entropy for Hfinger is also higher than that of FATT and Mercury; however, only one feature set achieves a higher level of this measure than p0f. Specifically designed to decrease the fingerprint generation level, feature set D achieves lower levels of collisions with malware and benign applications when compared to FATT and Mercury but with similar or only 2 pp higher fingerprint generation level. It also provides similar or higher fingerprint entropy. This feature set produces 1.5 pp more malware collisions than p0f and achieves lower fingerprint entropy and has lower levels of fingerprint generation and collisions with benign applications.

Regarding the above analysis, feature set C has been chosen as a default reporting mode for the Hfinger. Firstly, it provides a similar malware collision level as sets A and E, lower than sets B and D. Secondly, its fingerprint generation level is similar to that of set A and lower than of set E. Thirdly, its fingerprint entropy level is lower than of set E but almost identical of set A. Lastly, feature set C provides information about a higher number of features than set A, giving a more complex overview of a request for analyst. Feature set E achieves this with a higher fingerprint generation level.

6. Practical Usage Scenarios and Limitations

In this section practical usage scenarios for Hfinger are presented along with the discussion on limitations of this tool.

6.1. Practical Usage Scenarios

Hfinger was designed to be used as a standard network fingerprinting tool, and its usage cases are no different from other tools. It is capable of reading pcap files, thus it can analyze network traffic originating from different sources, for example, malware sandbox systems, honeypots but also enterprise networks. Hfinger can be used directly to analyze network traffic or it can be used as a subsystem, whose output is ingested by other analysis systems. When used as a standalone tool, it can help the analyst in network forensic objectives. While used as a subsystem, it can feed data into network monitoring or event logging systems, for example, SIEM (Security Information and Event Management) solutions.

Fingerprints created by Hfinger can also be used to identify and track malware in different scenarios. For example, when analyzing the network traffic of unknown malware, fingerprints created by Hfinger can be used to identify requests that were previously labeled as belonging to a particular malware family. Moreover, if the analyzed network traffic consists of multiple HTTP requests, fingerprints can help in grouping them, giving a basis for further analysis of the purpose of the requests—whether it was a connectivity check, C&C server check-in, or some other malicious activity. Additionally, Hfinger can extend and complement IDS systems by using alerts to search for requests that were not reported but have the same fingerprint as those identified. Another application is to identify similarities between different malware families when similar fingerprints for both are discovered; however, this can be achieved when using Hfinger with techniques other than default exact match search. These were not analyzed in this paper but can be used as a starting point for future work.

Hfinger cannot analyze HTTP network traffic secured with HTTPS protocol on its own. However, in many environments and architectures, HTTPS traffic can be inspected, for example, by using TLS keys in sandbox systems or TLS inspection systems in corporate networks.

6.2. Limitations

The presented research and the proposed tool, apart from the auspicious results as outlined in the previous sections, have their limitations that will be discussed below.

Firstly, the authors put maximum effort into the correct preparation of the data sets; however, not all biases could be eliminated. The malware labeling process involved the usage of the ET Pro IDS ruleset. It is a well-known, industry-tested intelligence source that both false positive and false negative errors could be present. That is, some benign HTTP requests were marked as malware, some malicious requests were not alerted, or the malware name provided by the rule was incorrect. Additionally, although malware families were carefully selected for the analysis, their distribution in terms of malware types might not reflect the actual distribution. These biases could influence the results of the analysis and Hfinger feature set selection process.

Secondly, Hfinger capabilities to analyze many features, including header values, can be less efficient for malware families that introduce many changes in the request structure. This can happen, for example, with malware used to perform DDoS attacks, where it is a common technique to change the value of *User-Agent* header with each request. In such a situation, the number of fingerprints created by Hfinger can increase. However, thanks to the fingerprint's modular structure, this issue can be addressed by ignoring the part of the fingerprint generating the higher level of noise.

Thirdly, some malware families tend to incorporate mimicking mechanisms to become similar to benign applications. Depending on the level of mimicry, Hfinger can help to uncover it. If the changes are simple, for example, altering the value of the *User-Agent* header to a benign one, the generated fingerprints will show only a change in this value. However, when the whole structure of a request is changed, then the issue is becoming harder to address. In the worst-case scenario, the request can be changed to such a degree that the malware fingerprint can be the same as of a benign application. Nevertheless, we believe that applying such a degree of mimicry mechanisms would require a lot of design effort and, from our experience, is not typical for malware developers.

Finally, fingerprints produced by Hfinger were designed to be used in exact match searches. Potentially, they can also be used to perform fuzzy searching, for example, by using only some parts of the basic fingerprint. However, we considered it is out of the scope of this paper. The main focus during Hfinger design was put on exact matching, which is supported by many security monitoring and logging tools, contrary to fuzzy search. Furthermore, fuzzy search functionality can be provided with different mechanisms, depending on the monitoring system, thus creating problems with interoperability and potential lack of support of some operations. These issues were analyzed during the design phase of Hfinger, and the decision was made to develop a solution that can be easily integrated into existing deployments of various systems and tools. Nevertheless, fuzzy searching or request clustering mechanisms can be treated as our future work.

7. Conclusions

This paper presents Hfinger, an HTTP request fingerprinting tool. Hfinger analyzes the network traffic and extracts information from different parts of the HTTP requests to provide a simple and interpretable for analyst representation of requests. The fingerprints provided by the tool can be used for exact match searches to identify similar requests between different pcap files and, as such, aid in threat hunting or as a step to identify unknown malware.

The results presented in this paper show that in the default Hfinger reporting mode, the generated fingerprints are 8–34 times more unique between malware families than in other three similar, community-proven, existing fingerprinting tools: FATT, Mercury, and p0f. In the default reporting mode, Hfinger introduces no collisions between malware and benign applications, contrary to the other tools. The number of generated fingerprints is at most about three times higher when compared to FATT and Mercury but 35% lower compared to p0f. Hfinger achieves higher levels of fingerprint entropy than FATT and Mercury but only slightly lower than p0f. In the authors' opinion, the three-fold increase in the number of fingerprints is justifiable by the significant (8–34 fold) increase of fingerprint uniqueness. Thus, this analysis confirms that Hfinger is an effective tool for malware HTTP request fingerprinting.

Hfinger can also operate in other reporting modes that can help achieve better fingerprint entropy levels, provide a lower number of fingerprints, or produce information about a broader set of request features. They offer better or at least comparable results for all measures defined in this paper compared to the other analyzed tools.

Future work will focus on enabling fuzzy searching. This includes, for example, capabilities for searching similar requests on the base of a fingerprint's substring, using a wildcard search or searching depending on the importance of fingerprint elements. Another direction is to use Hfinger as a basis for request clustering mechanisms, which can help to uncover new relations between requests.

Author Contributions: Conceptualization, P.B. and W.M.; Data curation, P.B.; Formal analysis, P.B. and W.M.; Funding acquisition, W.M.; Investigation, P.B. and W.M.; Methodology, P.B. and W.M.; Project administration, P.B. and W.M.; Resources, P.B.; Software, P.B.; Supervision, W.M.; Validation, P.B. and W.M.; Visualization, P.B. and W.M.; Writing—original draft, P.B. and W.M.; Writing—review & editing, P.B. and W.M. All authors have read and agreed to the published version of the manuscript.

Funding: This research received no external funding.

Institutional Review Board Statement: Not applicable.

Informed Consent Statement: Not applicable.

Data Availability Statement: Publicly available pcap file data set of Malware Capture Facility Project was analyzed in this study. It can be found here: https://www.stratosphereips.org/datasets-malware, accessed on 26 March 2021. The pcap file data set provided by CERT Polska cannot be shared due to legal, confidentiality, and privacy issues.

Acknowledgments: The authors would like to thank Paweł Pawliński (CERT Polska/NASK) for his insightful comments and suggestions that allowed for the improvement of this work.

Conflicts of Interest: The authors declare no conflict of interest. The funders had no role in the design of the study; in the collection, analyses, or interpretation of data; in the writing of the manuscript, or in the decision to publish the results.

Appendix A. List of Source Code Versions of the Analyzed Fingerprinting Tools

FATT:
https://github.com/0x4D31/fatt/commit/314cd1ff7873b5a145a51ec4e85f6107828a2c79, accessed on 26 March 2021
p0f3plus:

https://github.com/FlUxIuS/p0f3plus/commit/748cc69cc996e830f258f3f3c7b95ca7a4a74a3e, accessed on 26 March 2021
Mercury:
https://github.com/cisco/mercury/commit/500f5b74a710c0f1c423b8cb370c667aae44a7e3, accessed on 26 March 2021

Appendix B. List of Installed Benign Applications Used in Experiments

- Adobe Reader
- Discord
- GIMP
- LibreOffice
- Mozilla Thunderbird
- Notepad++
- Microsoft Office 2019 Home and Business
- Skype
- Slack
- Spotify
- Steam
- VLC media player
- Zoom
- µtorrent

Appendix C. Malware Families of the Final Malware Data Set Sorted by the Number of HTTP Requests

Table A1. Malware families of the final malware data set sorted by the number of HTTP requests in the final data set.

Malware Family Name	Number of Requests	Malware Family Name	Number of Requests
Upatre	62,257	KeyBase	141
Simda	57,730	STOP	139
Locky	44,498	Nessfi	136
Dridex	30,070	Jaff	136
Arkei	22,057	GrayBird	136
DirtJumper	18,486	Cannibal	130
Chthonic	14,410	1ms0rry	129
Vflooder	14,252	IcedID	122
Ursnif	11,756	Wannacry	113
Arid Viper APT	10,063	Adylkuzz	111
Emotet	9662	Amadey	103
Nemucod	8857	ArtraDownloader	99
Houdini	7583	Zeprox	96
Miuref/Boaxxe	7501	PowershellEmpire	88
Pushdo.S	7012	MegalodonHTTP	88
SmokeLoader	6523	BlackshadesRAT	82
Andromeda	5839	Banload	80
Nymaim	5590	GrandSteal	76
Matsnu	5522	Mokes	73
LokiBot	4415	EightRed	73
Kovter.B	4332	ZeroHTTP	70
Tinba	4004	Sakula	67
Formbook	3496	NetSupport	65
AgentTesla	3052	Legion	62
Gaudox	2880	FindPOS	60
BlackNET	2822	DDI.Bot	59
AZORult	2057	Agent.ZJL	57
Mydoom	1833	Adware.Liuliangbao.A	55
Htbot	1730	DCRS	54
Neutrino	1697	Dalexis	52

Table A1. *Cont.*

Malware Family Name	Number of Requests	Malware Family Name	Number of Requests
Kronos	1692	FTCode	50
PUP.Linkury	1481	MSIL.adv	47
Trickbot	1255	Maze	46
Necurs	1158	KPOT	45
Sage	1145	Sality	41
Hancitor	1034	Madness	41
CryptoWall	613	Dimnie	38
Pony	607	Instagram Like Bot	37
Wizzcaster	567	H1N1	36
QuantLoader	538	Panda	35
TVRat	436	Ratankba	34
Kelihos.F	406	Zeroaccess	33
MedusaHTTP	403	DownloadGuide	33
Karmen	397	Betabot	31
GuLoader	383	Alina.POS	31
KINS	351	SocStealer	30
Tofsee.AX	338	Sezin	30
Predator The Thief	286	Scarab	30
InstallCapital	274	Golroted.B	30
Terdot	256	Agima.o	30
TinyNuke	250	CobaltStrike	29
ColorFish	242	Philadelphia	28
HawkEye	234	Dapato	27
Sarwent	229	Mole	26
GandCrab	229	TorrentLocker	24
DustySky	200	FusionCore	23
Phorpiex	190	Qadars	20
DirCrypt	174	KrugBOT	20
Alphacrypt	174	JakyllHyde	20
Donvibs	168	HPDefender.B	20
DiamondFox	153		

References

1. Miller, S.; Smith, P. *Rise of Legitimate Services for Backdoor Command and Control*; Technical Report; Anomali: Redwood City, CA, USA, 2017. Available online: https://www.anomali.com/files/anomali-labs-reports/legit-services.pdf (accessed on 27 July 2020).
2. Broder, A.Z. Some applications of Rabin's fingerprinting method. In *Sequences II*; Capocelli, R., De Santis, A., Vaccaro, U., Eds.; Springer: New York, NY, USA, 1993; pp. 143–152.
3. Li, K.; Chen, R.; Gu, L.; Liu, C.; Yin, J. A Method Based on Statistical Characteristics for Detection Malware Requests in Network Traffic. In Proceedings of the 2018 IEEE Third International Conference on Data Science in Cyberspace (DSC), Guangzhou, China, 18–21 June 2018; pp. 527–532.
4. Perdisci, R.; Lee, W.; Feamster, N. Behavioral Clustering of HTTP-Based Malware and Signature Generation Using Malicious Network Traces. In Proceedings of the NSDI '10: 7th USENIX Symposium on Networked Systems Design and Implementation, San Jose, CA, USA, 28–30 April 2010; USENIX Association: San Jose, CA, USA, 2010; pp. 391–404.
5. GmbH, A.T. *AV-TEST Security Report 2019/2020*; Technical Report; AV-TEST Institute: Magdeburg, Germany, 2020. Available online: https://www.av-test.org/fileadmin/pdf/security_report/AV-TEST_Security_Report_2019-2020.pdf (accessed on 15 October 2020).
6. Laperdrix, P.; Bielova, N.; Baudry, B.; Avoine, G. Browser Fingerprinting: A survey. *arXiv* **2019**, arXiv:1905.01051.
7. Bortolameotti, R.; van Ede, T.; Caselli, M.; Everts, M.H.; Hartel, P.; Hofstede, R.; Jonker, W.; Peter, A. DECANTeR: DEteCtion of Anomalous OutbouNd HTTP TRaffic by Passive Application Fingerprinting. In Proceedings of the 33rd Annual Computer Security Applications Conference (ACSAC 2017), Orlando, FL, USA, 4–8 December 2017; Association for Computing Machinery: New York, NY, USA, 2017; pp. 373–386. [CrossRef]
8. Bortolameotti, R.; van Ede, T.; Continella, A.; Hupperich, T.; Everts, M.H.; Rafati, R.; Jonker, W.; Hartel, P.; Peter, A. HeadPrint: Detecting Anomalous Communications through Header-Based Application Fingerprinting. In Proceedings of the 35th Annual ACM Symposium on Applied Computing (SAC '20), Brno, Czech Republic, 30 March–3 April 2020; Association for Computing Machinery: New York, NY, USA, 2020; pp. 1696–1705. [CrossRef]
9. Shbair, W.M.; Cholez, T.; Francois, J.; Chrisment, I. A Survey of HTTPS Traffic and Services Identification Approaches. *arXiv* **2020**, arXiv:2008.08339.
10. Stringhini, G.; Egele, M.; Zarras, A.; Holz, T.; Kruegel, C.; Vigna, G. B@bel: Leveraging Email Delivery for Spam Mitigation. In Proceedings of the 21st USENIX Security Symposium (USENIX Security 12), Bellevue, WA, USA, 8–10 August 2012; USENIX Association: San Diego, CA, USA, 2012; pp. 16–32.

11. Bazydło, P.; Lasota, K.; Kozakiewicz, A. Botnet Fingerprinting: Anomaly Detection in SMTP Conversations. *IEEE Secur. Priv.* **2017**, *15*, 25–32. [CrossRef]
12. Fachkha, C.; Bou-Harb, E.; Debbabi, M. Fingerprinting Internet DNS Amplification DDoS Activities. In Proceedings of the 2014 6th International Conference on New Technologies, Mobility and Security (NTMS), Dubai, United Arab Emirates, 30 March–2 April 2014; pp. 1–5. [CrossRef]
13. Kim, T.; Ju, H. Effective DNS server fingerprinting method. In Proceedings of the 2011 13th Asia-Pacific Network Operations and Management Symposium, Taipei, Taiwan, 21–23 September 2011; pp. 1–4. [CrossRef]
14. Blaise, A.; Bouet, M.; Conan, V.; Secci, S. Botnet Fingerprinting: A Frequency Distributions Scheme for Lightweight Bot Detection. *IEEE Trans. Netw. Serv. Manag.* **2020**, *17*, 1701–1714. [CrossRef]
15. Ory, S.; Aharon, F.; Elad, S. *Passive Fingerprinting of HTTP/2 Clients;* Technical Report; Akamai: Cambridge, MA, USA, 2017.
16. Holland, J.; Schmitt, P.; Feamster, N.; Mittal, P. nPrint: Standard Packet-level Network Traffic Analysis. *arXiv* **2020**, arXiv:2008.02695.
17. Białczak, P.; Mazurczyk, W. Characterizing Anomalies in Malware-Generated HTTP Traffic. *Secur. Commun. Netw.* **2020**, *2020*, 8848863. [CrossRef]
18. Sahoo, D.; Liu, C.; Hoi, S.C.H. Malicious URL Detection using Machine Learning: A Survey. *arXiv* **2019**, arXiv:1701.07179.
19. Fowler, G.; Noll, L.C.; Vo, K.P.; Eastlake, D.; Hansen, T. The FNV Non-Cryptographic Hash Algorithm. Available online: https://tools.ietf.org/html/draft-eastlake-fnv (accessed on 27 July 2020).
20. Windows 10 Has an Undocumented Certificate Pinning Feature. Available online: https://web.archive.org/web/20170501183238/http://hexatomium.github.io/2016/09/24/hidden-w10-pins/ (accessed on 27 July 2020).
21. Windows Update—Interception. Available online: https://security.stackexchange.com/questions/31861/windows-update-interception (accessed on 27 July 2020).
22. Arora, J.S. (Ed.) *Introduction to Optimum Design*, 3rd ed.; Academic Press: Boston, MA, USA, 2012.

entropy

MDPI

Article

How to Effectively Collect and Process Network Data for Intrusion Detection?

Mikołaj Komisarek [1,2], Marek Pawlicki [1,*], Rafał Kozik [1], Witold Hołubowicz [2] and Michał Choraś [3]

1 ITTI Sp. z o.o., Rubież 46, 61-612 Poznań, Poland; mikolaj.komisarek@itti.com.pl or mikolaj.komisarek@pbs.edu.pl (M.K.); rafal.kozik@itti.com.pl (R.K.)
2 Institute of Telecommunications and Computer Science, Bydgoszcz University of Science and Technology, 85-796 Bydgoszcz, Poland; witold.holubowicz@pbs.edu.pl
3 Faculty of Mathematics and Computer Science, FernUniversität in Hagen, Universitatsstrasse 11, 58097 Hagen, Germany; mchoras@itti.com.pl
* Correspondence: marek.pawlicki@itti.com.pl

Abstract: The number of security breaches in the cyberspace is on the rise. This threat is met with intensive work in the intrusion detection research community. To keep the defensive mechanisms up to date and relevant, realistic network traffic datasets are needed. The use of flow-based data for machine-learning-based network intrusion detection is a promising direction for intrusion detection systems. However, many contemporary benchmark datasets do not contain features that are usable in the wild. The main contribution of this work is to cover the research gap related to identifying and investigating valuable features in the NetFlow schema that allow for effective, machine-learning-based network intrusion detection in the real world. To achieve this goal, several feature selection techniques have been applied on five flow-based network intrusion detection datasets, establishing an informative flow-based feature set. The authors' experience with the deployment of this kind of system shows that to close the research-to-market gap, and to perform actual real-world application of machine-learning-based intrusion detection, a set of labeled data from the end-user has to be collected. This research aims at establishing the appropriate, minimal amount of data that is sufficient to effectively train machine learning algorithms in intrusion detection. The results show that a set of 10 features and a small amount of data is enough for the final model to perform very well.

Keywords: NetFlow; network intrusion detection; network behavior analysis; data quality; feature selection

Citation: Komisarek, M.; Pawlicki, M.; Kozik, R.; Hołubowicz, W.; Choraś, M. How to Effectively Collect and Process Network Data for Intrusion Detection? *Entropy* **2021**, *23*, 1532. https://doi.org/10.3390/e23111532

Academic Editor: Ernestina Menasalvas

Received: 7 October 2021
Accepted: 15 November 2021
Published: 18 November 2021

1. Introduction

With the list of known network threats expanding every year, researchers and cybersecurity experts are constantly working on new safeguards and new tools of protection. Cybercriminals keep trying to pull newer and more sophisticated tricks to steal sensitive or personal data or cause damage to private businesses or government organizations [1,2].

To facilitate the use of machine learning to streamline network intrusion detection, good quality labeled data need to be collected. This enables the use of highly-accurate supervized learning techniques. The data-dependent algorithms are only as good as the data used to train them.

Motivation, Methodology and Main Objectives

One of the prevailing problems of research in the domain of intrusion detection is the changing characteristics of both network traffic and the contemporary threat landscape. The pace of changes in the field is tightly connected to the intensity of the cyber-arms-race. The constant change in the threat landscape causes the benchmark datasets to lose relevance. The privacy issues and the acquisition costs make the telecom companies reluctant to provide new, labeled data, which in turn causes a constant, high demand for new, relevant intrusion detection datasets.

The motivation and contribution of this paper stems from the realization that there is a discrepancy between the datasets available to the intrusion detection researchers and the types of data that are usable in a real-world deployment of real-time ML-based NIDS. This realization comes directly from the authors' experience in building ML-based NIDS, including the detection component for the H2020 SIMARGL project [3–8].

First and foremost, the data need to be in a flow-type format, in contrast to the packed-based IDS methods, as flow-type network characteristics offer a couple of important advantages. Flow type data can describe the activity on the network much more efficiently [9,10], and the flow-type standards are proper for high-speed networks, as the aggregations allow for significant cutback in the size of data. Additionally, the standards such as NetFlow, IPFIX or sFlow are widely adapted and recognizable in the security community [11,12]. However, not all the fields available in the IPFIX or other flow-based schemas are of value for intrusion detection. In fact, many fields do not contain any relevant information, or the information is redundant. This is reflected in the benchmark datasets, where many features have to be filtered out in the feature selection process of ML-based IDS. While feature selection is a process widely explored in the ML research community and is an important step in the formulation of any ML model, the reverse of the feature selection problem can be a major issue: not including important and informative features in the collected dataset. Having to perform feature selection is also an upfront computational cost, which has to be paid in the training phase. This is not necessary if a standard, informative set of flow-based features is established.

On top of that, many fields in the benchmark datasets are unusable from the perspective of a real-time NIDS, as they can only be calculated having collected a significant amount of flows, such as in [13]. Additionally, some widely used benchmark datasets contain custom-made features: for example, the amount of certain indicators (such as creating programs or entering certain directories on the host machine) in the actions of users in the case of benchmark datasets from the KDD family [14]. These kinds of features are realistically unobtainable in the wild for a number of reasons, starting from the inconvenience of providing these characteristics in a real-time scenario and ending with the issues of privacy.

In a real-world scenario, the deployment of an NIDS requires the collection of a sample labeled dataset from the target network due to deployment shift [15]. This is a costly and inconvenient process, so acquisition of the minimum amount of data is common sense. This paper explores the notion of minimal sets of data required for effective detection. One more important contribution stems from the fact that establishing a standardized set of features effective for NIDS helps with the use of transfer learning techniques across multiple datasets. Based on the popular flow-based data schemas, the research process presented in this paper addresses a research gap related to the verification of a list of features that contribute to network intrusion detection. In addition, the research aims to answer the question of what minimum amount of data is sufficient to effectively and efficiently train a machine learning model for threat detection. Thus, the main objectives of the paper are as follows:

1. To establish and verify an optimal set of flow-based features usable for network intrusion detection,
2. To establish the minimal amount of labeled data necessary to train a machine-learning-based NIDS for effective deployment,
3. To clear the path for anyone wishing to collect an NIDS dataset.

To validate the findings of the paper, a set of four commonly used machine learning algorithms is trained using the established flow-based features on five benchmark datasets.

To summarise and clearly state the major contributions, the paper does the following:

1. Establishes a set of effective and usable flow-based features based on five recent benchmark datasets.
2. Establishes a minimum amount of data that allows the training of an ML classifier in IDS.

3. Validates the findings by training a set of different ML models and reports the results.

As illustrated in Figure 1, five steps can be distinguished between network traffic flowing through a certain point in a network and actually publishing the dataset for the research community to use. The traffic has to be collected and reliably labeled, then an adequate and usable set of features needs to be extracted, then the collected dataset should go through evaluation and validation procedures to check its usefulness for ML procedures, and then the dataset can be published for the community to use. The work contained in this paper revolves around improving the feature extraction and selection phase, and the validation of this work is also provided.

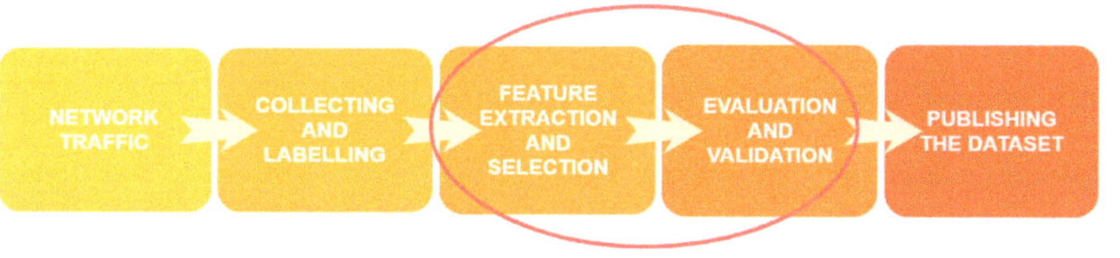

Figure 1. The steps required to go from network traffic to publishing of a dataset suitable for ML methods. The red ellipse indicates the focus of this paper.

Our previous research has focused on analyzing network traffic based on the NetFlow data format. In [3], we have proposed a new dataset derived from a real-world, in-the-wild network. The dataset is collected, described and now published. The dataset has 44 features and contains labeled data. Its use for intrusion detection has been verified and validated by using the following algorithms: random forest classifier, gradient boosting classifier, and a neural network. The detection efficiency oscillated around 99%. After the publication of the dataset, the focus of our work shifted to further improving the quality of the data collection process, for the next iteration of the dataset. This paper contains the results of this work.

The paper is structured as follows: in Section 2, a brief overview of existing approaches both in feature selection and in the way network data are prepared is presented, along with a review of articles regarding data quality. Sections 3 and 4 describe how the datasets were collected and present the datasets used in this research article. Section 5 describes the methodology used with respect to feature selection and verification of the amount of data needed to effectively train the algorithms. Finally, Section 6 presents the results and evaluates the impact of the amount of data on the training of ML models. The paper closes with conclusions and future plans.

2. Related Works

In the 1990s, a group of researchers led by Professor Richard Y. Wang conducted a study in which they formulated the concept of "data quality", which equals "fitness for use" [16]. Within this research, the definition of "data quality dimension" was also introduced, which referred to a set of attributes that defines the construct of data quality.

Since the 1990s, the evolution of the Internet has caused the approach to the field of data quality to change dramatically. Li Cai and Yangyong Zhu present the current challenges in the era of big data in their research [17]. They identified several problems that arise in this day and age, the first being the variety and complexity of data sources and types. This phenomenon appears in the literature, e.g., in [18] or [19]. Another challenge is the huge amount of data coming from all directions and the fact that it is difficult to assess the quality of these data in a short period of time. Moreover, the change and validity of these data is very short, which makes data processing even more important nowadays. In the rest of the paper, the authors proposed the establishment and a hierarchical structure of a data quality framework and presented a process for assessing the quality of large amounts of data.

The authors of [20] provide an extensive analysis and review of currently existing data quality approaches to big data. At the very beginning, they cite the concepts of big data and present the life cycle of such data. The initial phase of collection starts with data generation or retrieval, then one enters the phase of data acquisition to move on to data storage in the next stage and then to data processing and analytics. The whole process ends with the data visualization aspect. In the next part of the article, the authors analyze the quality of data by presenting a wide range of articles that show the problems of maintaining quality in large datasets. They conclude that there is no complete reference model for data quality and management in big data.

In addition to the quality of the data itself, the size of the dataset has quite a significant impact on the machine learning process. The authors of the research paper [21], who base their research on the medical domain, examine the impact of the size of the learning set on model quality and performance. For this purpose, they conducted a set of experiments on six popular machine learning models using medical datasets. To measure the effect of data volume on model performance, they prepared a set of three subsets of different sizes and a series of metrics to compare performance. The authors of the paper emphasize that their study shows that it is not the size that affects the performance of the classifier but the degree of the dataset that represents the original distribution. Another conclusion from the research is that for a limited dataset, the AdaBoost and Naïve Bayes classifiers perform best and the decision tree classifier performs worst.

The authors of [22] used three distributed algorithms—extreme learning machines (ELM), distributed random forest, and distributed random boosted-trees—to detect botnet attacks. In the research paper, they presented the concepts and architecture of the system, which was based on big query data processing. Network data analysis in the form of NetFlow was used as a use case. The results provided in the conclusion show that this is a very promising work.

In [23], the authors have addressed the topic of data characterization, namely the problem associated with the imbalance of infected samples from normal traffic. For this reason, the authors presented a number of studies regarding data balancing and its impact on various machine learning algorithms.

The authors of [11] analyze data in the form of NetFlow and IPFIX with respect to network traffic monitoring. They point out at the very beginning of the paper that these protocols are used for scaled fast network flow export. The article itself introduces the reader to the history of these two protocols, and outlines the fundamental differences between them. An example architecture of a flow monitoring system is presented by the authors. In addition, a comprehensive comparison of network traffic collection tools is provided.

A similar analysis is performed by the authors of [24]. They rely on a review of machine learning and data mining methods used in cyber analytics to support intrusion detection.

The effect of data volume on machine learning effectiveness has also been examined in [25], in which researchers use data from Tweets to test algorithms such as decision trees, naïve Bayes, nearest neighbor and radial basis function network. Based on the results, the developers suggest that increasing the data size improves performance but the effect of this improvement decreases as the sizes of the datasets increase. They also note that it is more important to add additional samples to small datasets than to larger ones. The best classifier proved to be naïve Bayes, which was also the fastest in the training process and achieved good results on the smallest datasets.

A slightly different area is explored by the authors of [26]. They focus on investigating the optimal size of the number of features in the random forest algorithm. The authors' conclusions emphasize that the hypothesis is true and that there is no functional relationship between the optimal size and the characteristics of the datasets being checked. They confirm this after using the out-of-bag error method and SearchSize using random forest.

This work is mainly concerned with good quality network data based on NetFlow and selecting the optimal number of features and set size with respect to intrusion detection. There are many research articles that base the detection of undesirable events in computer networks by verifying NetFlow data.

In [27], the authors use NetFlow-based network traffic. K-means and genetic algorithms were compared and tested. These two approaches were used to find the undesirable parts of the network traffic. Ultimately, the results that were obtained in this paper prove that the genetic algorithm is better suited to calculate the so-called survival curves.

The authors of [28] perform intrusion detection in network traffic. However, the proposed method is based on identifying anomalous end-user nodes and their network traffic patterns. The authors point out that frequently changing IP addresses make their method ineffective. It can be concluded that when identifying anomalies, IP addresses and ports should not be used.

The network traffic intrusion detection architecture proposed in [29] is based on the use of a time series clustering algorithm. The authors show that the algorithm is able to detect anomalies in live data without any prior knowledge of the data.

In [4,5], the authors focused on presenting the concept of architecture and software, the task of which was to analyze the traffic in real-time from the data provided by the stream. The authors use the scalable Apache Kafka environment, Apache Spark and the Elasticsearch database for this purpose. For efficient intrusion detection, network flows in the data stream are grouped by source IP address into one-minute time windows.

The authors of [30] propose a moderate architecture of a convolutional neural network (CNN) to facilitate a decrease in the resources consumed by computations in large-scale intrusion samples, attempting to better the classification metrics.

In [31], the researchers present two approaches grounded in wavelets to effectively mine and analyze network security log databases. Using wavelets allows the extraction of adequate frequency components. The authors conclude that using wavelet transforms grants the ability to de-noise the data, which in turn permits faster querying.

The authors of [32] introduce multiscale Hebbian learning to tackle the challenge of inadequately labeled data in network intrusion detection. The experiment conducted on the UNSW-NB15 dataset shows that the approach can spot overlapping classification boundaries.

A mixed wavelet-based neural network model for cyber security situation prediction is evaluated in [33]. The approach shows significant improvements over the state-of-the-art.

In [34], a comprehensive survey of machine learning and deep learning approaches to intrusion detection can be found. The study taxonomizes the IDS systems by detection method and source of data, then lists the common learning algorithms employed for IDS, including both the shallow and deep learning models.

3. Machine Learning over NetFlow Data

In this work, the main focus of the research is on finding a suitable data scheme to detect intrusions in network traffic in an effective manner. This paper showcases the process of evaluating flow-based features in the task of network intrusion detection. This work also establishes the minimum amount of data needed to train the various machine learning algorithms.

The most popular versions of NetFlow are versions 5 and 9, thus this work will focus on fields obtainable using this format. The topic of network intrusion detection with the use of NetFlow is a well established approach in the research community. Multiple papers have been published proving that it is possible to build a working ML IDS based on flow-based data [35].

3.1. Collecting Data

Data collection is of crucial importance in the entire process of building a machine learning model. Providing good quality and sufficient amount of training data allows the algorithm to be trained effectively. NetFlow collection can be achieved by tools such as Nprobe or Ntop [36]. The ability to collect traffic is also possible using a configured ElasticStack environment. Nprobe [37] is a tool that includes both a probe and a collector. It grants the ability to export traffic as NetFlow v5/v9/IPFIX. The authors of this tool point out the following as the main advantages of this software: the full support for IPv4 and IPv6, and the ability to automatically export to an SQL database and to a Kafka stream, as well as low CPU and RAM resource consumption.

Nowadays, with the size of network traffic growing at a staggering pace, there is a need for scalable solutions that can handle the increasing amount of data. In [5], an in-depth description of the software and its architecture is presented. The detector relies on collecting data using a NetFlow collector, and the data are transferred directly to the Kafka stream. The data are then processed by the solution, which performs intrusion detection, and the result along with the features of the sample are recorded to the Elasticsearch database. This solution architecture is highly scalable and tools such as Apache Kafka guarantee redundancy.

3.2. Datasets

Five datasets were selected for testing on NetFlow-based collections. They represent recent network traffic and therefore they reflect relatively current network behavior. The authors of [38] converted five popular datasets to a strict NetFlow format. A detailed description of the individual features available within the sets can be found in Table 1. The datasets provide 33 numeric parameters and four text parameters.

- UNSW-NB15 [39]—The dataset was created in 2015, with the IXIA PerfectStorm tool. Using this software, clean traffic and various types of network anomalies were generated. Approximately 100 GB of data stored as PCAP files was collected and thanks to the developers at The Cyber Range Lab of the Australian Centre for Cyber Security (ACCS), the collection has been made public as part of further research into improving network security. The structure of the collection originally contained 49 features and encompassed 2,218,761 samples of clean traffic, which is about 87.35% of the whole collection. The rest, i.e., 321,283 network frames, is made up of executed attacks.
- BoT-IoT [40]—Developers in Australia (ACCS) also created this dataset in 2018. In this case, a network flow taking place in a real network environment was recorded. This collection estimates about 69 GB of data in PCAP format and contains 42 features. The diversity of traffic in this collection is very uneven as it contains only 477 frames and there are 3,668,045 flows of the infected traffic. This results in normal traffic of only 0.01%.
- ToN-IoT [41]—this data collection is very similar to the BoT-IoT collection, as it also contains very many attacks and very little normal traffic. The collection comes from the IoT network, more precisely from service telemetry data, and was recorded in 2020. The number of infected frames equals 21,542,641 samples while normal traffic is only 796,380 flows. This represents a percentage of 96.44% for the infected samples and 3.56% for normal traffic, respectively.
- CSE-CIC-IDS2018 [42]—in 2018, another dataset made available through a collaboration between two organizations: Communications Security Establishment (CSE) and the Canadian Institute for Cybersecurity (CIC), was released. This is a very realistic set, as the scenario was designed using the infrastructure of five large organizations and server rooms. Normal traffic was generated by human users and several different machines were used to attack these networks. The whole collection contains 73 features and consists of a large amount of data amounting to 16,232,943 flows. The attacks in this collection represent 2,748,235 samples and the normal traffic represents 13,484,708 flows.
- UQ-NIDS [43]—a dataset that was created by combining the four previously presented datasets. It represents the advantages of shared datasets, where it is possible to combine multiple smaller datasets, which leads to a larger and more versatile NIDS

dataset containing flows from multiple network configurations and different attack settings. This network dataset contains 11,994,893 flows, of which 9,208,048 (76.77%) are benign flows and 2,786,845 (23.23%) are attacks.

Table 1. Description of the original features found in the datasets: UNSW-NB15, BoT-IoT, ToN-IoT, CSE-CIC-IDS2018, UQ-NIDS.

Feature	Description
IPV4_SRC_ADDR	IPv4 source address
IPV4_DST_ADDR	IPv4 destination address
L4_SRC_PORT	IPv4 source port number
L4_DST_PORT	IPv4 destination port number
PROTOCOL	IP protocol identifier byte
L7_PROTO	Layer 7 protocol (numeric)
IN_BYTES	Incoming number of bytes
OUT_BYTES	Outgoing number of bytes
IN_PKTS	Incoming number of packets
OUT_PKTS	Outgoing number of packets
FLOW_DURATION_MILLISECONDS	Flow duration in milliseconds
TCP_FLAGS	Cumulative of all TCP flags
CLIENT_TCP_FLAGS	Cumulative of all client TCP flags
SERVER_TCP_FLAGS	Cumulative of all server TCP flags
DURATION_IN	Client to Server stream duration (msec)
DURATION_OUT	Client to Server stream duration (msec)
MIN_TTL	Min flow TTL
MAX_TTL	Max flow TTL
LONGEST_FLOW_PKT	Longest packet (bytes) of the flow
SHORTEST_FLOW_PKT	Shortest packet (bytes) of the flow
MIN_IP_PKT_LEN	Len of the smallest flow IP packet observed
MAX_IP_PKT_LEN	Len of the largest flow IP packet observed
SRC_TO_DST_SECOND_BYTES	Src to dst Bytes/sec
DST_TO_SRC_SECOND_BYTES	Dst to src Bytes/sec
RETRANSMITTED_IN_BYTES	No. of r-d TCP flow bytes (src->dst)
RETRANSMITTED_IN_PKTS	No. of r-d TCP flow packets (src->dst)
RETRANSMITTED_OUT_BYTES	No. of r-d TCP flow bytes (dst->src)
RETRANSMITTED_OUT_PKTS	No. of r-d TCP flow packets (dst->src)
SRC_TO_DST_AVG_THROUGHPUT	Src to dst average thpt (bps)
DST_TO_SRC_AVG_THROUGHPUT	Dst to src average thpt (bps)
NUM_PKTS_UP_TO_128_BYTES	Packets whose IP size $\leq$ 128
NUM_PKTS_128_TO_256_BYTES	Packets whose IP size > 128 and $\leq$256
NUM_PKTS_256_TO_512_BYTES	Packets whose IP size > 256 and $\leq$512
NUM_PKTS_512_TO_1024_BYTES	Packets whose IP size > 512 and $\leq$1024
NUM_PKTS_1024_TO_1514_BYTES	Packets whose IP size > 1024 and $\leq$1514
TCP_WIN_MAX_IN	Max TCP Window (src-dst)
TCP_WIN_MAX_OUT	Max TCP Window (dst-src)
ICMP_TYPE	ICMP Type * 256 + ICMP code
ICMP_IPV4_TYPE	ICMP Type
DNS_QUERY_ID	DNS query transaction Id
DNS_QUERY_TYPE	DNS query type (e.g., 1 = A, 2 = NS.)
DNS_TTL_ANSWER	TTL of the first A record (if any)
FTP_COMMAND_RET_CODE	FTP client command return code

3.3. Feature Selection

The first stage of the research will be to sift and select features from the NetFlow datasets and thus establish the most relevant flow-based set of features to feed the ML algorithms. Only those features will be extracted that show the best performance in the intrusion detection process and positively affect the detection metrics of the model. Several selection techniques will be applied to maintain the validity results and provide the most relevant set of features.

In a 2020 survey of feature selection techniques [44], the authors describe the different variants and possibilities to correctly pick features. The authors of [44,45] indicate that feature reduction leads to reduced complexity, which translates into reduced computation time. The feature selection techniques that will be used in the following experiments can be divided into four categories [46–48] :

- Filter methods,
- Wrapper methods,
- Embedded methods,
- Hybrid methods.

The categorical values that are present in the datasets to be subjected to the feature selection technique must be preprocessed by functions such as ordinal encoder, one hot encoder, or other methods [49]. The list of features found in the five analyzed datasets in this paper has several categorical values, such as IP addresses or destination and source ports. These features are not involved in the research conducted in this article and are omitted, so only the non-categorical values remain in the dataset, although an overview of all methodologies that can be used when categorical values are present in the data can be found in [50]. The rejection of these features is dictated by the fact that IP addresses and ports change dynamically from network to network and relying on them limits the capabilities of the algorithm.

The first technique that was used to study the number and effectiveness of features is LASSO regularization [51,52], which is an embedded method and combines both wrapping and filtering methods. The term regularization refers to the concept the application of which is to prevent data overfitting. The way this method works is based on adding penalties to the parameters to reduce the freedom of the models. There are two main regularization techniques, namely ridge regression and LASSO regression [52]. In the LASSO technique, shrinkage is used. This is where the data values are shrunk towards the center point as the mean. The LASSO regularization technique itself seeks to create simple models with a reduced list of parameters. This type of regression is used for models that exhibit high levels of collinearity, or to automate variable selection/parameter elimination. When the model uses an L1 regularization technique then it is called LASSO regression. If it uses an L2 regularization technique, then it is called ridge regression. Due to the fact that this paper will use the L1 approach, a penalty equal to the absolute value of the coefficient size is added in this technique.

The choice of the L1 technique in this research was dictated by a key difference between the L1 and L2 techniques. Namely, LASSO reduces the coefficient of a less important feature to zero, completely eliminates it from the dataset, which works perfectly for feature selection. In turn, L2 is mainly used to avoid the over-fitting problem.

Some coefficients may become zero and be eliminated from the model. Larger penalties result in coefficient values closer to zero (ideal for creating simpler models). Because it is a linear model type, a penalty is imposed on the coefficient that participates in the multiplication of each predictor. The mathematical notation of this technique is shown below:

$$\frac{1}{2n}\sum_{i=1}^{n}(y_{real}-y_{pred})^2+\alpha\sum_{j=1}^{p}\mid a_j\mid \tag{1}$$

In Equation (1), it is assumed that the dataset has n instances and p features. The y_{real} and y_{pred} parameters define for us the predicted value result and the real value result. The parameter a is considered a hyperparameter in this formula. The purpose of regression is to reduce the values of the coefficients to exclude useless features. When a is 0, it reverts to the original linear regression. If α is too large, it neglects the first part of the cost function and the results are unreliable. The LASSO regression concept ultimately leads to the optimization of the cost function. This is achieved by reducing the absolute value of the coefficients. This technique is only likely to work if the features are normalized.

Random Forest Importance [53] is another method that was used to compare the effect of parameters on model quality. Calculating the importance of features found in a dataset using the random forest technique answers the question of what features will be appropriate in subsequent classification, as well as regression. In his 2001 paper, Breiman [54] presented a technique called out-of-bag (OOB) importance score. Its operation is based on calculating the difference between the original mean error and the randomly permuted mean error in the OOB samples. All feature values are stochastically changed for each tree and using this operation using random forest model to predict this permuted feature and obtain the mean error. If the average error decreases drastically, it means that the feature is strongly correlated [55].

Another way that shows the importance of a feature is to rank the tree due to the decrease in impurity (Gini impurity) relative to all trees [56]. According to the principle of the algorithm, the most impure trees are at the beginning and the least impure trees are at the end. With this division, a set of most important features can be easily created. The mathematical formulation of this procedure can be represented as follows. For each node A for the decision tree, the partitioning is performed after decreasing the impurity of the node $R(A)$. The impurity of a node is represented by a Gini index. The determination of the Gini Index can be defined as subtracting the sum of squared probabilities of each class from one. If the samples of class Z are contained in a subset of A then the impurity Gini(A) can be defined as seen in Equation (2)

$$R(Z) = 1 - \sum_{j=1}^{Z} (P_j)^2 \tag{2}$$

where P_j is the relative frequency of class j in Z, or in other words, it is the probability of an element being classified in another class. After splitting Z into two different nodes Z_1 and Z_2 with two different data sizes N_1 and N_2, the Gini index can be defined with the formula in Equation (3):

$$Gini_{split}(Z) = \frac{N_1}{N} Gini(Z_1) + \frac{N_2}{N} Gini(Z_2) \tag{3}$$

The splitting of a given node occurs when $Gini_{split}(z)$ is the smallest value. The importance value of feature X_j in a single tree T_k is defined by the formula in Equation (4):

$$S_k(X_j) = \sum_{t \in T_k} \Delta Gini_{split}(t) \tag{4}$$

Finally, this formula is applied to every tree in the set. In the above Equation (4), t represents a single node belonging to a single tree T_k, for each node, the Gini split method is calculated. This is how the best influencing features on the model are selected using the RF importance method.

The final feature selection method used in this paper is the chi-square method [57]. It is based on the calculation of chi-square between each feature and the target value.

$$chi^2 = \sum_{e_t \in \{0,1\}} \sum_{e_c \in \{0,1\}} \frac{(N_{e_t e_c} - E_{e_t e_c})^2}{E_{e_t e_c}} \tag{5}$$

In Equation (5) N is the observed value of w, and E the expected value. e_t takes the value of 1 if the document contains the term t, and 0 otherwise. e_c takes the value 1 if the document belongs to class c, and 0 otherwise.

3.4. The Effect of Training Data Size on the Model

As part of the development of the area of network traffic intrusion detection, the next objective of this paper was to conduct a study related to finding a sufficient amount of learning data that can be used to effectively teach machine learning or neural network algorithms, taking into account the aspect of no loss of efficiency of a given algorithm. The process of testing such an assumption started by dividing the sets into smaller parts. Given five different datasets based on the same scheme, nine subsets containing different numbers of samples were extracted from each of them. To maintain the greatest reliability between data iterations of testing within each larger subset, data from smaller subsets were included. The process of the algorithm was shown in the pseudocode format in Algorithm 1. The preparation of this stage of the research began with loading all five sets into memory, then within each iteration/set the nine smaller subsets were extracted. For each subset, four algorithms were trained and their performance was tested on the test data by generating ROC curve plots and a set of metrics. The ROC curve is a graphical representation on a graph on the Y axis of a value relating to specificity, and on the X axis of a value relating to a false-positive rate. The range of values of this curve is represented from 0 to 1. Values closer to 1 indicate better performance.

Algorithm 1 The process of extracting and splitting a dataset.

```
Require: datasets[D1, D2, D3, D4, D5]
  trainDataset, testDataset = dividing_the_set_in_a_ratio_of_70_to_30
  subDataset = [10, 100, 500, 1000, 2000, 10000, 20000, 30000, N]
  tempDataset = []
  for each d ∈ trainDataset do
      for each size ∈ subDataset do
          tempDataset = d[size] + tempDataset
          d = d − tempDataset
      end for
  end for
```

The final phase of testing was the preparation of a summary set, which contained a set of metrics with respect to the trained model. In this phase, the standard split of training to test data was 70 to 30. Several metrics were used to evaluate the performance and correctness of the model to give a good summary of the results. The models were evaluated [58,59] using: Accuracy (ACC-Equation: (6)), Precision (Pr-Equation: (7)), Recall (Re-Equation: (8)), F1-Score (Equation: (9)), Balanced accuracy (BCC-Equation: (10)) [60] and the Matthews correlation coefficient (MCC-Equation: (11)) [61]. Within the listed metrics that describe the performance of given algorithms in this study, values such as True Positive (*TP*), True Negative (*TN*), False Positive (*FP*), False Negative (*FN*) were needed to calculate them.

Presented below are the individual formulas that were considered in the process of evaluating the performance of the algorithm.

$$Accuracy = \frac{TP + TN}{TP + FP + FN + TN} \tag{6}$$

$$Precision = \frac{TP}{TP + FP} \tag{7}$$

$$Recall = \frac{TP}{TP + FN} \tag{8}$$

$$F1 = 2 * \frac{Recall * Precision}{Recall + Precision} \tag{9}$$

$$BCC = \frac{\frac{TP}{TP+FN} + \frac{TN}{TN+FP}}{2} \tag{10}$$

$$MCC = \frac{TN * TP - FN * FP}{\sqrt{(TP + FP)(TP + FN)(TN + FP)(TN + FN)}} \tag{11}$$

3.5. Classification Models

To validate the effectiveness of the established set of features, four different widely used ML algorithms were trained and tested. The algorithms were chosen to cover different paradigms of machine learning: AdaBoost for tree-based based algorithms, ANN for gradient-based algorithms, naïve Bayes for Bayes theorem-based algorithms. The random forest was included, as in our experience it is a good fit for flow-based IDS. The performance of the four algorithms was tested for optimal dataset size and network traffic intrusion detection.

The first reported algorithm in the set is random forest. As the name suggests, this algorithm consists of multiple decision trees that form a forest. This tree-based algorithm has been utilized in many IDS and research papers; some examples of its use and effectiveness can be found, e.g., in [62–64]. The model is trained using bagging (bootstrap aggregation) techniques. The outcome of the algorithm is determined by the average scores from each tree. The random forest eliminates the limitations of the decision tree algorithm. It reduces the overfitting to the datasets and increases precision [65]. Entropy is worth mentioning within this algorithm. Entropy [66–68] is a measure of disorder or, in other words, uncertainty. It is expressed by the Formula (12):

$$E(x) = \sum_{i=1}^{c} -p(x_i) \log_2 p(x_i) \tag{12}$$

In this formula, $p(x_i)$ expresses the measure of the probability of the frequency of the occurrence of element/class "i" in the data. When using machine learning algorithms, the goal of the data scientist is to reduce the disorder. The metric for reducing this disorder is expressed by the following Formula (13):

$$IG(Y, X) = E(Y) - E(Y|X) \tag{13}$$

The operation of this metric is to subtract the entropy of Y from a given X from the entropy of Y itself, given the additional information X has about Y. This process is called information amplification. As there is more uncertainty reduction, more information about Y is obtained from X. Entropy as a metric is involved in the decision tree process. During the construction of decision trees, data partitioning is calculated using information gain (IG). IG is a measure that defines how much "information" a feature gives us about a class. The attribute with the highest information gain will be split first in the tree construction process. The number of trees hyperparameter was set to 100.

The second algorithm used for intrusion detection in this paper is the adaptive boosting classifier (AdaBoost). The algorithm itself has its effectiveness proven in the scientific literature on intrusion detection [69–71]. The technique involves adding an element of boosting and adaptively adjusting the weights when a misclassification is made. In this process, weak weights are converted to strong weights. Boosting is used to reduce bias as well as variance for supervized learning. The algorithm used in this research had the learning rate set to 1 and the N_components hyperparameter to 100.

The third model used in the intrusion detection study was based on a neural network. A simple neural network was designed, which consisted of two hidden layers and two dropout layers. The first hidden layer contains 32 neurons and the second layer contains 16 neurons, the two abandonment layers were set to 0.01, while the activation layer in the hidden layers was set to the Rectified Linear Unit (ReLU). The last layer contained the number of neurons equal to the number of classes and used a softmax activation function. The loss function was set to the "categorical_crossentropy" method while the chosen optimization algorithm was adaptive momentum [72]. Early stopping stopped the learning process at 16 epochs, and the batch size was set to 20.

The final algorithm used in our study is naïve Bayes. This classifier is based on Bayes' Theorem with the assumption of independence of predictors. It states that the presence of a feature in a class is not dependent on any other feature. The naïve Bayes model is easy to build and is particularly useful for very large datasets [73]. The default hyperparameters were used for this classifier.

All types of models that have been selected for this research have characteristics that make them promising for an in-the-wild application of IDS. The artificial neural network model was used because of the fact that ANNs continue to learn even when other methods reach their full potential. Using more data for training this algorithm can improve the detection performance results. While there are a myriad of ANN-based algorithms that could be applied here—convolutional neural networks, recurrent neural networks, etc.—there already exists plenty of research that deals with the specifics of certain deep neural network paradigms in IDS. The focus of the use of ANNs in this paper is only on validation of the feature set and the minimal amount of data. The second model to be discussed is the AdaBoost. This algorithm is fast, easy to use and does not require extensive tuning of hyperparameters. Random forest has proven itself in many network attack studies and its performance has always been high and the results satisfactory; the authors found promising results of using this algorithm in previous work [3–5]. The last chosen algorithm is the naïve Bayes classifier. Its performance is based on a strong assumption of independence, and in a literal sense, it refers to the statement that the probability of one attribute does not affect the probability of another. One of the most important aspects of why this algorithm was selected for validation of this research is that it can perform better than other algorithms in situations with little training data.

In summary, parameter tuning was applied for each model to obtain the best possible results. For this purpose, the GridSearch technique was used. The effects of parameter tuning for each model can be observed in Table 2. The table contains two columns that represent the name of the parameter and its final value. A 10-fold cross validation was used during the GridSearch process.

Table 2. The final result of tuning hyperparameters by using the GridSearch technique.

Model	Parameter	Value
Random Forest	n_estimators	200
	max_features	auto
	max_depth	8
	criterion	entropy
AdaBoost	n_estimators	230
	learning_rate	0.05
Naïve BAYES	var_smoothing	10^{-9}
ANN	epochs	16
	batch size	20
	loss function	categorical_crossentropy

4. Experiments and Results

The first stage of research in this paper was to select an appropriate number of features from five network datasets based on the NetFlow format. Each set was subjected to a feature selection process. The correlations of features with each other in the dataset were evaluated. The following values were eliminated from the set due to high correlation with other features: OUT_PKTS, CLIENT_TCP_FLAGS, MIN_TTL, SHORTEST_FLOW_PKT,

DST_TO_SRC_SECOND_BYTES. A summary of the exploration of the feature space offered by the NetFlow schema culminates in Figure 2, showing the feature correlation map.

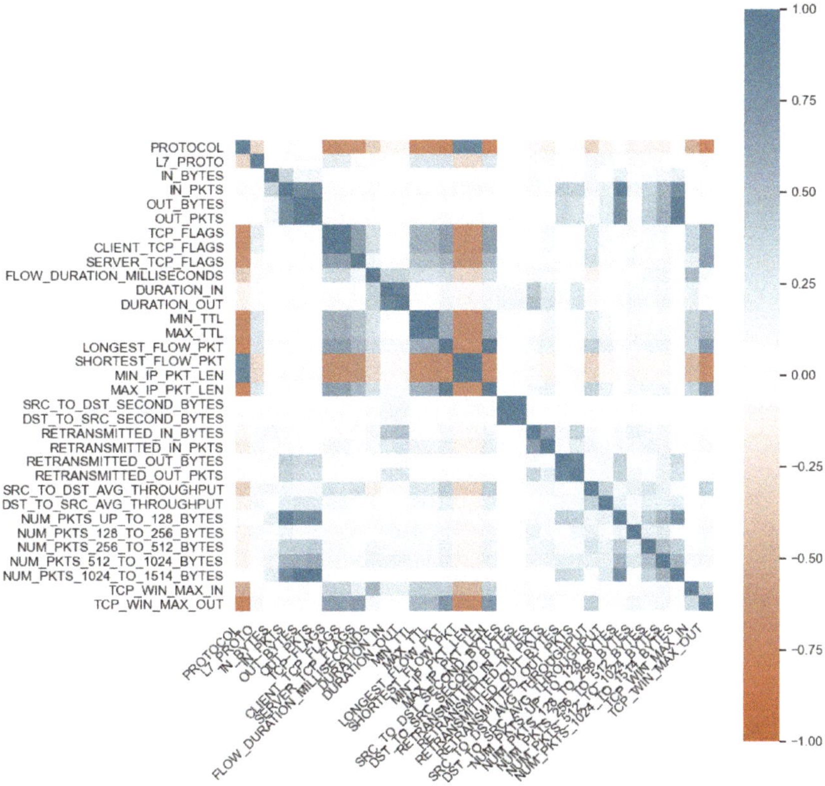

Figure 2. The correlations of features in the datasets: NF-UNSW-NB15, NF-BoT-IoT, NF-CSE-CIC-IDS2018, NF-UQ-NIDS, NF-ToN-IoT.

In the next stage of the testing, three separate feature verifications were conducted for each of the five sets, with the goal of selecting the 10 features that were most useful for the entire set. For this purpose, the methods used for the tests were: Chi2, random forest importance, and LASSO L1. The results of these tests for each dataset can be found in Figures 3–7. From the above results of examining the ability of individual parameters, it can be clearly concluded that the set of features worth considering is not 33. To achieve similar results with such a list of anomalies, it is sufficient to select 10 features.

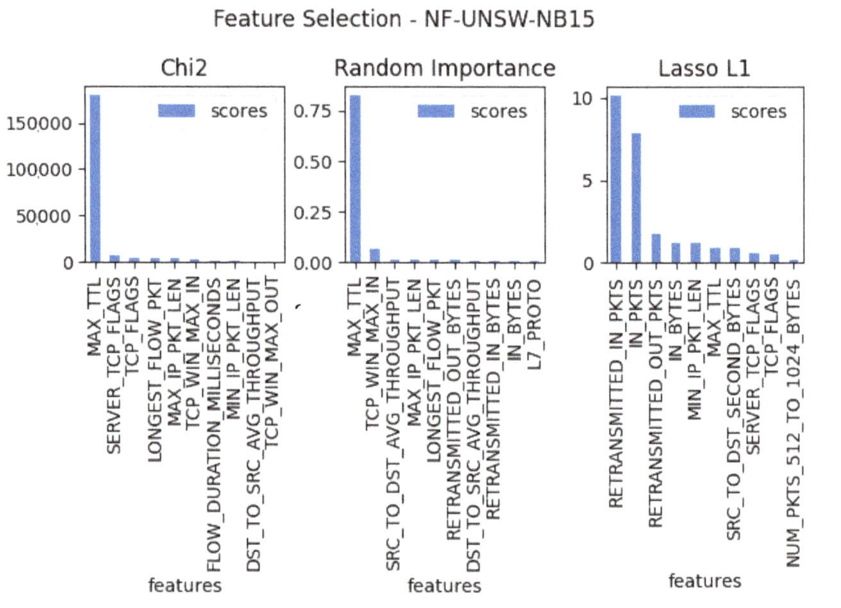

Figure 3. The top 10 best features determined by using three algorithms: LASSO L1, random forest Importance, and Chi2 for dataset nf-unsw-nb15.

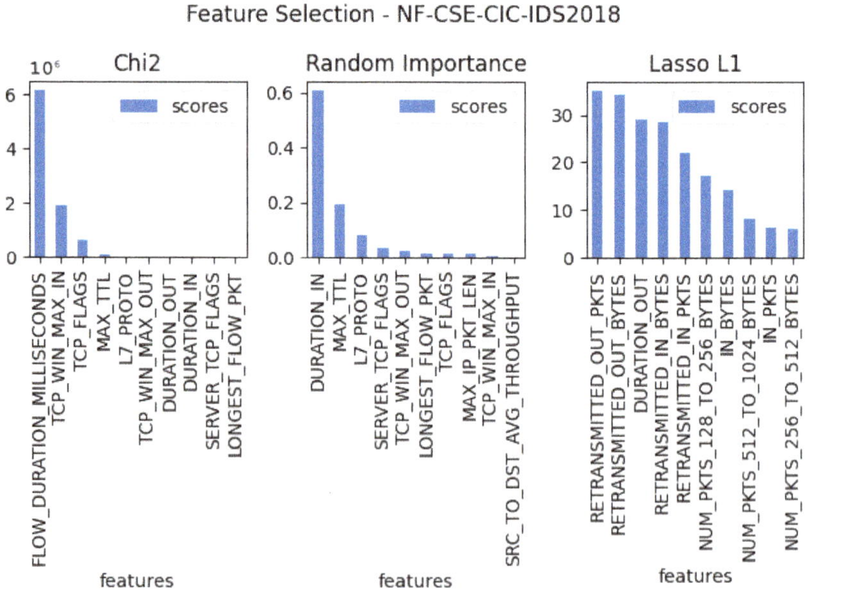

Figure 4. The top 10 best features determined by using three algorithms: LASSO L1, random forest importance, and Chi2 for dataset nf-cse-cic-ids.

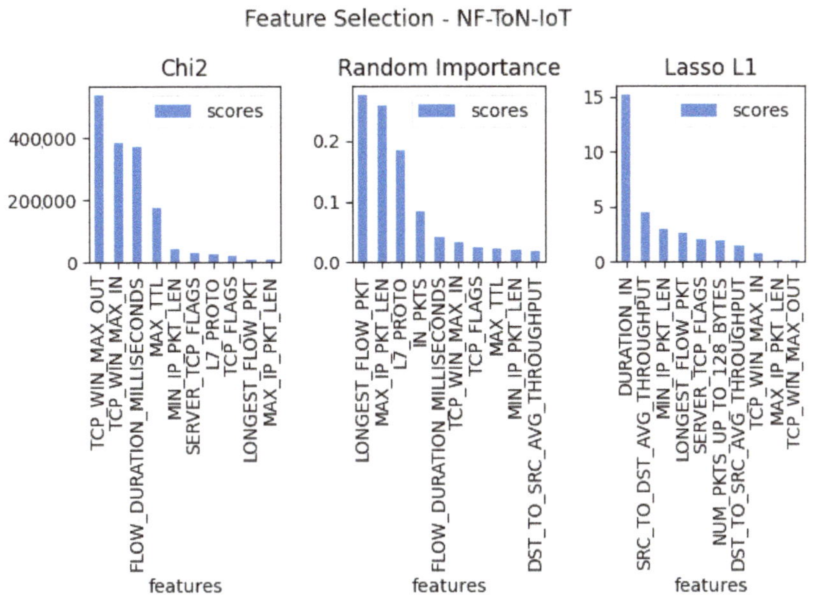

Figure 5. The top 10 best features determined by using three algorithms: LASSO L1, random forest importance, and Chi2 for dataset nf-ton-iot.

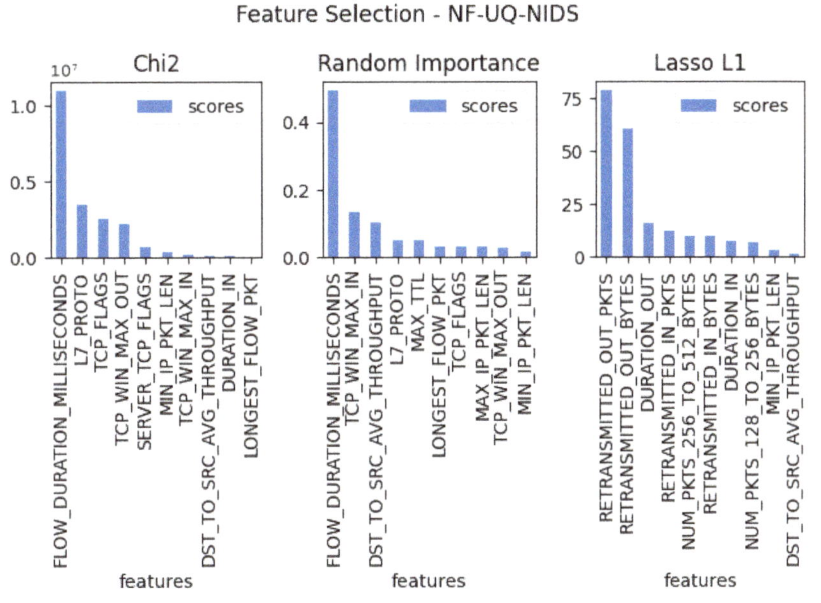

Figure 6. The top 10 best features determined by using three algorithms: LASSO L1, random forest importance, and Chi2 for dataset nf-uq-nids.

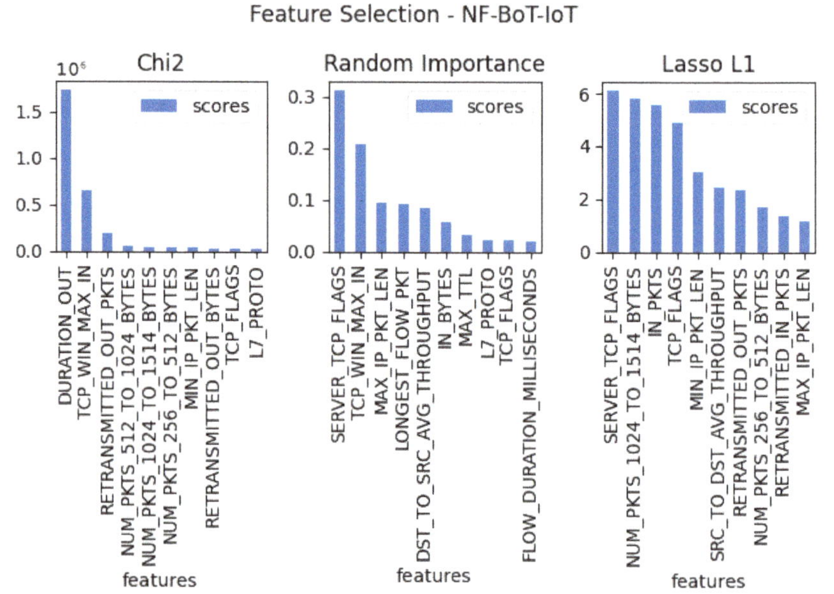

Figure 7. The top 10 best features determined by using three algorithms: LASSO L1, random forest importance, and Chi2 for dataset nf-bot-iot.

The second stage of the research was conducted to verify the effect of the data volume on intrusion detection in network traffic. Using five NetFlow datasets further divided into subsets with appropriate number of samples, the results in the form of a ROC curve plot were collected. Nine smaller subsets were separated from each of the five sets. The sizes of these subsets were as follows: 50, 100, 500, 1000, 2000, 10,000, 20,000, 30,000, and the last subset was equal to 30% of each total dataset. Another important note is that 50% of each subset consists of infected traffic, and 50% of clean traffic. The first algorithm that was trained on all the subset datasets used in this paper was random forest. The configuration of this model was default, so the parameter selection was not modified. The result for all the datasets and the random forest algorithm is shown in Figure 8.

It appears that for such anomalies, it only takes about 500 samples, i.e., 250 attack and 250 benign traffic samples, to perform detection with the effectiveness similar to much larger subsets. The second algorithm used to study the size of data needed for learning is AdaBoost. The data splitting is identical to the previous algorithm and the results for all the datasets and the AdaBoost algorithm have been shown in Figure 9.

The third model from which the results were collected is naïve Bayes. The performance of naïve Bayes was not on par with the other methods. The results are presented in Figure 10.

The efficiency of the neural network on individual datasets is the last test conducted as part of the research contained in this paper. The neural network was built based on the configuration that was mentioned in the earlier section "Classification Models". The entire list of studies for this set and the ANN can be found in Figure 11.

The research in this paper concludes with a comprehensive summary of the performance results of each algorithm on the five network datasets. The results of this summary can be found in Table 3. A number of metrics were used to correctly evaluate the effectiveness of a given model to verify the intrusion detection performance. It can be observed that the AdaBoost and random forest algorithms perform very well in detecting anomalies in network traffic.

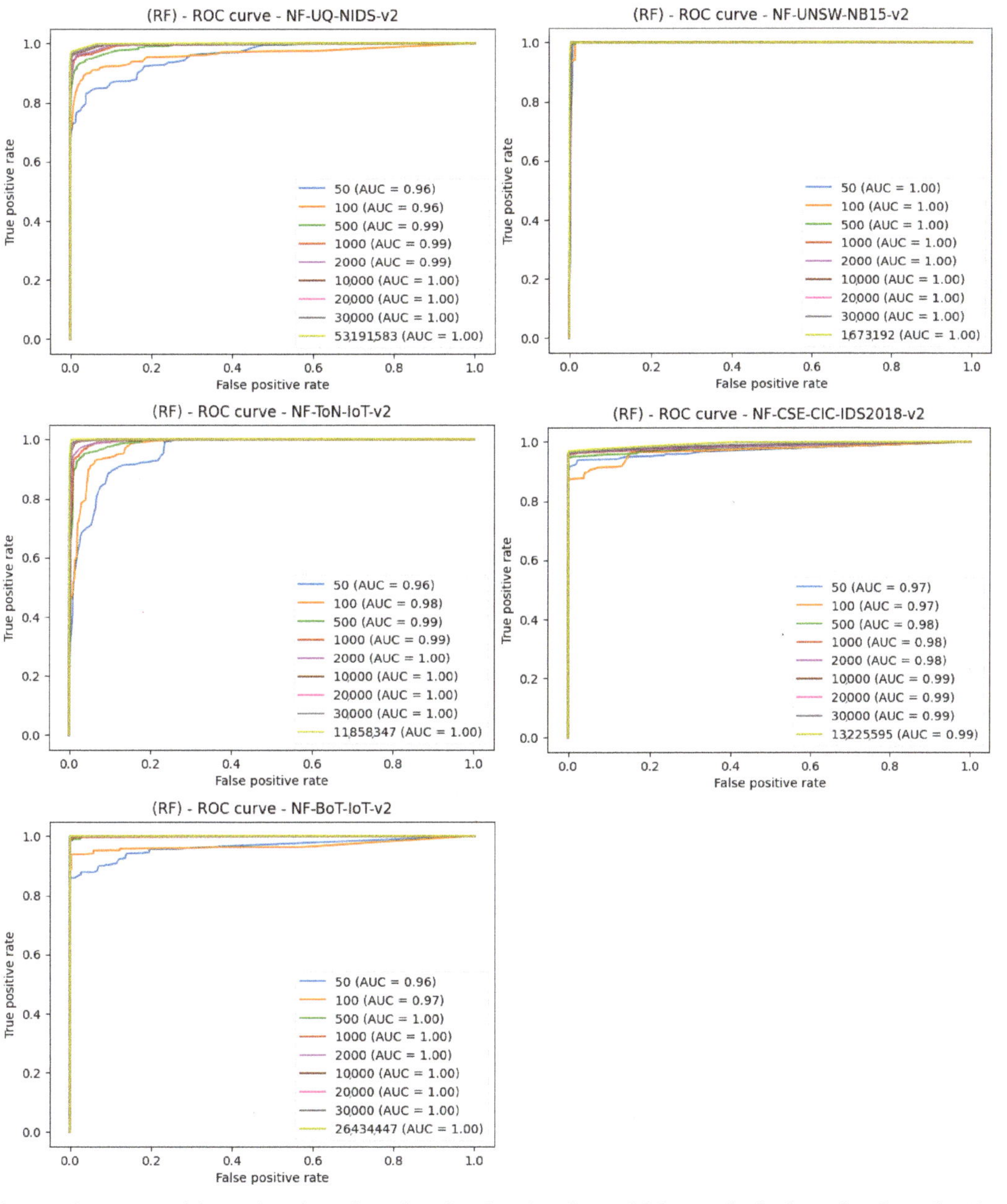

Figure 8. Comparison of the number of samples and machine learning effects—ROC curve plot for the random forest algorithm.

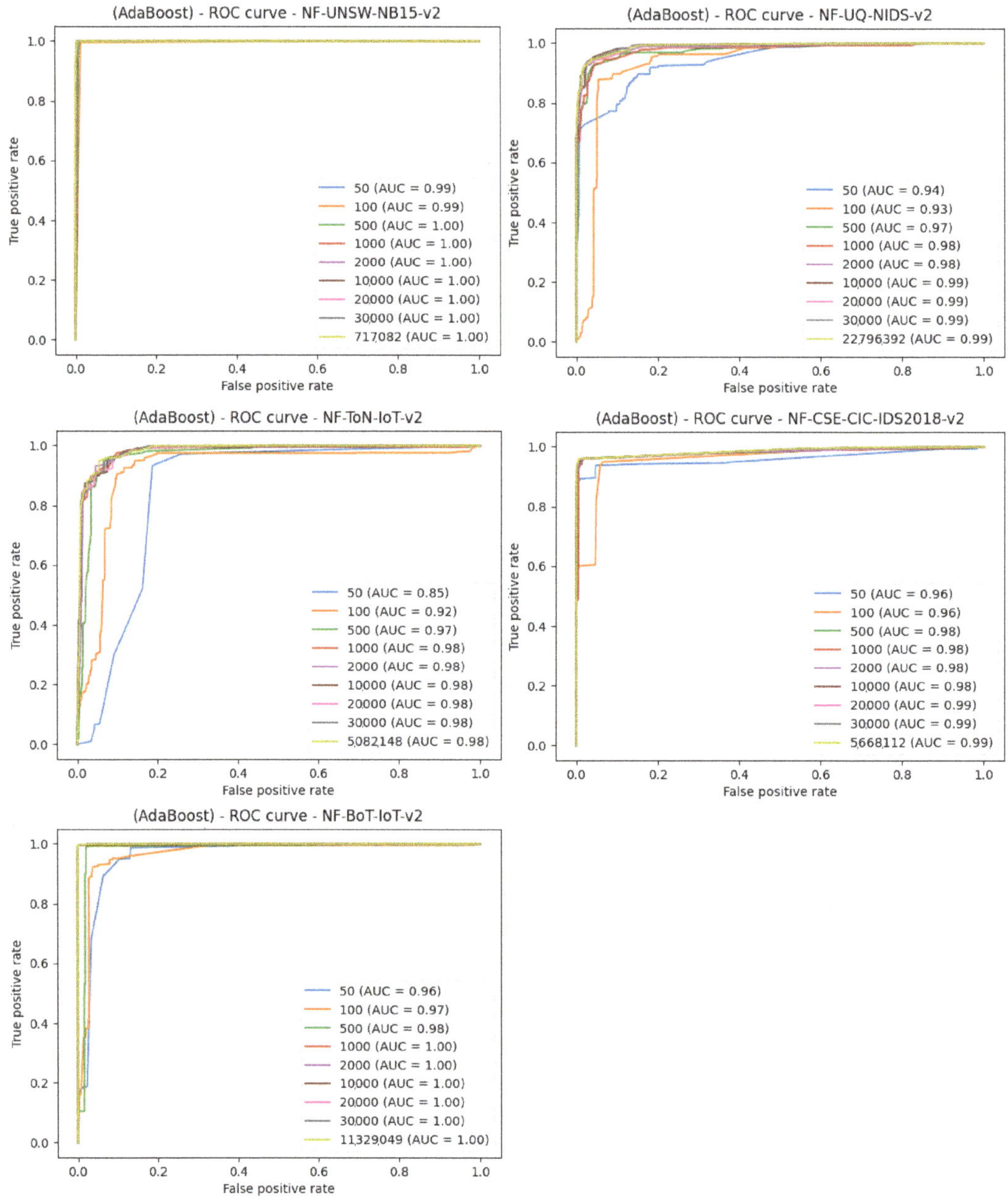

Figure 9. Comparison of the number of samples and machine learning effects—ROC curve plot for the AdaBoost algorithm.

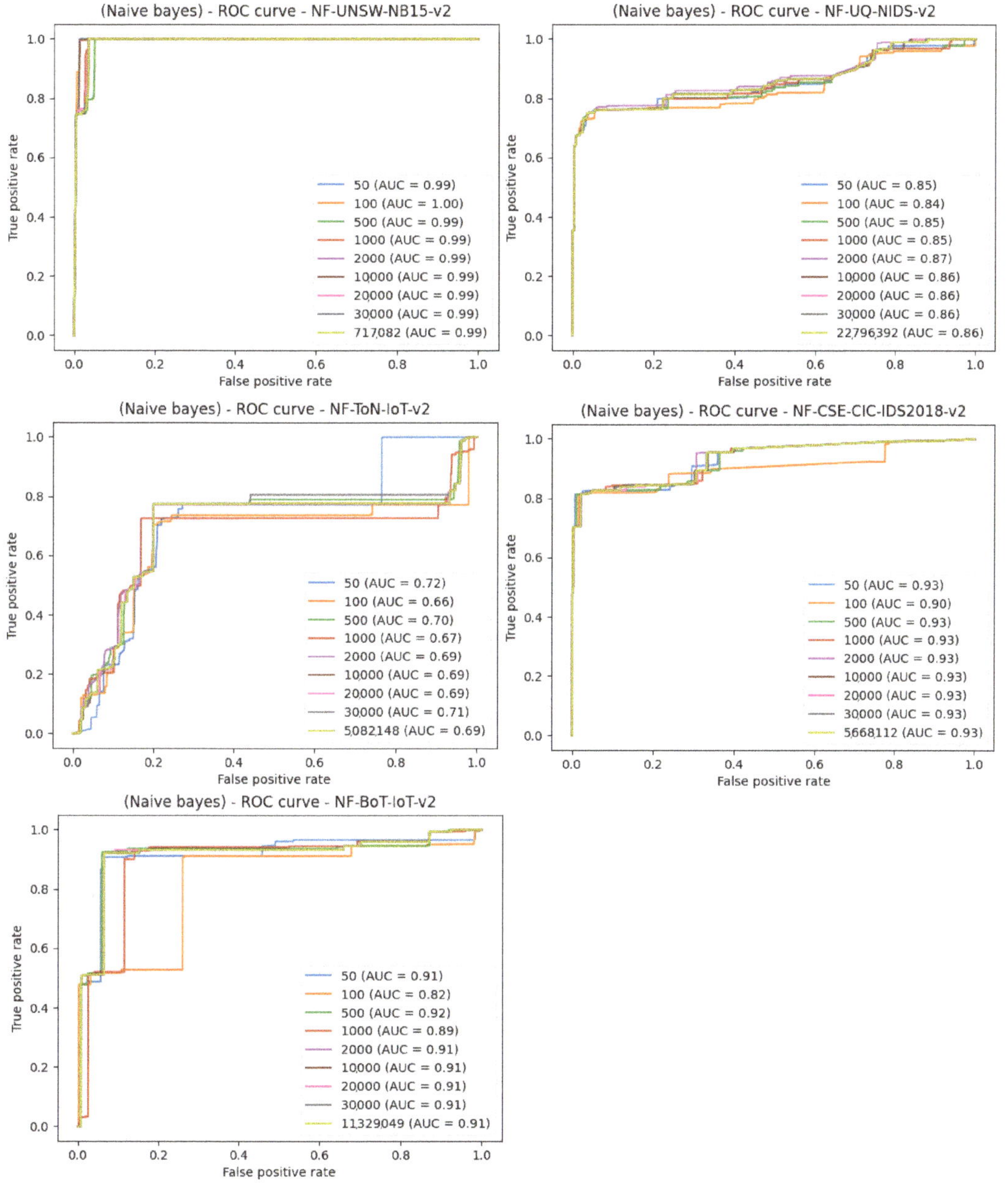

Figure 10. Comparison of the number of samples and machine learning effects—ROC curve plot for the naïve Bayes classifier.

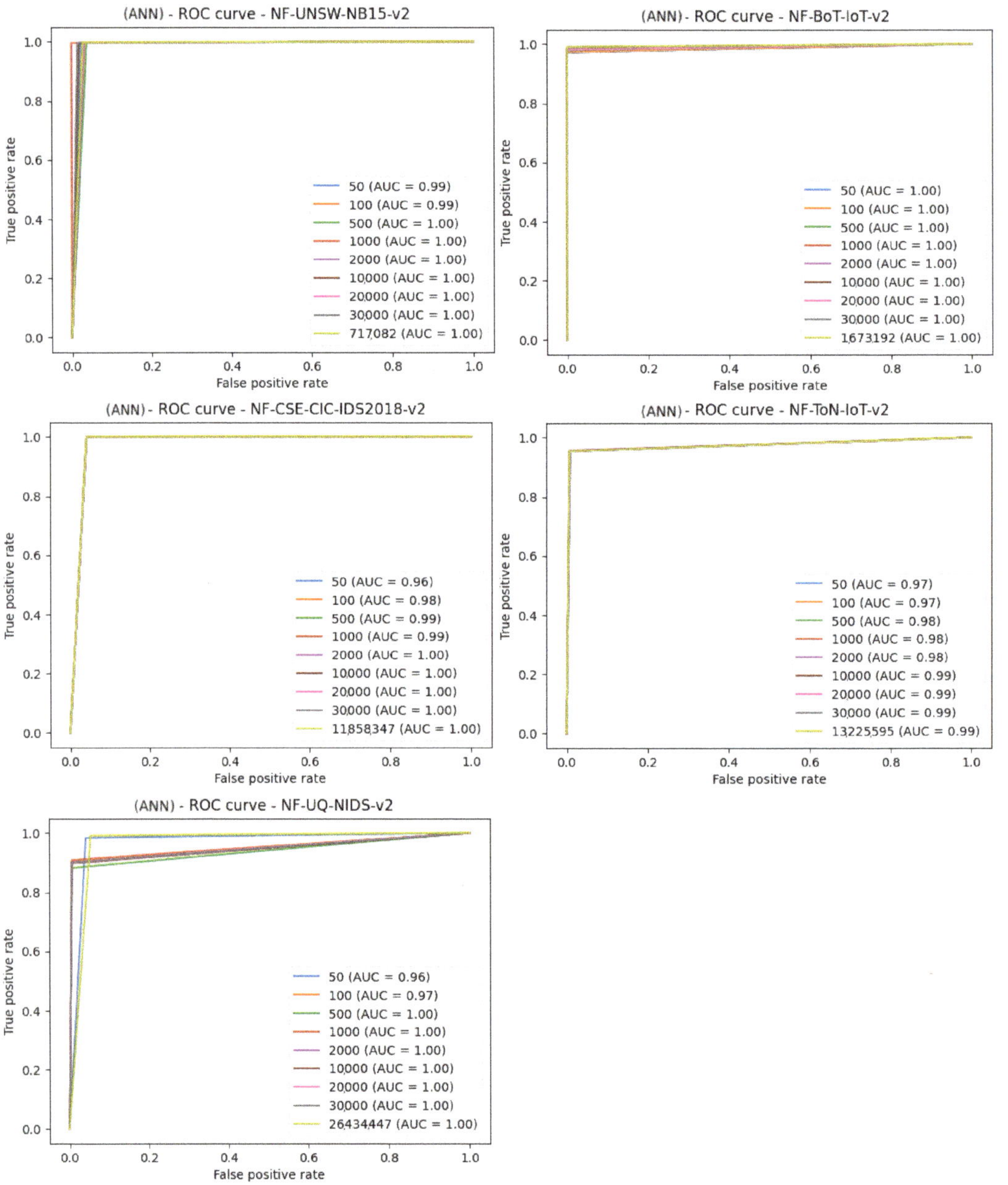

Figure 11. Comparison of the number of samples and machine learning effects—ROC curve plot for ANN.

Table 3. A summary of the effectiveness of the algorithms on each dataset using the listed metrics and established feature set.

Dataset	ACC	Precision	Recall	F1	BCC	MCC	AUC_ROC
			Random Forest				
UNSW-NB15	1	1	1	1	0.9858	0.9653	0.9653
BoT-IoT	1	1	1	1	0.9989	0.9970	0.9989
CSE-CIC-IDS	1	1	1	1	0.9828	0.9800	0.9828
UQ-NIDS	0.98	0.98	0.98	0.98	0.9837	0.9557	0.9837
ToN-IoT	1	1	1	1	0.9962	0.9933	0.9962
			ADABOOST				
UNSW-NB15	1	1	0.95	0.97	0.9943	0.9406	0.9943
BoT-IoT	1	1	1	1	0.9828	0.9800	0.9828
CSE-CIC-IDS	0.99	0.99	0.99	0.99	0.9666	0.9574	0.9666
UQ-NIDS	0.95	0.95	0.95	0.95	0.9540	0.8928	0.9540
ToN-IoT	0.94	0.94	0.94	0.94	0.9321	0.8690	0.9321
			Naïve BAYES				
UNSW-NB15	0.98	0.98	0.98	0.98	0.8706	0.7902	0.8706
BoT-IoT	0.94	0.99	0.94	0.97	0.6275	0.0667	0.6275
CSE-CIC-IDS	0.94	0.94	0.94	0.94	0.8869	0.7303	0.8869
UQ-NIDS	0.82	0.86	0.82	0.83	0.8524	0.6644	0.8524
ToN-IoT	0.64	0.72	0.64	0.64	0.6807	0.3538	0.6807
			ANN				
UNSW-NB15	1	1	1	1	0.9930	0.9164	0.9930
BoT-IoT	1	1	1	1	0.9034	0.8653	0.9034
CSE-CIC-IDS	0.99	0.99	0.99	0.99	0.9791	0.9755	0.9791
UQ-NIDS	0.97	0.97	0.97	0.97	0.9701	0.9290	0.9701
ToN-IoT	0.98	0.98	0.98	0.98	0.9681	0.9461	0.9681

5. Discussion

The main motivation for the research was to find the minimal set of features and the minimum size of data for intrusion detection based on the NetFlow scheme. Using five recent benchmark datasets, in the first step feature selection was performed with three methods.

The experiments showed that across all the datasets, the methods indicated common features to be the most informative. The number of selected features that do not cause a decrease in the machine learning model capability oscillates around 10 out of 33 features available in the sets. As can be observed in the figures, the importance of features drops sharply in most instances after just a few values.

The final set of features is provided by the findings coming from the use of the feature selection methods, which are presented in Figures 3–7. Some features provided higher importance scores than others depending on the dataset and measuring method. However, as can be observed from the figures, many of the features are repeated, which means that these features have the greatest impact on machine learning performance, regardless of the dataset and selection method. The final list of features was compiled from the results of feature selection and compared with the results of the correlation coefficient of the features, and the features that were strongly correlated with one another were removed.

The following is the list of features extracted from the evaluated datasets using the feature selection methods. The list is validated as input data by training four different ML algorithms.

1. FLOW_DURATION_MILLISECONDS
2. TCP_WIN_MAX_IN
3. DURATION_OUT
4. MAX_TTL
5. L7_PROTO
6. SRC_TO_DST_AVG_THROUGHPUT
7. SHORTEST_FLOW_PKT
8. MIN_IP_PKT_LEN
9. TCP_WIN_MAX_OUT
10. OUT_BYTES

In the second round of research, a study was conducted on the effect of data volume on intrusion detection in network traffic. In the first stage, the available datasets were divided into smaller sets, four algorithms were trained on them, and the results were collected in the form of ROC curve plots. Following [74], the future direction is to focus on the explainability of IDS.

6. Conclusions and Threats to Validity

The important contribution from this study is the conclusion that large amounts of data are not needed for effective intrusion detection. The algorithms did not show much more efficiency and effectiveness after exceeding 2000 samples, which included 1000 samples of normal traffic and 1000 samples of infected traffic. Another important finding comes in the fact that not all the NetFlow fields available as features are informative features for ML classifiers.

The nature of network traffic changes over time, with new services, new equipment and novel threats all being reflected in the traffic characteristics. Thus, the datasets used for network intrusion detection have to convey those changes and the relevant, current phenomena. The rapid pace of those changes causes the datasets to become obsolete with the passage of time. This paper serves as a set of general guidelines for the collection of relevant network intrusion detection datasets adequate for stream processing solutions.

Author Contributions: Conceptualization, W.H. and M.C.; Formal analysis, M.K., M.P., R.K. and W.H.; Funding acquisition, W.H. and M.C.; Investigation, M.K. and R.K.; Methodology, M.P.; Resources, M.C.; Software, M.K.; Supervision, R.K. and W.H.; Validation, R.K.; Visualization, M.K.; Writing—original draft, M.K. and M.P.; Writing—review and editing, M.P. and R.K. All authors have read and agreed to the published version of the manuscript.

Funding: This work is funded under the SIMARGL Project—Secure Intelligent Methods for Advanced RecoGnition of malware and Stegomalware, with the support of the European Commission and the Horizon 2020 Program, under Grant Agreement No. 833042.

Institutional Review Board Statement: Not applicable.

Informed Consent Statement: Not applicable.

Data Availability Statement: The data supporting the reported results are publicly available benchmarks and can be found at: https://staff.itee.uq.edu.au/marius/NIDS_datasets/ (accessed on 14 November 2021).

Conflicts of Interest: The authors declare no conflict of interest.

References

1. Kaur, J.; Ramachandran, R. The Recent Trends in CyberSecurity: A Review. *J. King Saud-Univ.-Comput. Inf. Sci.* **2021**. [CrossRef]
2. Pawlicka, A.; Jaroszewska-Choras, D.; Choras, M.; Pawlicki, M. Guidelines for Stego/Malware Detection Tools: Achieving GDPR Compliance. *IEEE Technol. Soc. Mag.* **2020**, *39*, 60–70. [CrossRef]

3. Mihailescu, M.E.; Mihai, D.; Carabas, M.; Komisarek, M.; Pawlicki, M.; Hołubowicz, W.; Kozik, R. The Proposition and Evaluation of the RoEduNet-SIMARGL2021 Network Intrusion Detection Dataset. *Sensors* **2021**, *21*, 4319. [CrossRef]
4. Komisarek, M.; Pawlicki, M.; Kozik, R.; Choras, M. Machine Learning Based Approach to Anomaly and Cyberattack Detection in Streamed Network Traffic Data. *J. Wirel. Mob. Netw. Ubiquitous Comput. Dependable Appl.* **2021**, *12*, 3–19. [CrossRef]
5. Komisarek, M.; Choras, M.; Kozik, R.; Pawlicki, M. Real-time stream processing tool for detecting suspicious network patterns using machine learning. In Proceedings of the ARES 2020: The 15th International Conference on Availability, Reliability and Security, Virtual Event, Ireland, 25–28 August 2020; Volkamer, M., Wressnegger, C., Eds.; ACM: New York, NY, USA, 2020; pp. 60:1–60:7. [CrossRef]
6. Komisarek, M.; Pawlicki, M.; Kowalski, M.; Marzecki, A.; Kozik, R.; Choraś, M. Network Intrusion Detection in the Wild-the Orange use case in the SIMARGL project. In Proceedings of the 16th International Conference on Availability, Reliability and Security, Vienna, Austria, 17–20 August 2021; pp. 1–7.
7. Szczepanski, M.; Komisarek, M.; Pawlicki, M.; Kozik, R.; Choraś, M. The Proposition of Balanced and Explainable Surrogate Method for Network Intrusion Detection in Streamed Real Difficult Data. In *International Conference on Computational Collective Intelligence*; Springer: Cham, Switzerland, 2021; pp. 241–252.
8. Choraś, M.; Pawlicki, M. Intrusion detection approach based on optimised artificial neural network. *Neurocomputing* **2021**, *452*, 705–715. [CrossRef]
9. Ring, M.; Wunderlich, S.; Scheuring, D.; Landes, D.; Hotho, A. A survey of network-based intrusion detection data sets. *Comput. Secur.* **2019**, *86*, 147–167. [CrossRef]
10. Ghafir, I.; Prenosil, V.; Svoboda, J.; Hammoudeh, M. A Survey on Network Security Monitoring Systems. In Proceedings of the 2016 IEEE 4th International Conference on Future Internet of Things and Cloud Workshops (FiCloudW), Vienna, Austria, 22–24 August 2016; pp. 77–82.
11. Hofstede, R.; Celeda, P.; Trammell, B.; Drago, I.; Sadre, R.; Sperotto, A.; Pras, A. Flow Monitoring Explained: From Packet Capture to Data Analysis with NetFlow and IPFIX. *IEEE Commun. Surv. Tutor.* **2014**, *16*, 2037–2064. [CrossRef]
12. Claise, B.; Bryant, S. *Specification of the IP Flow Information Export (IPFIX) Protocol for the Exchange of IP Traffic Flow Information;* Technical Report, RFC 5101; IETF: Wilmington, DE, USA, January 2008.
13. Sharafaldin, I.; Lashkari, A.H.; Ghorbani, A.A. Toward generating a new intrusion detection dataset and intrusion traffic characterization. *ICISSp* **2018**, *1*, 108–116.
14. Dhanabal, L.; Shantharajah, S. A study on NSL-KDD dataset for intrusion detection system based on classification algorithms. *Int. J. Adv. Res. Comput. Commun. Eng.* **2015**, *4*, 446–452.
15. Subbaswamy, A.; Saria, S. From development to deployment: Dataset shift, causality, and shift-stable models in health AI. *Biostatistics* **2020**, *21*, 345–352. [CrossRef] [PubMed]
16. Wang, R.Y.; Strong, D.M. Beyond Accuracy: What Data Quality Means to Data Consumers. *J. Manag. Inf. Syst.* **1996**, *12*, 5–33. [CrossRef]
17. Cai, L.; Zhu, Y. The Challenges of Data Quality and Data Quality Assessment in the Big Data Era. *Data Sci. J.* **2015**, *14*, 2. [CrossRef]
18. Demchenko, Y.; Membrey, P.; Grosso, P.; Laat, C. Addressing Big Data Issues in Scientific Data Infrastructure. In Proceedings of the 2013 International Conference on Collaboration Technologies and Systems (CTS), San Diego, CA, USA, 20–24 May 2013. [CrossRef]
19. Becker, D.; King, T.D.; McMullen, B. Big data, big data quality problem. In Proceedings of the 2015 IEEE International Conference on Big Data (Big Data), 2015, Santa Clara, CA, USA, 29 October–1 November 2015; pp. 2644–2653. [CrossRef]
20. Taleb, I.; Serhani, M.A.; Dssouli, R. Big Data Quality: A Survey. In Proceedings of the 2018 IEEE International Congress on Big Data (BigData Congress), San Francisco, CA, USA, 2–7 July 2018; pp. 166–173. [CrossRef]
21. Althnian, A.; AlSaeed, D.; Al-Baity, H.; Samha, A.; Dris, A.B.; Alzakari, N.; Abou Elwafa, A.; Kurdi, H. Impact of Dataset Size on Classification Performance: An Empirical Evaluation in the Medical Domain. *Appl. Sci.* **2021**, *11*, 796. [CrossRef]
22. Kozik, R.; Pawlicki, M.; Choraś, M. Cost-Sensitive Distributed Machine Learning for NetFlow-Based Botnet Activity Detection. *Secur. Commun. Netw.* **2018**, *2018*, 8753870. [CrossRef]
23. Pawlicki, M.; Choraś, M.; Kozik, R.; Holubowicz, W. On the Impact of Network Data Balancing in Cybersecurity Applications. In *International Conference on Computational Science;* Springer: Cham, Switzerland, 2020; Volume 12140, pp. 196–210.
24. Buczak, A.L.; Guven, E. A Survey of Data Mining and Machine Learning Methods for Cyber Security Intrusion Detection. *IEEE Commun. Surv. Tutor.* **2016**, *18*, 1153–1176. [CrossRef]
25. Prusa, J.; Khoshgoftaar, T.M.; Seliya, N. The Effect of Dataset Size on Training Tweet Sentiment Classifiers. In Proceedings of the 2015 IEEE 14th International Conference on Machine Learning and Applications (ICMLA), Miami, FL, USA, 9–11 December 2015; pp. 96–102. [CrossRef]
26. Han, S.; Kim, H. On the Optimal Size of Candidate Feature Set in Random forest. *Appl. Sci.* **2019**, *9*, 898. [CrossRef]
27. Oujezsky, V.; Horvath, T. Traffic Similarity Observation Using a Genetic Algorithm and Clustering. *Technologies* **2018**, *6*, 103. [CrossRef]
28. Vaarandi, R.; Pihelgas, M. NetFlow Based Framework for Identifying Anomalous End User Nodes. In Proceedings of the 15th International Conference on Cyber Warfare and Security: ICCWS 2020, Norfolk, VA, USA, 12–13 March 2020.

29. Flanagan, K.; Fallon, E.; Connolly, P.; Awad, A. NetFlow Anomaly Detection Though Parallel Cluster Density Analysis in Continuous Time-Series. In *Wired/Wireless Internet Communications*; Koucheryavy, Y., Mamatas, L., Matta, I., Ometov, A., Papadimitriou, P., Eds.; Springer International Publishing: Cham, Switzerland, 2017; pp. 221–232.
30. Mohamed, S.; Ejbali, R. Deep Learning with Moderate Architecture for Network Intrusion Detection System. In *Intelligent Systems Design and Applications*; Abraham, A., Piuri, V., Gandhi, N., Siarry, P., Kaklauskas, A., Madureira, A., Eds.; Springer International Publishing: Cham, Switzerland, 2021; pp. 774–783.
31. Liu, W.; Duan, H.X.; Ren, P.; Li, X.; Wu, J.P. Wavelet based data mining and querying in network security databases. In Proceedings of the 2003 International Conference on Machine Learning and Cybernetics (IEEE Cat. No. 03EX693), Xi'an, China, 5 November 2003; Volume 1, pp. 178–182. [CrossRef]
32. Siddiqui, S.; Khan, M.S.; Ferens, K. Multiscale Hebbian neural network for cyber threat detection. In Proceedings of the 2017 International Joint Conference on Neural Networks (IJCNN), Anchorage, AK, USA, 14–19 May 2017; pp. 1427–1434. [CrossRef]
33. He, F.; Zhang, Y.; Liu, D.; Dong, Y.; Liu, C.; Wu, C. Mixed Wavelet-Based Neural Network Model for Cyber Security Situation Prediction Using MODWT and Hurst Exponent Analysis. In *Network and System Security*; Yan, Z., Molva, R., Mazurczyk, W., Kantola, R., Eds.; Springer International Publishing: Cham, Switzerland, 2017; pp. 99–111.
34. Liu, H.; Lang, B. Machine Learning and Deep Learning Methods for Intrusion Detection Systems: A Survey. *Appl. Sci.* **2019**, *9*, 4396. [CrossRef]
35. Fejrskov, M.; Pedersen, J.M.; Vasilomanolakis, E. Cyber-security research by ISPs: A NetFlow and DNS Anonymization Policy. In Proceedings of the 2020 International Conference on Cyber Security and Protection of Digital Services (Cyber Security), Dublin, Ireland, 15 June 2020; pp. 1–8. [CrossRef]
36. Qureshi, S.; Menghwar, G.; Tunio, S.; Ullah, F.; Nazir, A.; Wajahat, A. Performance Analysis of Open Source Solution -ntop I I for Active and Passive Packet Analysis Relating to Application and Transport Layer. *Int. J. Adv. Comput. Sci. Appl.* **2019**, *10*, 4. [CrossRef]
37. Deri, L. nProbe: An Open Source NetFlow Probe for Gigabit Networks. In Proceedings of the TERENA Networking Conference 2003, Zagreb, Croatia, 19–22 May 2003.
38. Sarhan, M.; Layeghy, S.; Moustafa, N.; Portmann, M. NetFlow Datasets for Machine Learning-based Network Intrusion Detection Systems. *arXiv* **2020**, arXiv:2011.09144. Available online: https://arxiv.org/abs/2011.09144 (accessed on 14 September 2021).
39. Moustafa, N.; Slay, J. UNSW-NB15: A comprehensive data set for network intrusion detection systems (UNSW-NB15 network data set). In Proceedings of the 2015 Military Communications and Information Systems Conference (MilCIS), Canberra, Australia, 10–12 November 2015; pp. 1–6. [CrossRef]
40. Koroniotis, N.; Moustafa, N.; Sitnikova, E.; Turnbull, B. Towards the development of realistic botnet dataset in the Internet of Things for network forensic analytics: Bot-IoT dataset. *Future Gener. Comput. Syst.* **2019**, *100*, 779–796. [CrossRef]
41. Alsaedi, A.; Moustafa, N.; Tari, Z.; Mahmood, A.; Anwar, A. TON_IoT Telemetry Dataset: A New Generation Dataset of IoT and IIoT for Data-Driven Intrusion Detection Systems. *IEEE Access* **2020**, *8*, 165130–165150. [CrossRef]
42. Sharafaldin, I.; Habibi Lashkari, A.; Ghorbani, A.A. Toward Generating a New Intrusion Detection Dataset and Intrusion Traffic Characterization. In Proceedings of the 4th International Conference on Information Systems Security and Privacy-ICISSP, INSTICC, Funchal, Portugal, 22–24 January 2018; SciTePress: Setúbal, Portugal, 2018; pp. 108–116. [CrossRef]
43. Sarhan, M.; Layeghy, S.; Portmann, M. Towards a Standard Feature Set for Network Intrusion Detection System Datasets. *arXiv* **2021**, arXiv:cs.NI/2101.11315.
44. Honest, N. A survey on Feature Selection Techniques. *GIS Sci. J.* **2020**, *7*, 353–358.
45. Khalid, S.; Khalil, T.; Nasreen, S. A survey of feature selection and feature extraction techniques in machine learning. In Proceedings of the 2014 Science and Information Conference, London, UK, 27–29 August 2014; pp. 372–378. [CrossRef]
46. Ferreira, A.J.; Figueiredo, M.A. Efficient feature selection filters for high-dimensional data. *Pattern Recognit. Lett.* **2012**, *33*, 1794–1804. [CrossRef]
47. Jović, A.; Brkić, K.; Bogunović, N. A review of feature selection methods with applications. In Proceedings of the 2015 38th International Convention on Information and Communication Technology, Electronics and Microelectronics (MIPRO), Opatija, Croatia, 25–29 May 2015; pp. 1200–1205. [CrossRef]
48. Mostert, W.; Malan, K.M.; Engelbrecht, A.P. A Feature Selection Algorithm Performance Metric for Comparative Analysis. *Algorithms* **2021**, *14*, 100. [CrossRef]
49. Potdar, K.; Pardawala, T.; Pai, C. A Comparative Study of Categorical Variable Encoding Techniques for Neural Network Classifiers. *Int. J. Comput. Appl.* **2017**, *175*, 7–9. [CrossRef]
50. Hancock, J.T.; Khoshgoftaar, T.M. Survey on categorical data for neural networks. *J. Big Data* **2020**, *7*, 28. [CrossRef]
51. Muthukrishnan, R.; Rohini, R. LASSO: A feature selection technique in predictive modeling for machine learning. In Proceedings of the 2016 IEEE International Conference on Advances in Computer Applications (ICACA), Coimbatore, India, 24 October 2016; pp. 18–20. [CrossRef]
52. Osman, H.; Ghafari, M.; Nierstrasz, O. Automatic Feature Selection by Regularization to Improve Bug Prediction Accuracy. In Proceedings of the 2017 IEEE Workshop on Machine Learning Techniques for Software Quality Evaluation (MaLTeSQuE), Klagenfurt, Austria, 21 February 2017. [CrossRef]
53. Strobl, C.; Boulesteix, A.L.; Zeileis, A.; Hothorn, T. Bias in random forest variable importance measures: Illustrations, sources and a solution. *BMC Bioinform.* **2007**, *8*, 25. [CrossRef]

54. Breiman, L. Random Forests. *Mach. Learn.* **2001**, *45*, 5–32. [CrossRef]
55. Nguyen, T.T.; Huang, J.Z.; Nguyen, T.T. Unbiased Feature Selection in Learning Random Forests for High-Dimensional Data. *Sci. World J.* **2015**, *2015*, 471371. [CrossRef] [PubMed]
56. Suryakanthi, T. Evaluating the Impact of GINI Index and Information Gain on Classification using Decision Tree Classifier Algorithm. *Int. J. Adv. Comput. Sci. Appl.* **2020**, *11*, 612–619. [CrossRef]
57. Nihan, S. Karl Pearsons chi-square tests. *Educ. Res. Rev.* **2020**, *15*, 575–580. [CrossRef]
58. Hossin, M.; Sulaiman, M.N. A Review on Evaluation Metrics for Data Classification Evaluations. *Int. J. Data Min. Knowl. Manag. Process* **2015**, *5*, 1–11. [CrossRef]
59. Fatourechi, M.; Ward, R.K.; Mason, S.G.; Huggins, J.; Schlögl, A.; Birch, G.E. Comparison of Evaluation Metrics in Classification Applications with Imbalanced Datasets. In Proceedings of the 2008 Seventh International Conference on Machine Learning and Applications, San Diego, CA, USA, 11–13 December 2008; pp. 777–782. [CrossRef]
60. Brodersen, K.H.; Ong, C.S.; Stephan, K.E.; Buhmann, J.M. The Balanced Accuracy and Its Posterior Distribution. In Proceedings of the 2010 20th International Conference on Pattern Recognition, Istanbul, Turkey, 23–26 August 2010; pp. 3121–3124. [CrossRef]
61. Chicco, D.; Jurman, G. The advantages of the Matthews correlation coefficient (MCC) over F1 score and accuracy in binary classification evaluation. *BMC Genom.* **2020**, *21*, 6. [CrossRef] [PubMed]
62. Primartha, R.; Adhi Tama, B. Anomaly detection using random forest: A performance revisited. In Proceedings of the 2017 International Conference on Data and Software Engineering (ICoDSE), Palembang, Indonesia, 1–2 November 2017; pp. 1–6. [CrossRef]
63. Huč, A.; Šalej, J.; Trebar, M. Analysis of Machine Learning Algorithms for Anomaly Detection on Edge Devices. *Sensors* **2021**, *21*, 4946. [CrossRef]
64. Biswas, P.; Samanta, T. Anomaly detection using ensemble random forest in wireless sensor network. *Int. J. Inf. Technol.* **2021**, *13*, 2043–2052. [CrossRef]
65. Seifert, S. Application of random forest based approaches to surface-enhanced Raman scattering data. *Sci. Rep.* **2020**, *10*, 5436. [CrossRef]
66. Gulati, P.; Sharma, A.; Gupta, M. Theoretical Study of Decision Tree Algorithms to Identify Pivotal Factors for Performance Improvement: A Review. *Int. J. Comput. Appl.* **2016**, *141*, 19–25. [CrossRef]
67. Yang, N.; Li, T.; Song, J. Construction of Decision Trees based Entropy and Rough Sets under Tolerance Relation. In *International Journal of Computational Intelligence Systems*; Atlantis Press: Paris, France, 2007. [CrossRef]
68. Zhang, H.; Zhou, R. The analysis and optimization of decision tree based on ID3 algorithm. In Proceedings of the 2017 9th International Conference on Modelling, Identification and Control (ICMIC), Kunming, China, 10–12 July 2017; pp. 924–928. [CrossRef]
69. Mazini, M.; Shirazi, B.; Mahdavi, I. Anomaly network-based intrusion detection system using a reliable hybrid artificial bee colony and AdaBoost algorithms. *J. King Saud Univ.-Comput. Inf. Sci.* **2019**, *31*, 541–553. [CrossRef]
70. Yuan, Y.; Kaklamanos, G.; Hogrefe, D. A Novel Semi-Supervised Adaboost Technique for Network Anomaly Detection. In Proceedings of the 19th ACM International Conference on Modeling, Analysis and Simulation of Wireless and Mobile Systems, Malta, 13–17 November 2016; pp. 111–114. [CrossRef]
71. Li, W.; Li, Q. Using Naive Bayes with AdaBoost to Enhance Network Anomaly Intrusion Detection. In Proceedings of the 2010 Third International Conference on Intelligent Networks and Intelligent Systems, Shenyang, China, 1–3 November 2010; pp. 486–489. [CrossRef]
72. Kingma, D.P.; Ba, J. Adam: A method for stochastic optimization. *arXiv* **2014**, arXiv:1412.6980.
73. Wibawa, A.; Kurniawan, A.; Murti, D.; Adiperkasa, R.P.; Putra, S.; Kurniawan, S.; Nugraha, Y. Naïve Bayes Classifier for Journal Quartile Classification. *Int. J. Recent Contrib. Eng. Sci. IT (IJES)* **2019**, *7*, 91. [CrossRef]
74. Szczepański, M.; Choraś, M.; Pawlicki, M.; Kozik, R. Achieving explainability of intrusion detection system by hybrid oracle-explainer approach. In Proceedings of the 2020 International Joint Conference on Neural Networks (IJCNN), Glasgow, UK, 19–24 July 2020; pp. 1–8.

MDPI
St. Alban-Anlage 66
4052 Basel
Switzerland
Tel. +41 61 683 77 34
Fax +41 61 302 89 18
www.mdpi.com

Entropy Editorial Office
E-mail: entropy@mdpi.com
www.mdpi.com/journal/entropy

www.ingramcontent.com/pod-product-compliance
Ingram Content Group UK Ltd.
Pitfield, Milton Keynes, MK11 3LW, UK
UKHW060123010826
14089UKWH00006B/128

9 783036 553139